# Remote Sensing Applications for Dryland Natural Resource Management



NIPA® GENX ELECTRONIC RESOURCES & SOLUTIONS P. LTD.
New Delhi-110 034

www.nipaers.com
Browse, Search, Read & Buy...
Print Books
eChapters
eBooks
Publishers
Forthcoming
Subject Catalogues
New Titles
Online Resources on:
✓ Current Affairs
✓ Reasoning, Logic & Aptitude
✓ Competitive Examinations
✓ English Language Lab
✓ Writing & Pronunciation Tools
✓ Personality Development
✓ Online Programmes for Professionals
✓ Interview Preparation
Pay Using
PhonePe
Google Pay
paytm
PayPal
UPI
CC Avenue

# Remote Sensing Applications for Dryland Natural Resource Management

*Edited by*
**Mahesh Kumar Gaur**

NIPA® GENX ELECTRONIC RESOURCES & SOLUTIONS P. LTD.
New Delhi-110 034

**NIPA® GENX ELECTRONIC RESOURCES & SOLUTIONS P. LTD.**

101,103, Vikas Surya Plaza, CU Block
L.S.C. Market, Pitam Pura, New Delhi-110 034
Ph : +91 11 27341616, 27341717, 27341718
E-mail: newindiapublishingagency@gmail.com
web: www.nipabooks.com

***For customer assistance, please contact***
Phone: + 91-11-27 34 17 17
Fax: + 91-11- 27 34 16 16
E-Mail: feedbacks@nipabooks.com



ISBN: 978-81-19254-28-6



Composed and Designed by NIPA®.

# LIST OF CONTRIBUTORS


| | |
|---|---|
| Agnihotri, Ila | Research Scientist, Regional Remote Sensing Centre (West), NRSC, ISRO, Jodhpur- 342 003. Rajasthan, India |
| Bera, A.K. | Scientist, Regional Remote Sensing Centre (West), NRSC, ISRO, Jodhpur-342 003. Rajasthan, India |
| Bhadra, BK | Scientist, Regional Remote Sensing Centre (West), NRSC, ISRO, Jodhpur-342 003. Rajasthan, India |
| Chuchra, Jyotsana | Research Scientist, Regional Remote Sensing Centre (West), NRSC. ISRO, Jodhpur - 342003. Rajasthan, India. |
| Dhyani, P. | Professor, BIT Mesra, Ranchi, Jharkhand, India |
| Gaur, Mahesh Kumar | Associate Professor, Government Bangur College, Pali – 306401. Rajasthan, India |
| Joseph, Manoj | Scientist, Regional Remote Sensing Centre (West), NRSC, ISRO, Jodhpur-342 003. Rajasthan, India |
| Kar, Amal | Principal Scientist & Head of Division, Division of Natural Resources & Environment, Central Arid Zone Research Institute, Rajasthan Jodhpur 342 003. India. |
| Mishra, Divya | Research Scientist, Regional Remote Sensing Centre (West), NRSC. ISRO, Jodhpur - 342003. Rajasthan, India. |
| Moharana, P.C. | Senior Scientist, Division of Natural Resources & Environment, Central Arid Zone Research Institute, Jodhpur 342 003. Rajasthan, India |

| | |
|---|---|
| Palaria, S | Professor & Head, Department of Remote Sensing, Maharishi Dayanand Saraswati University, Ajmer - Rajasthan, India |
| Punia, M. P. | Professor, BIT, Jaipur, Rajasthan India |
| Ram, Balak | Ex Principal Scientist, Central Arid Zone Research Institute, Jodhpur, Rajasthan India |
| S. Rama Subramoniam | Scientist, Regional Remote Sensing Centre (West), NRSC, ISRO, Jodhpur-342 003. Rajasthan, India |
| Sharma, Aditi | BIT Mesra, Ranchi, Jharkhand, India |
| Sharma, Chilka | Assistant Professor, Banasthali University, Tonk - 304022, Rajasthan, India |
| Sharma, J.R. | Chief General Manager, Regional Remote Sensing Centres, NRSC, ISRO, Hyderabad, Andhra Pradesh |
| Sharma, Narvada | PGT Geography, KV No. 1. AFS, Jodhpur Rajasthan, India. |
| Sharma, Trapti | Maharishi Dayanad Saraswati University, Ajmer - Rajasthan, India |
| Singh, Himani | Research Scientist, Regional Remote Sensing Centre (West), NRSC, ISRO, Jodhpur- 342 003. Rajasthan, India |
| Shirohy, Sneha | Research Scientist, Regional Remote Sensing Centre (West), NRSC, ISRO, Jodhpur- 342 003. Rajasthan, India |
| Tyagi, Mani | BIT Mesra, Ranchi, Jharkhand, India. |
| Yadav, Kamini | Research Scientist, Regional Remote Sensing Centre (West), NRSC, ISRO, Jodhpur- 342 003. Rajasthan, India |

**DR. J.R. SHARMA**

Chief General Manager, Regional Centres
National Remote Sensing Centre
ISRO, Department of Space,
HYDERABAD. (Andhra Pradesh)

# FOREWORD

Sustaining population growth in Drylands is a big challenge for Research & Development Institutions working for Natural Resource Management in this region. Besides harsh, climatic conditions, poor soil conditions and poor water availability are major challenges and constraints for development. Among many tools developed during last 4-5 decades, Remote Sensing & GIS technologies offers great potentials for baseline survey, monitoring and management of natural resource, besides integration of socio-economic factors with natural resource data in GIS environment.

India, one of the few leading nations in the world having state of art satellites with variety of spatial and temporal resolution offers great opportunities to use Remote Sensing based inputs for preparation of development plans. There have been number of studies conducted in last few years on food security, sustainable livelihood, activities for employment generation and overall sustainable development using space based inputs and there integration with traditional data sets.

This book titled "Remote Sensing Applications for Dryland Natural Resource Management edited " by Dr. Mahesh Kumar Gaur is a collection of wide variety of articles using satellite data as well as new methods of Natural Resource Management in Drylands of India.

There are 19 studies addressing desertification paradigms and measurement in Indian context, drought and data mining, drought prediction based on rainfall and NDVI data sets, ground water analysis, micro level & village level and modeling, land degradation & poverty nexus in dry lands, desert hydrology & fluvial process and current demography occupational structures of desert region.

I feel, this book will help the students, researchers and policy makers to adopt some of the ideas based on micro level studies for overall Natural Resource Management in the region and development of the area.

**J.R. SHARMA**

ISRO, Department of Space

Hyderabad (A.P.)

# PREFACE

The land surface of our earth is home to humankind as well as most of the world's flora and fauna. Though dwarfed by the sea it still covers a vast area: if all Earth's land was put together without the oceans it would form a planet only slightly smaller than Mars.

Amassing land cover data enables researchers to see how it changes over time. Analysis of archived data helps accurately chart the rates that deserts are growing and forests are cleared, a coastline erodes or urban settlements grow. On a more rapid timescale, comparing satellite images can swiftly identify the damage done by disasters such as earthquakes, fires or floods.

Earth observation is essential for monitoring, reporting and verification. Satellite navigation is an important space tool also. Combining radar imagery with a technique called interferometry can identify tiny changes in land surface down to a scale of millimetres: this can be used to monitor urban subsidence as well as the slow movement of tectonic plates and the 'breathing' of volcanoes.

Since satellites are expected to become an increasingly important means of obtaining observations nationally for comparing climate variability and change over different parts of the Earth. And space technology can play fundamental role in addressing climate challenges, now and in the future. With the availability of a number of satellites, India can try to achieve an environmental and climate monitoring system comparable to what we achieved for meteorology, some 25 years ago. These satellites will provide the data necessary to make annually updated country-wide land-cover maps. These maps will be essential for the country to fulfill UN's mandate on Reducing Emissions from Deforestation and Forest Degradation program especially in the arid and semi-arid areas which are resource-rich.

The severe droughts and accelerated land degradation process of recent years underline India's continuing vulnerability to desertification

- threatening the livelihoods of millions of countrymen. Desertification and land degradation is a common problem across our country, because of circumstances in common: similar historical backgrounds, climatic conditions, land use patterns, cultural characteristics and vegetation types. The country needs to develop a common, harmonized information system available to all states affected by desertification and land degradation processes, making it possible to permanently access the data needed to monitor changes as they occur. The World Watch Institute estimates the Earth's landmasses lose up to 24 thousand million tonnes of topsoil annually. Since, our economy is highly dependent on agriculture, so soil is to be considered one of our major natural resources to be efficiently protected so a maximum level of productivity need to be sustained. And our varied climate is such that large parts of the country are sensitive to wild fires.

The good news is that dryland desertification can be remedied or even reversed, using appropriate land management techniques. Forecasting is needed of areas most at risk, and this is what satellites can provide. As well as highlighting any relevant land use change, the view from space can reveal increased surface reflectivity, temperature, dryness and dustiness. Infrared sensors can be used to detect vegetation stress from environmental changes.

With climate change a priority, India needs to develop a joint strategy on space and climate. As part of this, Editor invited experts from the space sector to provide research inputs for contribution. In order to effectively respond to climate change challenges, decision-makers need scientifically based climate knowledge inputs.

Arid and semi-arid India is now facing a threefold holistic crisis: economic, food, and climate. What has emerged from these crises is the vital importance of inter-linkages among them on the one hand, and the missed opportunities in putting these pieces together on the other. This book has tried to explore these challenges though in-depth discussions of the individual. It is anticipated to inspire a forward looking debate that looks at the lessons from the past and points to actions for the future. Expert views have been shared by scientists and persons of eminence on the national and state level challenges with futuristic remedial approaches.

**Dr. Mahesh Kumar GAUR**

**Editor**

# CONTENTS

# Chapter - 1

# Geomorphology in Quantification of Aeolian Dynamism in Thar Desert at Local to Regional Scales

**Amal Kar**

## Introduction

Geomorphology is the study of earth's landscapes over time and space, and involves an analysis of both its surface features as well as the processes responsible. Historically, the subject evolved through monumental studies on supposed evolution of the landscape, based on intuitive knowledge of the researchers on landscape response to endogenic and exogenic forces, as well as visions of possible changes over a time scale of centuries to millennia or more. Too much emphasis on the study of possible landscape evolution in an era of little technological back-up facilities meant that the discourses were largely philosophical in nature, with fewer measurable parameters to substantiate the concepts or the reasoning, although reflecting deep understanding of the researchers involved. As older concepts began to be confronted with newer ones, and applied aspects started receiving greater attention, a need was felt for quantification of the features and processes. This was more so for an understanding of the fluvial landscape, although aeolian and other process-based studies also gradually incorporated quantitative analysis as a means of understanding the landscape. Despite the awareness, lack of adequate equipment hindered the progress of quantification in the pre-World War II period.

During the past half a century geomorphology has undergone a radical transformation, especially with the advent of digital revolution.

Analysis based on quantification of parameters is now fast replacing the classical geomorphology that was based largely on power of observation, intimate knowledge of forms and processes and intuitive reasoning, yet having fewer measurements. Analysis of earth's land surface dynamics is now more data-demanding, and hypotheses on land surface evolution are required to be tested on the basis of quantifiable parameters. This calls for instrumentation at critical locations and time, as well as simulation studies through laboratory modelling. Since geomorphic feature dynamics are related to changes in weather and climate, geology, structural framework, soils, vegetation, land uses, etc., an understanding of a number of allied research fields like climatology, geology, sedimentology, pedology, hydrology, plant ecology, etc., has become a necessity for the study of geomorphology. As humans have become a major driving force in accelerating the geomorphic processes faster than the geological rates, and in changing the geomorphic features in a way never before, the rates of changes in processes and forms have also become very important considerations in geomorphological studies. This is true for the arid geomorphological studies also, where responses of the aeolian landscape to induced changes are far more complex than under a fluvial regime, and current understanding of the phenomena is not very perfect.

Here we propose to discuss the issues involved in quantification of the aeolian process-form interactions in the Thar Desert with some examples.

## Quantification of the Potentials of Aeolian Dynamics

Sand dunes and other sandy landforms constitute the dominant landscape of Thar Desert (also called the Great Indian Sand Desert). Wind-blown sand is ubiquitous here, so much so that hardly any kilometer-grid can be identified to the west of the Aravalli Ranges that does not have any aeolian sand on the surface or the subsurface. Yet, there is not much study to measure or to model the phenomena of aeolian dynamics, especially the aeolian process-form interactions. The major difficulties in continuing with satisfactory measurement of aeolian processes have so far been the paucity of right kind of field instruments, lack logistics for installation and maintenance of the sophisticated instruments under extreme heat and dust in open fields, and difficulties in simultaneous measurement of geomorphic and atmospheric parameters (from ground level to about 3 m height) during the hot summer months, especially during the periods of sand storms (mostly in May and June). Despite the drawbacks some studies have been carried out by us at periodic intervals to quantify sand mobility and bedform migration in the Thar Desert, using both field methods and the application of fundamental principles to the gathered information on morphometric, sedimentological and atmospheric parameters. After working with the different approaches I have a growing feeling that the idea of generating adequate spatio-temporal data using field methods alone is utopian in this vast desert, given the complexity of the terrain and logistic problems. Compared to this, the application of fundamental principles using few required input parameters is a better choice due not only to its strong scientific rooting in causes and consequences, but also due to the gradual availability of the required inputs from many different sources, including ground-based, aerial and satellite sensors. Provided the input parameters are correctly quantified, there is very little chance that the derived results are incorrect. One needs only to carry out a few sample field measurements to test the validity of the results.

The dominant sand and dust raising wind in Thar Desert start from March onwards when the westerlies create few dust storms, especially in the northwestern part. This is followed by the SW wind that builds up from April onwards, peaks in May-June when several episodes of sand/dust storms take place across the desert, and dissipates with the onset of monsoon rains by end of June to middle of July. Generally during the normal monsoon years, sand reactivation along the eastern fringe of the desert stops by the end of June, while in the western part it ends by early to middle July. If the rains fail the strong wind continues to drift sand till the middle of August, or even September.

**Erosivity of wind :** For a better appreciation of the spatial pattern of wind strength to erode sediments, we measured the erosivity of sand-shifting wind through a climatic wind erosion index (WEI), using a modified Chepil et al. (1962) formulae (Yallon and Ganor, 1966):

$$WEI = (100^{*}V^{3})/(2.9\ (PE)^{2}),$$

where V is the mean wind speed at 10 m height, PE is Thornthwaite's measure of precipitation effectiveness, and 2.9 is the average climatic index at Garden City, Kansas (USA), which was retained only to compare the calculated values with those measured by the quoted authors in USA, Israel and elsewhere, but this is not mandatory. We replaced the annual wind summaries that were suggested by Chepil et al. (1962), with the wind data of March to July for all stations within Thar Desert, both in India and Pakistan, and classified the calculated WEI into six groups, ranging from very low to extremely high (Kar, 1993; Table 1).

**Table 1 :** Wind erosion index for meteorological stations in Thar Desert

| Wind erosion index (%) | Category | Station |
|---|---|---|
| 480 and above | Extremely high | Jaisalmer, Chhor |
| 120-479 | Very high | Phalodi |
| 60-119 | High | - |
| 30-59 | Moderate | Bikaner, Jodhpur, Barmer, Pachpadra, Hyderabad, Sukkur, Khanpur |
| 15-29 | Low | Ganganagar, Churu, Nagaur, Jaipur, Badin |
| 1-14 | Very low | Hisar, Sikar, Sambhar, Ajmer, Erinpura, Bahawalpur |

According to the above calculation, the maximum potential wind erosion is found to occur roughly to the southwest of Jaisalmer in Rajasthan, and adjoining parts of Pakistan. It decreases gradually towards east and northeast, concurrent with the increasing rainfall and decreasing wind speed. When we matched the pattern of this potential wind erosion with the observed spatial pattern of present-day dune mobility, we found that under a natural set up the high megabarchanoids occur in the extremely high wind erosion index zone (WEI= >480), and rarely beyond it, while clusters of barchans occur within very high erosion index zone (WEI= 120-479). In areas having <120 index values barchans have been noticed mostly under very high human pressures on the sandy landscape.

**Erodibility of sediments :** The most vulnerable landforms to wind erosion/ deposition are the sand dunes and other sandy landforms. Using the percent stable aggregates in a typical soil as the criteria for erodibility (Skidmore, 1988) we then determined the broad wind erodibility classes for western Rajasthan (Kar and Joshi, 1995; Table 2). With the help of wind erosivity and wind erodibility maps it is now possible to identify the major areas vulnerable to wind erosion (Kar, 1996).

**Table 2 :** Wind erodibility classes in western Rajasthan

| Soil character | Average stable aggregate (%) | Erodibility class |
|---|---|---|
| Very deep, fine sand of dune complex | 1-5 | Very severe |
| Deep, loamy fine sand and fine sand of sandy plains and some dunes | 5-10 | Severe |
| Shallow loamy fine sand on sandy plains | 5-8 | Severe |
| Loamy medium sand of small dunes on flood plains | 7-10 | Moderate |
| Deep brown sandy loam and loamy fine sand on flat alluvial plains | 10-15 | Moderate |
| Deep flood plain soils, fine sand, sandy loam and loam | 30-40 | Slight |
| Seirozem soils with sandy loam, loam, clay loam and silty clay loam | 20-50 | Slight |
| Medium to fine textured gray brown soil | 40-50 | Negligible |

## Quantification of Aeolian Dynamics at Local Scale

**Potential sand transport :** Once the broad spatial pattern was quantified, we calculated the average sand transport rate (Q; in t per m width) in the driest and the windiest westernmost area of the Thar, i.e., Jaisalmer district, using Bagnold's (1941) formulae:

$$Q = \alpha\, C \sqrt{(d/D)} * (\rho/g) * (v - V_t)^3,$$

where $\alpha$ is a constant, C is a sorting coefficient (1.8 for graded dune sand), d is average particle diameter of the locality, D is standard particle diameter (0.25 mm), $\rho$ is air density, g is gravitational constant, v is average wind velocity, and Vt is threshold wind velocity.

Particle size analysis of a large number of sandy landforms in Jaisalmer district, including sand dunes, interdune plains, other sandy plains and low sandy hummocks suggested that d could be taken as 0.22 mm (Kar, 1994, 2002). Following Bagnold (1941) we kept $V_t$ as 4 m $sec^{-1}$, although several other values are also quoted (Lettau and Lettau, 1969; 1978; Wasson and Nanninga, 1986; Cooke *et al.*, 1993).

Considering the above, and taking v as 39 km $h^{-1}$ for a typical sandstorm continuing for 6 h or more (as measured during the mid to late 1980s), we found a sand transport rate of 195 kg/ m width/ h, or 4.68 t/ m width/ day on a bare flat surface. Using the long-term average wind speed for the sand-shifting summer months the average sand transportation rate for the rocky terrain of Jaisalmer-Pokaran was then calculated as 35.16 t per m width per year, provided there is no vegetative cover. The values provided the basis for a conservative estimate of the mean deflation rate (Es) on a typically bare, deep, dry aeolian plain, using Wilson's (1971) formulae: Es = Q/x, where x is the saturation distance (i.e., 5 m over dry loose sand). We estimated the annual potential deflation as ~7 t $m^{-2}$, and the theoretical rate for lowering of the sandy surface as 3.38 m $y^{-1}$.

**Sand movement on a vegetated surface:** Apart from wind strength, mobility of sand is also influenced by the type, density and arrangement of plants, including its height, crown cover and basal cover. Buckley (1987) used the Bagnold (1941) formulae with plant cover data to quantify the sand transport rate (q) as:

$$q = [\{0.174/\log(z/k')\}^{3}*C\,v(d/D)\rho/g]*\ [\{V(1-kP)-V_t\}^{3}],$$

where k is a constant, and P is plant cover. We found that a 4-8% cover of erect and spreading plant type reduces sand transport by 30-50% of that on bare sand, assuming a wind speed of 10 m $s^{-1}$ (i.e., 36 km $h^{-1}$). A 17% plant cover reduces sand transport rate by 84%. With an average wind velocity of 15 m $s^{-1}$, the 17% plant cover can reduce sand transport to 22% of that under bare sand condition.

Using the Buckley equation for bare sand (P= 0) and a wind speed of 39 km $h^{-1}$ for 20-days we calculated the potential soil loss from a bare sandy plain near Jaisalmer as ~187000 t $ha^{-1}$, or a deflation of 9.50 m. If the deflation is restricted to 20 cm depth, the estimated sediment loss according to this calculation becomes 3941 t $h^{-1}$. The value agrees well with our field measurement near Jaisalmer during a sandstorm when a deflation of 15-20 cm took place, resulting in the loss of approximately 3754 t $ha^{-1}$ sandy soil. Further deflation was restricted by a hard layer of calcium carbonate nodules (kankar pan) at that depth.

Now, considering a 10 m deep sandy plain with a uniform thickness of aeolian sand, and a *Lasiurus sindicus* grass cover of 17%, the estimated sand transport for 20 days is 12.78 t per m width. Theoretically this should lead to a soil loss of 25554 t $ha^{-1}$ in 20 days, as against the calculated 187218 t $ha^{-1}$ from a barren, un-vegetated sandy plain. In other words, the 17% grass cover protects sand by more than 86% of that under bare sand situation. A *L. sindicus* cover of 4-8%, on the other hand, may protect the surface by 30-54% of that under bare sand (Kar, 2005 a). Thus, a 4-8% cover of erect and spreading plant type reduces sand transport by 30-50% of that on bare sand, assuming a wind speed of 10 m $s^{-1}$ (i.e., 36 km $h^{-1}$). A 17% plant cover reduces sand transport rate by >80%. With an average wind velocity of 15 m $s^{-1}$, the 17% plant cover can reduce sand transport to 22% of that under bare sand condition.

**Sand transport through migrating barchans :** Barchans are the fastest moving sand dunes in the desert, affecting roads, canals, settlements and other infrastructures. An effort was made to measure their rates of movement, especially on the shallow rocky terrain near Pokaran where these dunes form and migrate in a SW to NE direction (Kar, 1993, 1998; Ramakrishna et al., 1994).

We monitored two barchans of 2.29 m and 2.24 m height, respectively, for three days in July 1985, when a sandstorm with an average wind speed of 32-36 km per hour was sweeping the Pokaran-Jaisalmer area. Theoretical calculation with v= 34 km $h^{-1}$ provided a sand transport rate of 2.367 t/ m width/ day on the pedimented surface. Thus, in three days the sand transport on the flat pedimented surface was 7.101 t/ m width. To find out the sand transportation rate along the inclined surface of the barchans we used the following equation of Bagnold (1941):

$$Q' = Q/ (\cos \theta * (\tan \propto + \tan \theta),$$

where Q′ is sand transport rate along an inclined plane, $\propto$ is dynamic friction angle for sand, and $\theta$ is slope of the surface.

The rate of advancement of the barchans was then calculated using the Bagnold formulae in Wasson and Nanninga (1986) as:

$$c = (QT)/ ((\cos\theta (\tan \propto + \tan\theta))/\rho h),$$

where, c is dune advancement rate, T is sandstorm duration in hour, h is the crest height of the barchan in meters, and $\rho$ is density of the loosely packed sand.

With an average wind speed of 34 km $h^{-1}$ and h of 2.29 m, the Q¢ was found to be 0.122 t/ m width/ hour (Kar, 1994). The advancement of the 2.29 m high barchan was calculated as 0.027 m $h^{-1}$, or 0.65 m in a day. At this rate we assumed that the dune would advance by 1.95 m in three days. The actual advancement was measured as 1.7 m. The calculated rate of Q′ on the 2.24 m high barchan was 1.99 m in three days, while the measured distance was 2.2 m. These small variations could be due to the use of daily wind speed data from a station ~50 kms away at Jaisalmer as a result of non-availability of hourly data at site, as well as lack of precision in distance and height measurement during the severe sandstorms. However, it demonstrates the utility of the fundamental equations in estimating sand transport and dune mobility.

Repeated measurement of the two barchans between June 1985 and June 1989 revealed that the 2.29 m barchan moved by 137.4 m, and lost 0.9 m height. When we plotted the crest height of the barchan against its travel distance between June 1985, and June 1989, and fitted a second order polynomial curve to it, we found that the dune might disappear at a distance of ~630 m from its June 1985 position. As the dune continues to travel this distance, it should also speed up, provided the wind speed remains constant. Barchans are known to have a phase of initial growth along its travel route and then a phase of decline. If what we observed in field was a phase of decline, then the phase of growth might have been somewhere upwind of the stretch we monitored.

**Bulk sand transport :** We then calculated the bulk sand transport along this barchan-infested strip of 2500 m length and 250 m width. We counted 16 barchans in this strip, which were almost regularly spaced (average 156.25 m). The average size of the barchans was 2.10 m. Detailed morphometric measurements on an average sized barchan helped to calculate its area and volume (i.e., 384.16 $m^2$ and 355.87 $m^3$, respectively). Since the average density of sand in the locality is 1970 kg $m^{-3}$, the mass of the barchan turned out to be 701.06 tonnes (Kar, 2009 a). Using the size information on this barchan we estimated the streamer transport rate and the time needed by the barchans for renewal of sand.

We found that the volume of sand contained in the barchans in the studied strip, when spread evenly over the plain on which these are moving, could cover an estimated 911.03 mm thickness of the plain. The average streamer length in the corridor (and presumably in the locality as well) was approximately 3.0 km, which imply that if a streamer of sand is released more than 3.0 km upwind of the barchan field, there is less chance of its being intercepted for sustaining the dunes here. The free streamer transport per m width per year was calculated as 1.03 $m^3$. Since an average barchan had a width of 21.0 m, the expected interception of sand by a single barchan in a year was estimated as 21.0*1.03 = 21.63 $m^3$, and that by all the sixteen barchans together in the strip as 346.08 $m^3$ (Kar, 2009 a).

In order to maintain equilibrium each barchan has to release an equal volume of sand from its horn tips. The process leads to a slow renewal of the sand load being moved along the dune. We calculated the time required by an average sized barchan in the area to renew its sand load as 16.45 years. To achieve this feat, the barchan needs to move a distance of 526.40 m, and to roll over its material 17 times over this distance, before sand renewal takes place. Since the total length of the barchan corridor was 5.0 km, an average sized barchan possibly renewed its sand load 10 times while crossing the corridor (Kar, 2009 a). Although our calculations were based on the configuration of an average dune in the area and some simplistic assumptions on aeolian processes, the outputs have implications for estimating the dune age, dune migration, and a host of other related questions, especially for the less accessible fields of megabarchans that are slow-moving as compared to the small barchans (Kar, 1990), as well as for sand control measures (Kar, 1996).

## Quantification of Aeolian Dynamics at Regional Scale

**Spatio-temporal variability of wind erosivity in Thar Desert and its surroundings:** We have shown earlier in this chapter how the pattern of WEI constructed from station-based meteorological data can be used to make realistic estimates of the spatial pattern of aeolian sand reactivation and bedform perturbation in the desert. We prepared the first map of wind erosivity in Thar Desert based on the mean station data for 1931-60 for the undivided India. Presently the meteorological data for the region are available from several global repositories in gridded form, while IMD provides the data for the Indian part of the desert. Data for the Pakistan part is difficult to access from their meteorological department.

Taking gridded data from the global repositories we tried to find out if the pattern of WEI and implied sand reactivation could be reconstructed for all the years from 1951 to 2000 in the Thar Desert and its former maximum extent (called the Megathar), match them with the atmospheric dust emission pattern, and then to predict the future patterns of sand under a global warming scenario. For a critical analysis of the pattern, we divided the Thar Desert into two sub-zones (east and west), and the Megathar into four sub-zones (east, north, south and west). Each of the two sub-zones of the Thar Desert was again divided into three areas of north, central and south.

We sourced the gridded datasets on precipitation, temperature, evaporation and wind speed from GPCC, VASClimo, IPCC, UK-HMO, NCEP-NCAR, DKRZ-ECHAM, EU-ERA40, Willmott and Matsuura, etc., and randomly tested their consistency with the available IMD data for Indian part. Additionally a number of satellite-derived products on rainfall, temperature, NDVI and dust emission (e.g., from AVHRR, MODIS, TOMS, MISR, TRMM, SPOT, etc.) were also perused. All the data were processed and integrated in a GIS environment for mapping and analysis. Meteorological data from GPCC-VASClimo, Willmott & Matsuura (0.5° spatial grid) and ECHAM5 simulations for the 20$^{th}$ century (~1.87° spatial grid) were found to be robust for the 1951-2000 period, and formed basic inputs for the analytical results discussed here.

Our results on decadal averages show that during the 1950s the spatial average of WEI was 30-50% higher than the mean for the 50-year period in all the sub-zones except in Megathar-east where it was almost at per with the 50-year mean (Kar, 2009 b). Since then Megathar-west has experienced less erosivity than the 50-year mean, while Megathar-south has seen 45% higher erosivity during the 1980s, but less than the 50-year mean in other decades. Other sub-zones experienced 2-18% higher erosivity than the 50-year mean during the

1960s, and 20-40% lower erosivity during the1970s. The 1980s, which saw the 4-year long drought throughout the region, experienced up to 20% higher erosivity, followed in the 1990s by 10-40% lower values than the 50-year mean. Erosivity during individual years had a broad inverse relationship with previous year's rainfall condition, although few cases of mismatches were also noticed. The mis-matches were essentially because the summer wind strength in the region depends on either or both of (a) the dominance of westerly wind that pushes back the moisture-laden monsoon current from the Bay of Bengal to create drought, and (b) a strong pre-monsoon wind from the Arabian Sea. Overall, the very high erosivity during the 1950s has not yet been matched in any of the sub-zones. Another interesting find is a positive relationship between land surface dryness and erosivity. It is found that despite global warming, both the eastern and the western Thar have experienced a gradual decline in the dryness during the last 50 years, which is also the case in the western and the northern Megathar. By contrast, the southern and the eastern Megathar have registered an increase. This has implication for vegetation cover maintenance and erosion potentials.

**Linkages with dust emission pattern :** We then attempted to link the dust emission signatures from TOMS satellite sensor (measured as aerosol index, AI) for 1979 to 2000 (with a data gap from 1993 to 1996) with the calculated wind erosion index in different sub-zones. It showed very good correspondence between the two variables till the mid-1990s, but not afterwards. A detailed analysis for the Thar as a whole, relating previous year's total rainfall and current year's June wind speed with the AI for June revealed almost synchronous changes in the curves for previous year's total rainfall, current year's June wind speed and current year's June AI till the late-1980s. Since then the rainfall showed an increasing trend during the 1990s, and AI the hints of a decline, before the instrument failed in late-1992. A new instrument since 1997 recorded steep increase in AI despite a falling wind strength and variable rainfall, and reflected a greater control of critical land use changes in the sandy terrain through tractor use and expansion of cropland (Kar, 2008).

The TOMS has been phased out by 2005, but the reliable data was produced till 2002. OMI, which operated since 2005, with sensor capabilities comparable to that of TOMS, worked well till 2009. Several other satellite sensors are also capturing the signatures of atmospheric dust and aerosols since the beginning of this Century. This includes MODIS, MISR, AATSR, GOME, SCIYAMACHY, MERIS, POLDER, CALIPSO, etc. Among these the MISR (since 2000) with nine view angles and four spectral bands has high spatial mapping capability. The CALIPSO (Cloud-Aerosol Lidar and Infrared Pathfinder Satellite

Observation) is very powerful in providing data since 2006 on the vertical profile of a dust column (up to 5 kms), as well as discriminating between different types of aerosol and clouds, but it does not provide spatial data in the manner other satellite sensors do. Also, because of its configurations it has a very limited swath, and the revisit time is too long to capture the properly the passage of any dust storm activity. Hopefully, a 2-D dust/aerosol mapping sensor in UV-spectrum region will be added in the A-Train satellite constellation in future to trail the CALIPSO for synchronous imaging of the land.

While the satellite-derived atmospheric dust and aerosol mapping has its own merits and applications, our concern remains mainly with proper understanding and quantification of sand reactivation from the desert and low-level dust/sand emission. The primary reasons are the negative impacts of the enhanced sand mobility on agricultural and other land uses, as well as on soil fertility and infrastructures like roads, railways, canals, etc., within the desert and its eastern fringe. Recent studies by Bollasina et al. (2008, 2011) have shown that the pre-monsoon aerosol build-up from the Thar Desert and adjoining western India, as well as anthropogenic aerosol loadings, have much wider ramifications as it leads to reduced cloudiness, higher downward shortwave radiation, higher 2-m temperature and lower precipitation in the eastern part of the Indian Subcontinent and adjoining parts of Southeast Asia. Bollasina et al. (2011) attributed the gradual drying of the east-central India, covering Bihar, eastern UP, Jharkhand and north Odisha, and areas to the east of this zone to the increasing aerosol level as it slows down the tropical meridional overturning circulation and weakens the South Asian monsoon in general. Such findings warrant more in-depth studies on the relevant geomorphic processes responsible for sand and dust emission and effective integration of the derived land surface data with the satellite-derived dust/aerosol and meteorological data.

**WEI in future climates :** For the Twenty-first Century (2001-2100) we chose to visualize the pattern of WEI in five-yearly averages and under A2 scenario (Kar, 2009 b). We first mapped the simulated precipitation and temperature data for 2001-2005 from some of the available models and matched them with the pattern from ground observation data. We found the ECHAM5-AR4 simulation data more appropriate. Additionally we calculated WEI for 5-yearly blocks from 1951 to 2000 using ECHAM5-20C-3M data, and regressed the spatial pattern on our previously calculated pattern, which revealed tolerable correlation between the two datasets, except for 1951-55, and strengthened our selection.

Our results show that during the present century mean WEI was at its lowest in all the subdivisions of the Thar and the Megathar during

2001-05, except in the Megathar-east that will possibly see its lowest mean during 2011-15. Between 1951 and 2100 the lower mean WEI values were recorded during 1991-95 and 1996-2000, except in Megathar-north where the trends in the mean values are mostly out of phase with the trends in other subdivisions. Between 2016 and 2030 mean WEI is likely to be very high in all the subdivisions of the Thar and the Megathar, equaling or surpassing the earlier peaks of 1961-65 and 1986-90. Although the erosivity is expected to decline slightly in 2031-35, it may gradually increase again from 2036-40, attaining a peak in the middle of the century, followed by a series of high-magnitude changes to the end of the century. In between, 2071-75 may see a major fall in mean WEI in all the subdivisions, tending to match the values of 2001-05. We expected a synchrony between WEI and surface dryness, but it was not always so, suggesting a complex relationship between the variables (Kar, 2009 b).

Wind strength is expected to increase steadily over the decades. Broadly, the above trends is expected to sharply increase the atmospheric dust load in the region from 2016-20, which may get accelerated if the present groundwater-dependent irrigation system collapses, because large-scale clearance of natural vegetation, expansion of the cropland and deep ploughing of sandy land took place in the wake of irrigation. This has already brought a large area under the potential threat of wind erosion. Much more effort in land conservation practices would, therefore, be required than at present to stabilize the aeolian bedforms, both in the Indian part and the Pakistan part of the desert (Kar, 2009 b).

**Digital mapping of wind erosion:** The WEI mapping demonstrated the potential of this digital technique in providing an estimated pattern of likely sand reactivation under natural condition. Superimposed on this pattern is the pattern due to human pressure on the sandy landscape. Taken together, one forms intuitively a scenario of wind erosion in the desert. Conventionally the wind erosion mapping is carried out from visual interpretation of signatures on FCCs of satellite imagery. This, however, leads to subjectivity and interpreter's bias. To minimise the interpreter's bias in wind erosion mapping, we tested a number of digital techniques in the southern part of Churu district in eastern Thar (Kar, 2005 b, 2011). We used the digital data of Landsat MSS and Landsat ETM+ for the purpose.

The radiometrically and geometrically corrected images were submitted to the following three processing techniques: (1) Principal Component Analysis (PCA), (2) Soil Brightness Index (SBI) classification from the Tasseled Cap Transformation coefficients, and (3) Colorimetric Brightness Index (CBI) classification. PCI was calculated using the procedures in ERDAS (1997).

SBI values were calculated using the following formulae (Kauth and Thomas, 1976, for MSS; Huang *et al.*, 2001, for ETM+):

$SBI\ (MSS) = 0.406MSS4 + 0.600MSS5 + 0.645MSS6 + 0.24MSS7$

$SBI\ (ETM) = 0.356ETM1 + 0.397ETM2 + 0.390ETM3 + 0.697ETM\ 4 + 0.229\ ETM5 + 0.160ETM7$

To convert the SBI values into sand reactivation categories, the DN ranges of some known sand-reactivated areas were measured from standard FCCs, and their corresponding SBI values were found out for calibration.

CBI was calculated using Mathieu *et al.* (1998) formulae:

$$CBI(MSS) = \sqrt{\frac{MSS4^2 + MSS5^2 + MSS6^2}{3}}$$

$$CBI(ETM) = \sqrt{\frac{ETM1^2 + ETM2^2 + ETM3^2}{3}}$$

The raw CBI values were then converted into sand reactivation categories using the protocol used for SBI.

Additionally, supervised classification of the scenes was performed, using maximum likelihood algorithm (Lillesand and Kiefer, 1987, for procedure), for which site-based information on the land cover characteristics, degree of reactivation, etc., from a number of localities were gathered through field campaigns. Results from PCI, SBI and CBI classifications were evaluated with respect to the output from supervised classification.

Digital image maps of surface brightness from PCI, SBI and CBI yielded broadly similar results. It was found that the PCA output, especially FCC of PC bands 1, 2 and 3, reasonably discriminated aeolian sand from fluvial sand, possibly because the aeolian sand rests in a matrix of fine silt and clay particles and often with a coating of silt/clay, which reflect less than the un-coated sand, while the fluvial sand gets washed out from the mixed deposits and appears brighter. PCA, however, did not help in recognition of sand reactivation categories.

Results from SBI and CBI classifications were almost the same (<5% difference between the two), but both suffered from a mixing of signatures with non-aeolian reactivation units. It was noticed that the almost overlapping radiance values from the following surface features influenced the accuracy of classification of the reactivated aeolian sand categories: riverine sand, dry gypsiferous surface, other dry salt-affected plains, and dry surface of salt lakes, all of which had higher brightness than the aeolian sand-reactivated areas. This pattern was also noticed during field campaign in several parts of the desert, when radiance from the variously reactivated aeolian sand surfaces, gypsiferous surfaces, salt-affected plains, natural vegetation at different states of degradation, agricultural fields with different crop varieties, etc., were measured at visible as well as near and middle infra-red wavelengths.

Although it is known that alkaline surfaces could be better differentiated from the saline and non-affected surfaces in the middle infrared band (10.4 to 12.5 μm), and that a standard FCC can improve separation of saline areas from non-saline ones, we decided to explore the possibility of density slicing of the SBI and CBI images at the critical limits of separation between aeolian-bright and other bright areas. This decision was based on knowledge of the tonal characteristics of a typical sand-reactivated area and a typical gypsiferous surface on standard FCC. A very thin pixel continuum was found between the highly reactivated sand category and the bright salt-affected surface, but the modal part of the two distributions could be separated appreciably. The brightness of dry surface of salt lakes were eliminated by first digitizing the lake boundaries in a vector environment, rasterization of the resultant map to fit it into the image, and then masking out of the brightness.

In spite of the near-similitude of results from SBI and CBI, the latter was preferred over the former because SBI calculation depends on calibration constants of the different wavelength bands in the sensor, which are difficult to obtain always. Also, when a satellite sensor quality degrades over time, one has to seek new calibration coefficients to calculate the values, and thus depend on results from repeated ground monitoring of some fixed objects, to be carried out by the satellite stations over a period of time. CBI calculation, on the other hand, does not require the satellite-dependent coefficients, and is based on the simple logic that aeolian sand surfaces undergo changes in colour with the degree of reactivation. This can be noticed during any visit to an arid sandy tract. Supervised classification needs accurate ground information for the training sites, and hence is ruled out for areas where ground information is scarce or, difficult to obtain. Misinformation from a few training sites may lead to large errors.

Using CBI classification we first mapped the sand reactivation pattern in January 1973 and January 2001, and then carried out a comparative study of sand reactivation pattern in the three tehsils during the period. It was found that the area experienced moderate to severe drought since the previous summer i.e., 1972 and 2000, with an average rainfall of 184 mm and 273 mm, respectively, against an annual average of 366 mm. Therefore both the scenes were expected to be similarly vulnerable to wind erosion. Image differencing for the reactivated sand categories did not, however, reveal much change in the area statistics, but significant spatial variability was noticed in sand reactivation categories from 1973 to 2001 (e.g., in Sujangarh tehsil). We co-evaluated the radiance images of the area with the CBI images to find out if there was any similarity in the changes in both the types. It was found that much of the eastern half, recording radiance values of <11 mW $cm^{-2}$ $Sr^{-1}$ $\mu m^{-1}$, had lower sand reactivation in 2001 than in 1973. This was due to more than a decade-long practice of groundwater-irrigation that had stabilized the landscape to a certain degree, despite the concomitant increase in cultivation pressure since 1973. The westernmost parts of Dungargarh and Sujangarh tehsils, on the other hand, showed large areas under moderate to severe sand reactivation. Yet, higher radiance (>11 mW $cm^{-2}$ $Sr^{-1}$ $\mu m^{-1}$) was especially concentrated in the north-west, where groundwater irrigation started late and deep ploughing of sandy soils using tractors had spread (Kar, 2011). Overall, 36% area did not show any change in sand reactivation category during the period, while 52% showed positive changes in reactivation and 12% negative changes (mainly in the eastern part).

The relative stability of the landscape in the eastern part as compared to the western part, despite groundwater irrigation and tractor ploughing in both the parts, may be interpreted in terms of lag effect of the landscape response to the human-induced changes. The sandy terrain in the eastern part experienced the changes in agricultural technology at least three decades ago, while such changes in the western part started taking place from the mid-1990s when a good source of groundwater was explored in parts of Sardarshahr, Dungargarh and Sujangarh tehsils. Vast sandy landscape of the interdune plains in the eastern part appears to have gradually adjusted to the changes brought in by irrigation and in its wake the tractor ploughing. Intensive cropping and application of water and fertilizer gradually helped to minimize sand reactivation from interdune plains, although pressure on dune slopes continued. Using this argument, and provided irrigation and tractorization continue to expand together at the same pace, and other pressures do not accelerate, it is likely that the sandy landscape in the western part will take at least another three decades to adjust to such a scenario of irrigated agriculture. Since the sandy terrain in the west has higher wind erosivity than in the east (Kar, 1993), the operations related to expansion of irrigated cropping there would most likely increase the vulnerability to wind erosion for a longer period of time before landscape stability sets in (Kar, 2011).

The brightness mapping, however, provides only an estimation of sand reactivation, especially as fresh sand deposition, but not deflation, which is also an important component of wind erosion. Quantum of net sand accumulation/deflation cannot be assessed properly until a very high-resolution digital elevation model (DEM) of the area is prepared and changes are monitored at certain time interval. This could be done either through rigorous ground-based data collection using differential GPS at some fixed benchmark sites (e.g., on a sand dune; Kar et al., 1998), or from high-resolution satellite remote sensing at periodic intervals.

There is another aspect to be resolved. While carrying out CBI mapping at regional scale, using AWIFS data we encountered the problem of higher brightness from gypsiferous and silt-rich sediments in large parts of north-western Thar. While small areas of higher brightness in salt ranns and similar surfaces can be masked, discriminating signatures of actual sand reactivation from non-such in silt-rich sediments within a large aeolian sand-covered area proved difficult. Hopefully we shall resolve this problem soon.

Yet another problem is the choosing of the scenes for mapping. So far we have used the images of dry cool period to map the phenomena, assuming that the scenes retain the residual of high sand reactivation that could not be stabilized during the preceding monsoon rains. Such mapping does not consider the actual scale of reactivation during a year, and hence, cannot be reasonably matched with the pattern of WEI. For understanding the intra- and inter-annual variability of wind erosion we need to find out a way to capture the essence of each year's sand reactivation pattern through time-series analysis.

**Other questions involved :** Quantification of regional-scale sand mobilization, as well as sand and dust emission and transport, need more data on several parameters. We have shown how the fundamental equations can be used to estimate the sand transport rates and bedform migration. There is scope for using such equations on gridded/pixel-based data. The modern era remote sensing is providing many of the required inputs directly or through interpretation of signatures. Global repositories of modelled earth science data are also generating numerous gridded spatial data. Geo-spatial modelling based on fundamental principles of sand reactivation and transport and linking them with the satellite signatures will be helpful in addressing the issues.

## Epilogue

Considering that the humans have become a major driving force in aeolian process acceleration and sand re-mobilization in the Thar Desert, at least since the second half of the last century, there has to be a major focus on understanding in quantitative terms the sensitivity of the different landscape elements to the changes, as well as types of responses provided by the aeolian system, and the likely fall out of the system's response. This calls for a detailed behavioral analysis of the inter-related parameters of the aeolian landscape under pressure, especially involving measurement of the rates of processes and sedimet flow, their behaviour under induced conditions, etc. The understanding should lead to proper modelling of the likely future scenarios under global climate change, and under population pressure. In the arid zone abrupt changes of climate are likely to cause more aberrant weathers, with increased frequencies of drought and flood, leading to an acceleration of both aeolian and fluvial processes that would impact the densely populated agrarian society. It is also time to monitor the changes in the surface topography and the fluxes, using both remote sensing and ground-based sensors to suggest how best the land users can deal with the changes to avoid catastrophic situations.

## References

Bagnold, R.A. 1941. *The Physics of Blown Sand and Desert Dunes*. Methuen, London, 265p.

Bollasina, M., Ming, Y. and Ramaswamy, V. 2011. Anthropogenic aerosols and the weakening of the South Asian summer monsoon. *Science*, 334: 502-505.

Bollasina, M., Nigam, S. and Lau, K.-M. 2008. Absorbing aerosols and summer monsoon evolution over South Asia: An observational portrayal. *Journal of Climate*, 21: 3221-3239.

Buckley, R.C. 1987. The effect of sparse vegetation on the transport of dune sand by wind. *Nature*, 325: 426-428.

Chepil, W.S., Siddoway, F.H. and Armbrust, D.V. 1962. Climatic factor for estimating wind erodibility of farm fields. *Journal of Soil and Water Conservation*, 17: 162-165.

Cooke, R.U., Warren, A. and Goudie, A.S. 1993. *Desert Geomorphology*. University of London Press, London, 526 p.

ERDAS. 1997. *ERDAS Field Guide*. ERDAS, Inc., Atlanta, 656 p.

Huang, C., Wylie, B., Yang, L., Homer, C. and Zylstra, G. 2001. Derivation of a Tasseled Cap Transformation based on Landsat-7 at-satellite reflectance. http://landcover.usgs.gov/pdf/tasseled. pdf. *U.S. Geological Survey*, 10 p.

Kar, A. 1990. The megabarchanoids of the Thar: Their environment, morphology and relationship with longitudinal dunes. *The Geographical Journal*, 156 (1): 51 61.

Kar, A. 1993. Aeolian processes and bedforms in the Thar desert. *Journal of Arid Environments*, 25: 83-96.

Kar, A. 1994. Sand dunes and their mobility in Jaisalmer district. In: *India - Geomorphological Diversity* (Eds., K.R. Dikshit, V.S. Kale and M.N. Kaul). Rawat Publishers, Jaipur, pp. 395-418.

Kar, A. 1996. Morphology and evolution of sand dunes in the Thar desert as key to sand control measures. *Indian Journal of Geomorphology*, 1: 177-206.

Kar, A. 1998. Present-day mobile crescentic dunes in the Thar Desert, India. In, *Quaternary Deserts and Climatic Change* (Eds., A.S. Alsharhan, K.W. Glennie, G.L. Whittle and C.G.St.C. Kendall). A.A. Balkema, Rotterdam (Netherlands) and Brookfield (USA); pp. 155-164.

Kar, A. 2002. Granulometry of aeolian bedforms in the extremely arid Jaisalmer region of Thar Desert. In: *Recent Advances in Geomorphology, Quaternary Geology and Environmental Geosciences: Indian Case Studies* (Eds., S.K. Tandon and B. Thakur). Manisha Publications, New Delhi; pp. 261-286.

Kar, A. 2005a. Role of shrubs in wind erosion control. In, P. Narain, M. Singh, M.S. Khan, and S. Kumar (eds.), *Shrubs of Indian Arid Zone*. Central Arid Zone Research Institute, Jodhpur, pp. 121-124.

Kar, A. 2005b. Digital remote sensing of wind erosion in the Thar Desert. *The Indian Journal of Geography*, 16: 1-6.

Kar, A. 2008. Climate change impact on the Thar Desert. *Geography and You*, September-October: 36-40.

Kar, A. 2009 a. An analysis of climate change through quantification of forms and processes in the Thar Desert. In, *Earth System Sciences, Volume 2* (Eds., A. Kumar, R.A. Kushwaha, and B. Thakur). Concept Publishing Co., New Delhi, pp. 363-381.

Kar, A. 2009 b. Wind erosion potentials in the Thar-Megathar region from AD 1951 to 2100: Patterns and implications. *Abstracts, International Conference on Nurturing Arid Zones for People and the Environment: Issues and Agenda for the 21st Century*. Central Arid Zone Research Institute, Jodhpur, pp. 24-25.

Kar, A. 2011. Geoinformatics in spatial and temporal analysis of wind erosion in Thar Desert. In, *Geoinformatics in Applied Geomorphology* (Eds., S. Anbazhagan, S.K. Subramanian and X. Yang). CRC Press, Boca Raton, USA, pp. 39-62.

Kar, A. and Joshi, D.C. 1995. Sand movement and control of aeolian hazard. In, A.K. Sen and Amal Kar (eds.), *Land Degradation and Desertification in Asia and the Pacific Region*. Scientific Publishers, Jodhpur; pp. 19 40.

Kar, A., Tsunekawa, A. and Miyazaki, T. 1998. Potentiality of global positioning system in sand dune measurement: A case study from the Thar Desert, India. In, *Quaternary Deserts and Climatic Change* (Eds., A.S. Alsharhan, K.W. Glennie, G.L. Whittle and C.G.St.C. Kendall). A.A. Balkema, Rotterdam (Netherlands) and Brookfield (USA), pp.

Kauth, R.J. and Thomas, G.S. 1976. The Tasseled Cap - a graphic description of the spectral-temporal development of agricultural crops as seen by Landsat. *Proceedings of the Symposium on Machine Processing of Remotely Sensed Data*. Purdue University, Indiana, pp. 4B41-4B51.

Lettau, H.H. and Lettau, K (eds.). 1978. *Exploring the World's Driest Climate*. IES Report 101, Center for Climatic Research, University of Wisconsin-Madison, Madison, 264p.

Lettau, K. and Lettau, H. 1969. Bulk transport of sand by the barchans of the Pampa de La Joya in southern Peru. *Zeitschrift fur Geomorphologie*, 13: 182-195.

Lillesand, T. M. and Kiefer, R. W. 1987. Remote Sensing and Image Interpretation. John Wiley & Sons, New York, 721 p.

Mathieu, R., Pouget, M., Cervelle, B. and Escadafal, R. 1998. Relationships between satellite-based radiometric indices simulated using laboratory reflectance data and typic soil color of an arid environment. *Remote Sensing of Environment*, 66: 17-28.

Ramakrishna, Y.S., Kar, A., Rao, A.S. and Singh, R.S. 1994. Micro-climate and mobility of a barchan dune in the Thar desert. *Annals of Arid Zone*, 33: 203-214.

Skidmore, E.L. 1988. Wind erosion. In: *Soil Erosion Research Methods* (Ed., R. Lal). Soil and Water Conservation Society of America, pp. 203-233.

Wasson, R.J. and Nanninga, P.M. 1986. Estimating wind transport of sand on vegetated surfaces. *Earth Surface Processes and Landforms*, 11: 505-514.

Wilson, I.G. 1971. Desert sandflow basins and a model for the development of ergs. *Geographical Journal*, 137: 180-199.

Yaalon, D.H. and Ganor, E. 1966. The climatic factor of wind erodibility and dust blowing in Israel. *Israel Journal of Earth Sciences*, 15: 27-32.

# Chapter - 2

# Water Management and Information System

**(India-WRIS Web GIS: Design and Development of Web Enabled Water Resources Information System of India)**

**J.R. Sharma**

## Abstract

*In the emerging era of wide spread use of IT tools, up-to-date information on water resources is the key to support economic development, conserve the nature as well as to improve the quality of life. Due to rapid rise in population and food demand, growing economy and improving living standards; the pressure on our water resources is increasing and it will become scarce in the coming decades if not managed at this stage in a integrated way. Water resources management requires a multi-disciplinary approach that combines a collection of technical tools and expertise along with stakeholders of diverse interests and priorities in planning and management and an operational water resources information system at national level is the first requirement.*

*The paper highlights the project India-WRIS WebGIS, jointly executed by ISRO and Central Water Commission and aims as 'Single Window' solution of all water resources & related data and information in a standardized GIS format to all concerned departments, organizations and stakeholders. The thin client scalable web enabled information system provides comprehensive, authoritative and consistent data of India's water resources along with allied natural resources data & information, web enabled tools to search, access, visualize, understand, look into context and study the spatial patterns. Based on the requirements and data availability, comprehensive information have been collected, thought fully categorized and arranged in GIS environment under 12 major and 30 sub information systems besides base sub information system having large number of attributes data of last 5 - 50 years. India-WRIS Web GIS has been designed using latest web 2.0 technologies which provides various GIS tools in a rich GUI assisted format for easy access and use.*

## 1.0 Introduction

In the emerging knowledge society and wide spread use to IT tools in different sectors, up-to-date information on water resources is the vital to support economic development, improve the quality of life as well as to conserve the nature and the environment. In this regard an operational water resources information system at national level is essential for planning and development of the country.

Decision-making processes in water resources management are characterized by a typical hierarchical structure and a high degree of complexity (Kaden et al., 1989). Water resources planning require a multi-disciplinary approach that brings together a collection of technical tools and expertise along with stakeholders of varied interests and priorities. Generally, the water management scenario is designed and influenced by a set of linked physical, biological, and socio-economic factors such as surface water hydrology, groundwater hydrology, climate, soils topography, land use, water quality, ecosystems, demographics, institutional arrangements and infrastructure (Biswas, 1981; Loucks, 1995;Bouwer, 2000; Zalewski, 2002). An information system is a set of data and functions, which is developed and disseminated to meet the needs of users. Key functions of an information system are viewing, presentation, interpretation, analysis and modeling of data (Maidment, 1997).

India-WRIS WebGIS portal aims a 'Single Window' solution of all water resources and related data & information in a standardized GIS format in national framework to all departments, organizations and stakeholders for water resource assessment, monitoring, water resource planning, development and integrated water resources management. It provides comprehensive, authoritative and consistent data of India's water resources along with allied natural resources data & information, web enabled tools to search, access, visualize, understand, look into context and study the spatial patterns. The WebGIS portal is designed and developed keeping in view multi-stakeholder users from all sections of society, varied and multi-source data input, current map policy, requirement of regular updates and near real time data accessibility, data security domains, scale of information and level of access of the portal as well as download of different GIS maps, data and value added products along with tool kit for further analysis and value addition under three user categories namely:

- All General Users (public domain fast track system)- Users are able to visit web portal and get the snapshots of the data sets on reduced scale of selected database and tools.
- Premium Users- Users are able to get the access to the India-WRIS web application detail datasets and tools by registration and password.

- CWC Intranet Users - These privileged users are able to get the full access to the India-WRIS web application and database. All the facilities developed are accessible by these users.

The information system on water resources has four key elements besides other facilities namely:

1. Data input/entry/collection system
2. Data storage, analysis, and transformation into 'user friendly' information
3. Interactive system for geo-visualization and temporal analysis and
4. Information dissemination system in public domain as downloads and further processing tools for value addition and customization.

Storage, processing, retrieval and dissemination of data constitute the most important aspects of the water resources management being multi stakeholder's involvement, people's participation and information sharing to increase transparency, public awareness, elevating the importance of water information and enlighten public involvement in water management. The thrust has been towards development of an open source user end web enabled information system. It provides adequate and contemporary information on the state of water resources which are must for planning and water resources management strategy. This, in turn, will ensure increase in public awareness about the crucial issues related with water and attract their participation in management, planning and development of water resources of the nation leading towards the holistic goal of water security.

## 2.0 Water Wealth of India

Water is one of the most important renewable natural resources for supporting life. With the increasing population of India as well as its all-round development, the utilization of water is also increasing at a fast pace. On an average, India receives annual precipitation (including snowfall) of about 4000 km$^3$. However, there exist considerable spatial and temporal variations in the distribution of rainfall and hence in availability of water in time and space across the country. It is estimated that out of the 4000 km$^3$ water, 1869 km$^3$ is Average annual potential flow in rivers available as water resource. Out of this total available water resource, only 1123 km$^3$ is utilizable (690 km$^3$ from surface water resources and 433 km$^3$ from ground water resources). The water demand in the year 2000 was 634 km$^3$ and it is likely to be 1093 km$^3$ by the year 2025. Due to rapid rise in population and growing economy of the country, there will be continuous increase in demand for water, and it will become scarce in the coming decades. (Table - 1)

According to the international norms, a country can be categorized as 'water stressed' when water availability is less than 1700 $m^3$ per capita per year whereas classified as 'water scarce' if it is less than 1000 $m^3$ per capita per year. In India, the availability of surface water in the years 1991 and 2001 were 2309 $m^3$ and 1902 $m^3$. However, it has been projected that per capita surface water availability is likely to be reduced to 1401 $m^3$ and 1191 $m^3$ by the years 2025 and 2050, respectively. The Per capita water availability in the year 2010 was 1588 $m^3$ against 5200 $m^3$ of the year 1951 in the country.

**Water Resources - India at a Glance**

- Area of the country as % of World Area : 2.4%
- Population as % of World Population : 17.1%
- Water as % of World Water : 4%
- Rank in per capita availability : 132
- Rank in water quality : 122
    - Average annual rainfall 1160 mm ( world average 1110 mm)
    - Range of distribution 150 -11690 mm
    - Range Rainy days 5-150, most rain 15 days in 100 hrs.
    - Range PET 1500 -3500 mm
    - Per capita water availability (2010) in $m^3$-1588

**Table - 1 :** India's water wealth

| Sl. No. | Water Resource at a Glance | Quantity ($km^3$) | Percentage |
|---|---|---|---|
| 1 | Annual precipitation (Including snowfall) | 4000 | 100 |
| 2 | Precipitation during monsoon | 3000 | 75 |
| 3 | Evaporation + Soil water | 2131 | 53.3 |
| 4 | Average annual potential flow in rivers | 1869 | 46.7 |
| 5 | Estimated utilizable water resources | 1123 | 28.1 |
| | Surface water | 690 | 17.3 |
| | Replenishable groundwater | 433 | 10.8 |
| | Current utilization of total | 634 | 15.85 |
| | Current utilization of utilizable water | 634 | 56.45 |
| | Storage created of utilizable water | 225 | 20.03 |
| | Storage (under construction) of utilizable water | 171 | 15.22 |
| 6 | Estimated water need in 2050 | 1450 | 129 |
| 7 | Estimated deficit | 327 | 29 |
| | Interlinking can give us | 200 | 17.8 |

*Source: Water Resources at a Glance 2011, CWC, New Delhi, (http://www.cwc.nic.in)*

## 3.0 Remote Sensing and GIS in Water Resources studies

The effective utilization of satellite remote sensing and GIS in water resources information generation and management can be broadly categorized as follows:

- Water Resources Assessment - It includes snow & glacier studies, surface water mapping and monitoring, wetlands mapping, runoff & hydrologic modeling and water balance studies.
- Water Resources Management - It comprises of irrigation water management, salinity and waterlogged area mapping & monitoring, monitoring new irrigation potential creation, evapo-transpiration studies, reservoir management, reservoir sedimentation and catchment area treatment
- Water Resources Development - It covers interlinking of rivers, ground water prospecting and recharge - systematic planning & development.
- Watershed Management - It contains water harvesting and soil erosion, watershed management and sustainable action plans for soil and water conservation.
- Flood Disaster Support - It includes flood disaster monitoring and management, flood forecasting, river engineering, urban flood management.
- Environmental Impact Assessment & Management - It is related to hydro-power development and water quality studies, identification & mapping of overexploited areas and augmentation of the resource.

## 4.0 Water Resources Management - Challenges / Issues in the Country

The major challenges/issues associated with the water resources management and development in the country are varied and complex and could be categorized as follows:

1. **Natural situation (Tropical Monsoon climate) -** Causes large scale spatial and temporal variation in water availability, recurring droughts and frequent floods.
2. **Human, Managerial and Developmental challenges -** There is increasing water demand and falling per capita availability, water use and energy efficiency, deterioration of water quality, reduction or deterioration of available resources (loss of surface storage), increasing competition/conflict within sectors, under and inefficient utilization of irrigation potential, over exploitation and depletion of ground water resources,

waterlogging and soil salinity in irrigated lands, fragmentation of management of water / management of shared resources, lack of spatial inventory for large number of water infrastructure in the country, currently used water resources potential estimates (CWC: 1988 & 1993 ; NCIWRD:1999) any, significant change in land use / land cover, demographic and utilization pattern in past few decades.

3. **Climate change impact** -Addressing the impact of climate change on water availability and economy. Analysis of scenarios for impacts on resources and use is required to evaluate water policies.

## 5.0 The Project India -WRIS

In view of the present status of water resources and growing demands of water for meeting the requirements of the fast increasing population as well as the difficulties that are likely to arise in future, a holistic, well planned long-term strategy is needed for sustainable water resources management in India. In order to address challenges the first requirement is reliable and updated information system and hence, the project "Generation of Database and Implementation of Web enabled Water resources Information System in the Country" (India-WRIS WebGIS) is being jointly executed by ISRO and Central Water Commission.

The objectives of the project are as follows:

- To collect available data from varied sources, generate database of country's water resources, organize in standardized GIS format and provide a thin client scalable web-enabled information system.
- To provide easier and faster access and sharing of nationally consistent and authentic water resources data through a centralized database and application server to all water resources departments / organizations as decided by CWC.
- To provide tools to create value added maps by way of multi-layer stacking of GIS database so as to provide integrated view to the water resources scenarios.
- To provide foundation for advanced modeling purposes and future Spatial Decision Support Systems (SDSS) including automated data collection system.

Based on the requirements and data availability, comprehensive information have been collected, thought fully categorized and arranged in GIS environment under 12 major and 30 sub information systems besides base sub information system having large number of attributes data of last 5 - 50 years. (Table-2)

**Table-2:** Main and sub Information Systems of India-WRIS

| Sl. No. | Main and Sub Information Systems |
|---|---|
| I. | BASE DATA INFO SYS<br>• Administrative, Infrastructure, Terrain (DEM) |
| II. | SURFACE WATER INFO SYS<br>1. Water Resource Region Info Sys<br>2. Basin Info Sys<br>3. Watershed Info Sys<br>4. River Info Sys<br>5. Surface Water Body<br>6. Water Resources Projects Info Sys<br>7. Command Area Info Sys<br>8. Minor Irrigation Info Sys<br>9. Canal Info Sys |
| III. | GROUND WATER INFO SYS<br>10. Aquifer / Litholog Info Sys<br>11. Ground Water Level Info Sys<br>12. Ground Water Potential Info Sys |
| IV. | HYDRO - MET INFO SYS<br>13. Meteorological Info Sys<br>14. Climate Info Sys<br>15. Hydro-Observation Info Sys |
| V. | WATER QUALITY INFO SYS<br>16. Surface Water Quality Info Sys<br>17. Ground Water Quality Info Sys |
| VI. | SNOW COVER / GLACIER INFO SYS<br>18. Snow Cover/Glacier Info Sys |
| VII. | INLAND NAVIGATION WATERWAYS INFO SYS<br>19. Inland Navigation Waterways Info Sys |
| VIII. | INTER-BASIN TRANSFER LINKS INFO SYS<br>20. Inter-basin Transfer Links Info Sys |
| IX. | HYDROLOGICAL EXTREMES INFO SYS<br>21. Flood Info Sys<br>22. Drought Info Sys<br>23. Extreme Events Info Sys |
| X. | LAND RESOURCES INFO SYS<br>24. Land Use / Land Cover Info Sys<br>25. Land Degradation Info Sys<br>26. Wasteland Info Sys<br>27. Soil Info Sys |
| XI. | WATER TOURISM INFO SYS<br>28. Water Tourism Info Sys |
| XII. | SOCIO - ECONOMIC INFO SYS<br>29. Rural Info Sys<br>30. Urban Info Sys |

## 6.0 India-WRIS Web GIS Application Architecture (Technologies & Tools)

The three components India-WRIS Web GIS Application are:

1. **Database Design & Generation :** The database for India-WRIS is highly complex with numerous sources involved. Much of the data is spatial in nature but the amount of associated data is very large and also having time series, and will further increase exponentially with the passage of time. The creation and management of such data is a colossal feat in itself and requires state of the art tools. The database standards and relationship have been developed for all type of data. The database generation software used have the capabilities of creating maps, viewing or exploring data, editing data, storing, conflation (integrating datasets from different sources), transforming (into different coordinates systems, different representations, re-sampling, resulting in new representation/ format of the same data), querying, analyzing etc.

   The database generation software used in India WRIS are AutoCAD Map, ERDAS IMAGINE, ESRI ArcGIS and IGIS. Along with these other utilities like MS Excel, Mat lab are used. Most of the dataset in India-WRIS is vector in nature, the software that has special tools to handle, edit and manipulate vector data is given preference. Since the data is massive in size, the software used should have inherent properties to handle the creation and maintenance of data in multi user environment. ESRI's ArcGIS suite of products is especially capable in this regard. ArcGIS also has capabilities to create Geo Databases which is a RDBMS file format for GIS data allows India-WRIS database to be manageable, scalable, compatible and easy to write queries for fast retrieval of results through web based applications over internet. (Table-3)

2. **Web application & User interface technology :** The major user requirement from the web portal is data dissemination; hence advance GIS data processing systems at the back end augmented with the best database connectivity over the internet is used so that the user is able to get intuitive and real time information.

   User has the facility for data visualization, analysis on the client side and use further to create customized reports. Adobe Flex is able to deliver Rich Internet Application (RIAs) across the enterprise and over the web efficiently. Using the Flex API, India-WRIS combines GIS based Web services from ArcGIS Server with other Web content, which are displayed in simple, dynamic mapping applications over the Web. All the published map services are compliant with OGC standards and the services can be accessed using WMS, WFS, WCS and KML standard formats. India-WRIS system is using Oracle 11g, relational database management system (RDBMS) which supports multi user system. ArcSDE as well as oracle together used to handle geospatial data and to create multiuser geo-database. (Figure-1)

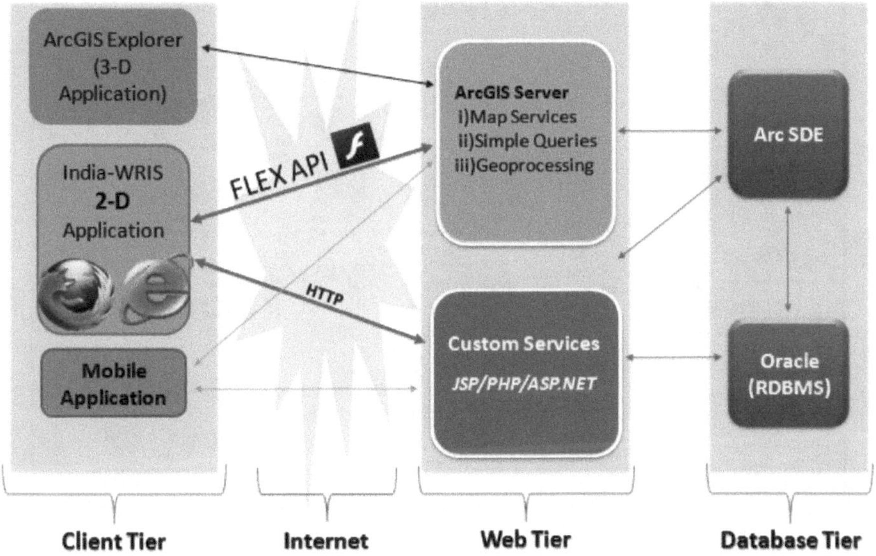


**Figure 1 :** Web Application Architecture

**Table-3:** Tasks and Software/ Technologies Used for India -WRIS

| Tasks | Software / Technologies | |
|---|---|---|
| WebGIS Front end | 2D | Adobe Flex, HTML, PHP |
| | 3D | ArcExplorer, .NET, ArcGlobe |
| | Mobile | .NET(Windows) / Flex(Android) / JAVA(Symbian) |
| | Meta Data | Visual Basic |
| Data Generation (Digital Image Processing / GIS Mapping) | ERDAS Imagine, ENVI, ArcMap, ArcCatalog, ArcSDE, AutoCAD 3D, IGIS, Map Window Library, GeodatabaseXML | |
| Publishing / Web Geodata Services | ArcGIS Server | |
| Geodatabase /Back end | Oracle 11g, MySQL | |

3. **Database storage & hosting :** In order to ensure reliable and secure, 24 x 7 availability of the WebGIS, a robust hosting architecture has been designed. The same has been replicated at three places namely, RRSC (West) - Jodhpur, the data generation and software development as lead centre; NRSC - Hyderabad for web hosting and CWC - New Delhi for intranet users and data validation & updation. (Figure-2)

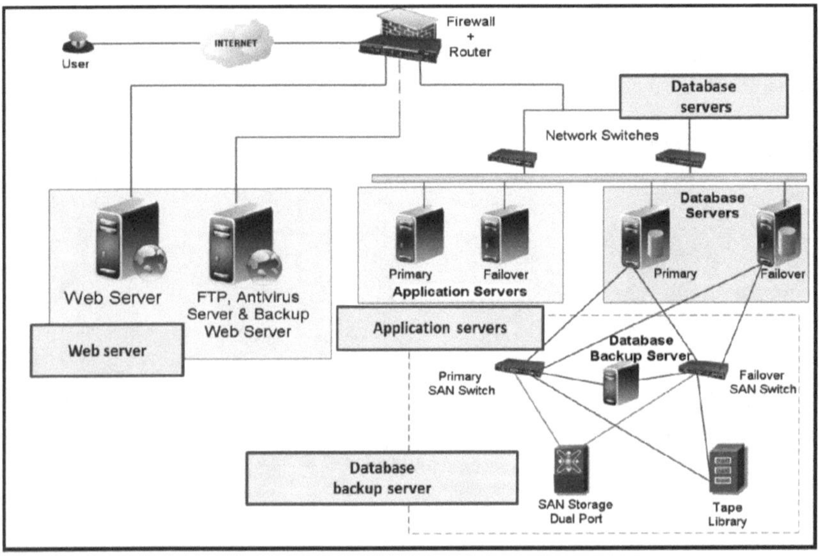


**Figure 2 :** Web hosting architecture

## 7.0 Designing the User Interface, Tools and Facilities in India-WRIS Web GIS

Considering large number of factors as; type and volume of data, large number of varied users, ease of handling, varied nature of internet connectivity in the country, information requirement by the users and available technologies; Utmost care has been taken by the WRIS team to design the user interface of the portal (Figure-3).

The home page is divided into four sections:

### 7.1 Universal toolbar

This universal toolbar is present at the top of the page. This has two sections:

First, the toolbar at the top of the page contains the links to pop-up window having information required by the user at any point and toolbar is visible at all times. The main links in this toolbar are:

- **About WRIS :** This page contains a brief overview of India-WRIS project including its history, scope, vision, goals, deliverables and time-frame for completion.
- **Accessibility :** This provides information for navigating through India-WRIS like screen resolution, keyboard shortcuts for easy navigation etc.
- **Tools :** Numerous tools along with symbols for easy access are described in this part. (Table -4)

**Figure 3 :** Home page and Front GUI of India-WRIS Web GIS

- **WRIS Mobile :** A precise version of India-WRIS has also been developed for mobile and handheld devices. This link provides more information and how to access.
- **Publications :** Various documents generated for India-WRIS are made available and reports being generated would be available to the users through this link. The documents are: 1. Software Design, System Architecture and Data Security; 2. Database Organization and Geo-database Standards; 3. Metadata Standards; 4. General Technical Guidelines - An Overview; 5. Theme-wise Database Generation Methodology; 6. Quality Assurance of Database - Standards & Guidelines; 7. Quality Assurance of Software - Standards & Guidelines.
- **WRIS Education :** All information about how to make use of India-WRIS information in the best possible ways.

- **FAQ :** This section contains answer to common questions and queries about the project and outcome.
- **Feedback :** Provides interface to post user suggestions & feedbacks.
- **Login :** For downloads and providing data inputs, login is provided based on user categories.
- **Join WRIS :** This section provides provision for new user to register and get connected to India-WRIS portal for updates.

Second, the advanced information toolbar is available right below the banner. It contains links to pages containing detailed information that a user requires when visiting the home page but may not require while exploring the other sections of the information system. The links available in this toolbar are:

- **Home :** This link leads to the main page of WRIS Portal.
- **Metadata :** This link leads the user to the Metadata Explorer which provides comprehensive information of the source of spatial and non-spatial data.
- **Report Generation :** This section features to automatically generate report of the user defined area / region containing the all data into tables and maps and allows Save As and Download in .pdf format.
- **Gallery :** This section presents the user with an image gallery of events related to development of India-WRIS. Like the website outlining the details of the project was launched by the Hon. Minister and Secretary, Ministry of Water Resources on December 10, 2009 during 5th Asian Regional Conference of ICID held at New Delhi and the β version of India-WRIS Information System (India-WRIS WebGIS) was launched on December 7, 2010 by Hon'ble Minister of Water Resources Shri. Pawan K. Bansal in a function organized at New Delhi.
- **WRIS Wiki :** Comprehensive information for the water resources assets and projects of the country is made available through WRIS-Wiki application.
- **Data Download :** Apart from viewing the available data, the user may also wish to take the data and perform analysis / do value addition. This link allows the download of GIS layers and associated attributes.
- **Help :** A comprehensive and universal help is documented in this section assisted with diagrams, screenshots and short videos.

- **Search :** Consolidated search is provided into the complete information system.

## 7.2 Main menu toolbar

This is the heart of India-WRIS information system where all the major links to the various Web GIS tools are provided in a rich GUI assisted format for easy access and use. The various sections accessible through this menu are:

### 7.2.1 WRIS Info Discovery

This section aids the user in discovering information contained in India-WRIS pertaining to a particular geographic area. The user can select his area of interest based on the Administrative units, Hydrological units and Constituency-wise and he is presented with a condensed list of all information available in India-WRIS for the area.

### 7.2.2 WRIS Explorer

This is the crux of India-WRIS WebGIS where all the data can be explored and interpreted using the various tools available for the purpose. The data in this section can be visualized using both 2D as well as 3D tools.

#### *7.2.2.1 WRIS Explorer - Geo Visualization*

2D Geo-Visualization -This section provides basic facility to visualize all the layers together in any combination by turning layers on and off as per the requirements.

3D Geo-Visualization - The 3D application is based on ArcGIS Explorer and its customized modules. The twin purpose of this application are 3D Visualization and 3D Analysis. The system consists of advanced features like: terrain view, fly around/fly to and measurements over terrain are provided.

#### *7.2.2.2 Sub-Information Systems*

There are 12 major information systems namely, base data, surface water, ground water, hydro-met, water quality, snow cover/glacier, inland navigation waterways, inter-basin transfer links, hydrological extremes, land resources, water tourism and socio-economic . These have been further divided into 30 sub-information systems. Each sub-information system is based on a particular theme. It contains relevant layers and specially created tools to make the best use of the data.

#### *7.2.2.3 Temporal Analyst*

A large amount of water resources and related data regarding hydrological, meteorological, pollution etc. are temporal in nature. In order to represent these datasets, a separate module has been created where facilities are provided to represent the time series data using suitable charts, animations and to compare the data across stations or years.

### 7.2.3 WRIS Connect

Meteorological information, flood forecasts and other water resources projects related data available in India-WRIS WebGIS could be very much helpful to the user for day to day life. WRIS Connect has been created with the purpose to keep the user up-to-date with this information. The main modules in WRIS Connect are :

#### *7.2.3.1 Hydro Observation Data*

This section enables the user to access the latest information from Hydro-Observation sites of CWC and meteorological information.

#### *7.2.3.2. AWS Data*

Automatic Weather Stations of ISRO and CWC is planned to be made available through this page.

#### *7.2.3.3 Query Interface*

User can have lot of queries answered directly through India-WRIS WebGIS Explorer and associated available tools. To explore more details, user can place his queries through Query Interface that contains set of fixed queries on various hydrological parameters. The answers are generated through different permutations and combinations of these fixed queries. The result of a query is displayed in spatial as well as non-spatial formats.

### 7.2.4 Input Data Builder

This module aims at keeping the data content of the various layers of India-WRIS up to date by providing facilities to the data providing sources to ingest the current attribute data directly into the relevant layers. The authorized users can enter the respective spatial and non-spatial data in the specified format into the information system through this facility. The three main modules of Input data builder are Spatial, Non-spatial and Metadata Input Builder.

### 7.2.5 Share Success Stories

The objective of this module is to connect people for water resources planning and management by providing platform to upload the success stories so that others can view, interact and practice.

### 7.2.6 Create Your WRIS and Hydrology Tools

This module provides facilities to the user to make further analysis of the downloaded data, adding new datasets using available general and hydrology tools.

## 7.3 General information toolbar

This toolbar is available in the lower section of the home page and provides links to general information about India-WRIS as:

- **Visitor Number :** Gives the total count of the users visited since the launch. This keeps a track of the users, their geographic location, time-duration spent on the page etc.
- **Disclaimer :** This document outlines the legal implications of downloading and using the data obtained from India-WRIS.
- **Sitemap :** Contains a graphical representation of the pages of India-WRIS along with their interlinking. This is to assist the user in easy navigation of various sections of the information system.
- **Links :** This page lists the useful links related to water, information systems and related documents spread across the World Wide Web that have been referred to during the development of India-WRIS.
- **Contact Us :** Address & contact details of the Project Director and Coordinators of the India-WRIS project are listed out in this section.
- **Last updated :** Displays the date of last modification or updation made in the current online version of India-WRIS Web GIS.

## 8.0 Conclusion

The hydrological processes are continuous as well as somewhat complex and therefore, a comprehensive, reliable and easily accessible Information System having time series data of the hydrological and meteorological observatories is a pre-requisite for effective management of water resources. The water resources Information System, India-WRIS, is likely to be fully developed by end of 2012; but after beta version launched in December 2010, further updated version would be launched in first quarter of 2012.

**Table - 4 :** Tools Inventory

| **Six Categories of Tools** |
|---|
| **I. Navigation Tools** |
| Zoom In: It zoom into a particular area on map that is selected by the user. |
| Zoom Out: Is zooms-out the map to come out of the detailing. |
| Pan: It allows user to Pan around the whole map |
| Full Extent: It allows viewing the map at the full extent. |
| Previous Extent: It allows going to previous extent when the map extentgedha |
| Next Extent: It allows returning from the previous extent. |
| Map Overview: Provides location of current view in context with larger map area. |
| Go To: Zoom to an area based on specified latitude and longitude |
| Select Area Zoom/Rubber Zoom: Smooth Zooming into a selected area |
| **II. Display Tools** |
| Swipe: It swipe the selected layer in the map to reveal underlying layers |
| Spotlight: It removes overlaid layer from the selected portion for better visualization |
| Magnifier: To view the zoom in layer details of selected portion only |
| Get Feature Info: On selecting a particular feature it displays summary info |
| Identify: identify the details of all the visible layers including the elevation details of the point |
| **III. Search and Query Tools** |
| Query Builder: create user defined queries |
| Search by Proximity |
| **IV. Personalization Tools** |
| Draw: Allows to draw a shape/line/point on map |
| Print: To print the current viewing area in land scape or portrait mode. |
| Save as Image |
| Bookmark: Allows bookmarking a specific location on the map for future reference. |
| Pin Mark: User can pin mark his location of interest and type his comment on same |
| **V. Advanced Tools** |
| Surface Profile: Generate the surface terrain height graph of selected points on map |
| Network Analysis/Route Tool: It specifies the defined route of road, rail and river network. |
| Tools to calculate parameters based on location and user input. |
| Linked View : Can View multiple view of different information in a single window. |
| **VI. Sharing Tools** |
| Share a Link: Share the current view of map with another user through mail. |
| iFrame: Sharing frame of India WRIS in other applications. |
| Links on twitter/Facebook/Google+ |

## Acknowledgements

We are grateful to Dr. V.K. Dadhwal, Director, NRSC/ISRO Hyderabad for guidance, support, keen interest in the project and constant encouragement. We are also grateful to Dr. G. Behera, DD (RS&GISAA) and Dr. Y.V.N. Krishnamurthy, DD (RCs), NRSC for overall supervision, guidance, timely help and encouragement.

Project Team wishes to acknowledge the help and support received from officials of CWC and from other Irrigation Departments / Water Resources Departments of all the states of the country for making available the large volume of time series data.

We gratefully thank all of the individuals for fruitful discussions, sharing their expertise and knowledge.

## References

Biswas, A. (1981). Integrated water management: Some international dimensions, *J. of Hydrology*, 51, (1-4) : 369-379

Bouwer, H. (2000). Integrated water management: Emerging issues and challenges, *Agricultural Water Management*, 45 (3) : 217-228.

Ganapathy, C. and Ernest, A. N. S., (2004). *Environ. Inf. Arch.*, 2 : 938-945.

Kaden, S., Becker, A. and Gnauck, A. (1989). Decision-support systems for water management. *IAHS Publ.* No. 180 : 11-21.

Kumar, R., Singh, R. D. and Sharma, K. D. (2005).Water resources of India, *Current Science*, 89 (5) : 794-811.

Lal, M. (2001).Climate change - Implications for India's water resources. *J. India Water Res. Soc.*, 21 : 101-119

Loucks, D. (1995). Developing and implementing decision support systems: A critique and a challenge, *Water Resource. Bulletin*, 31 (4) : 571-582

Maidment, D. (1997). Opportunities for the development of a global water information system. In : Land and Water Resources Information Systems - FAO Land and Water Bulletin 7 (Proceedings of a Technical Consultation Rome, Italy, 15-17 December 1997), pp. 115-120.

O'Hagan, RG, Robinson, B, Swan, G and Finny, D, 2008. Web-based visualisation of water information: an overview. CSIRO: Water for a Healthy Country, National Research Flagship.

Zalewski, M. (2002). Ecohydrology- the use of ecological and hydrological processes for sustainable management of water resources, *Hydrological Sciences Journal*, 47 (5) : 823

# Chapter - 3

# Fluvial Process Measurement of Arid Ephemeral Channels in Western Rajasthan

**P.C. Moharana and Amal Kar**

## Introduction

Arid western part of Rajasthan which contains the Indian segment of the Thar Desert, accounts for ~62% of the country's hot arid area and is a region of low rainfall. Under such physico-climatic situation, fluvial processes function much less frequently and more episodically than the aeolian processes, but their impacts on landforms and society are large, as noticed on several occasions in arid regions. There is a distinct east-west gradient of rainfall and temperature in the region. Along the eastern margin bordering Aravallies, the mean annual rainfall is 500 mm while in the westernmost part at Jaisalmer, it is less than 150 mm. The rainfall is largely monsoon driven which brings 80 - 90 per cent of total rainfall received during the year. There is large inter-annual rainfall variability of rainfall which leads to climatic extremes of drought and floods of different magnitudes. For example on an average, the Thar Desert in India experiences flood-causing rainfall in 2-3 years in a decade, but a maximum of 5 flood causing rainfall years in a decade was recorded during the 1950s, 1970s and the 1990s (Rao et al. 2006). In most cases the floods in Thar Desert are caused when rainfall of ~300 mm or more is recorded in 2-3 consecutive days (Kar, et al. 2007) and the region fall mostly in agro-climate zone; Western Dry Region. This zone is further divided into; transitional plains of inland drainage and transitional plains of Luni basin. Heavy rains during the first week of July 1990 and then in the first week of August 1990 also caused flood damages in the Luni basin (Kar, 1994). In 2006, several stations in the northern part of Barmer district and adjoining part of Jaisalmer district received rainfall of 300-400 mm during 17-24 August, causing flood. During 4-6 July 2007, some areas of Jodhpur and Pali district received a short burst of high intensity rain. The dry Jasawant Sagar dam got filled up in a single day on the 6th July and breached the next morning. In the upper catchment of Luni River, the Pichiak Dam developed cracks due to huge water accumulation in the reservoir and collapsed at two places causing huge flush-out of water and flood.

## Evidences of Fluvial Process in Arid Rajasthan

Landforms and the land forming processes in the Thar are influenced by high moisture deficiency that creates conditions favorable for the Aeolian processes to function more efficiently than the fluvial processes. But many areas within the desert with thick fluvial deposits could not have been formed by processes solely under dry climate conditions. Several lines of evidences confirm the existence of a high-energy fluvial regime in western Rajasthan during the late-Quaternary period. Geological evidence about the presence of a well drained fluvial system comes from the riverine pre-history of all the saline lakes in western Rajasthan (Anonymous, 2002). Other notable evidences of fuvilal processes are the discovery of a number of palaeochannels in the Thar Desert through the use of satellite remote sensing data (Ghose et al., 1979; Kar and Ghose, 1984, Kar, 1993, 1999a, Sahai et al., 1993). These palaeochannels are presumed to be the relics of a river system that drained western Rajasthan at different time period. Presence of dry valleys (the Ghaghar, for example) in NW Rajasthan, Haryana and also in Sindh province in Pakisthan is supportive evidence to the presence of an erstwhile fluvial regime in the Thar Desert region. Kale, 2000, based on analysis of the slack water flood deposits along some small tributary valleys of the Luni revealed at least 17 extreme flood events that had taken place in Luni River during the past 1000 years. Kar 1994a and Bajpai 2004 found evidences of role of lineaments over quaternary sedimentation in Luni basin and indicated how, palaeochannels in this tract sometimes get revived during periodic floods as happened during 1979 flood and after the flood. Scientists from CAZRI also interpreted satellite images to reveal some hitherto unknown, NE to SW flowing buried courses of Saraswati-Drishadvati river system through the Thar Desert (Ghose et al., 1979, Kar and Ghose, 1984; Kar 1999c). Based on the borewell lithologs and Quaternary stratigraphy along the courses, they suggested that these courses were buried within the sand dunes and sandy plains of the Thar and most of these courses could be older than 40,000 years. They also suggested that the Luni and its major tributaries drained into this Himalayan system.

## Quantitative Geomorphology of Arid Drainage Basins

The above studies on fluvial processes prompted researches later on for carrying out quantitative geomorphological studies in arid

Rajasthan (Ghose et al., 1967, 1969, Singh et al 1973). Their studies based on bi-variant and multi-variants analyses, indicated that the basins on phyllites schist and rhyolites have greater values of bifurcation ratio, stream length ratio, drainage density etc than the basins on granite or sandy alluvial plains. Therefore, such basins can be expected to generate higher runoff volumes that can flow out of the basin faster with minimum infiltration losses. In the international scenario, the most notable but earlier contributions began with the dynamic work from geomorphological team from USGS led by L.B.Leopold (1964) or by Columbia University led by A.N.Strahler between 1950-1960, what today, is known in the history of worldwide geographical thinking as "Quantitative geomorphology". The laws of drainage morphometry by Horton (1945), modified by Strahler (1952,1964) and Smart et al (1967) were the contemporary pioneering works which prompted researches on dynamic aspects of fluvial processes ; Leopold & Miller (1970) on measurements of morphological changes in river channels, Schuum (1960) on the aspects of sediment type and shape factor to modern fluvial deposits and Schick (1987) on tentative sediment budget of arid watershed in Negev, Couch 1987 on the relationship of gully sidewall shape to sediment production, vertical exchange of coarse bedload in desert streams by Schick et al. 1987 and geomorphological transformation functions calculated from DEM, Moussa (1997).

*DEMs, Remote Sensing and GIS based Studies:* Digital elevation models (DEMs) have been commonly used in a geographic information system (GIS) for representing topography and for extracting topographical and hydrological features. DEMs can be generated by several techniques which include ground surveys, topographical maps, stereo photogrammetry, laser scanning and InSAR. The most widely used method of obtaining a DEM, is to interpolate it from vectorized contour lines digitized from regional topographical maps. Shuttle Radar Topography Mission (SRTM) and Advanced Spaceborne Thermal Emission Reflection Radiometer (ASTER), are now the most commonly available DEMs. SRTM collected data in 2000 using InSAR technique which has been processed to generate DEMs with 1 arcsecond (approximately 30 m) and 3 arcsecond (approximately 90 m) spatial resolutions whereas, SRTM is an advanced multispectral imager that generates DEMs with 15-m spatial resolution. Many researchers have used Digital Terrain Models (DTM) to study the hydrologic and geomorphic responses of rainfall. Some of the important models include the Areal Nonpoint Source Watershed Environment Response Simulation model (ANSWERS; Beasley & Huggins, 1982), the Agricultural Nonpoint Source Pollution model (AGNPS; Young et al., 1989) and Systeme Hydrologique European model (SHE; Abbot et al, 1986). The Geomorphic Instantaneous Unit Hydrograph (GUIH) adopted by Gupta et al (1980) and others, is one of the most commonly used methods for determining rainfall – run-off response. The physically based models, such as WEPP (Water Erosion Prediction Project) are process-based models for runoff and soil erosion prediction (Laflen et al., 1997). The universal soil loss equation (USLE) and the revised USLE (RUSLE) are most widely used to model agricultural land. Moharana & Kar, 2002 used sparse topographic information to create a DEM and delineated watersheds in sandy terrain and Jenson, 1991, worked on applications of hydrologic information automatically extracted from digital elevation models.

In the following paragraphs, we present findings of two of the studies carried out in Central Thar relating to (1) Quantitative measurement of erosion processes in ephemeral channels (2) GIS based automated delineation of watersheds in sandy terrain.

## Erosion Process Measurements of Arid Ephemeral Channels

There are quite a few number of studies related to ephemeral flow modeling in the dry and shallow channels in arid Rajasthan; ephemeral flow modeling in arid region, Sharma et al (1996), sediment transport modeling in stream channels in arid zone, Sharma et al, (1994). These studies showed how sharply hydrographs lead to flash floods peaked and how high transmission losses due to a dominant sandy terrain lead to poor connectivity between channel segments. Yet studies involving quantitative information on the catchment processes in ephemeral channel of arid region in India have been felt for a long time, especially to model the impacts and to harness the benefits of water and sediment flows. We carried out a field based work to monitor the upland processes during the infrequent monsoon rainfall in Central Thar and their impact on sediment movement, especially the bed load. In the following, we discuss some of the initial results from the measurement of erosion from this upland stream catchment located to the north-east of Jodhpur in the central part of Thar Desert. The study involved detailed field measurements of cross profiles in 17 permanent observation points, longitudinal profile measurements in over 110 points, and experiments to correlate particle size – their movement in response to run-offs.

For the period considered, the mean annual rainfall was 180 mm with highest annual rainfall of 330 mm in 2008 and lowest of 120 mm in 2006. There have been 7 high intensity rain storms (40 – 119 mm) which was able to generate run-offs in a range of 3-38 %.

*The catchment*: The studied catchment has developed on a hill - rocky/gravelly pediment - colluvial plain sequence on rhyolite near Agolai, a small town in Jodhpur district of western Rajasthan. Two ephemeral channels (av. depth 1.0 - 1.5 m), originating from low rhyolite hills (relative relief ~18 m), join together on the pediment to form a second order shallow gully that carries the water and sediments from the catchment to a sink (a pond) about 1 km downstream of the source. We used a range of field methods (Miller, 1970, Graf, 1980) to measure erosion from the hill-slopes, gullied channels and stream banks, as well as to monitor the particle movement after a measured rainfall event. Only two significant run-off generating events took place during 2007 that moved bed loads to measurable distances.

*Long profile and channel hypsometry*: A standard Theodolite was used for measuring height along and across the main channel before and after every run-off-generating rainfall event, which helped to construct hypsometric curves at different time intervals. Long profile of the main channel was measured at 104 fixed locations with reference to a benchmark, both before and after a run-off generating event. The variations in the long profile in response to major rainfall events during 2008 are presented in Fig. 1-3.

*Channel Cross profile measurements:* Simultaneously, channel cross profiles were measured at 19 locations, for which colored wooden stacks were placed at fixed end points on either side of the channel and height was measured at 40 cm interval. Cross profile of channel in the lower, middle and upper section of the stream measured during 4 major rainfall events at Agolai in 2008. Analysis of profiles showed an average of 0.5 cm deposits and 1-2 cm of erosion in the upper reach, 4-10 cm of incision and localized deposition (10-12 cm) in the middle reach and 10-13 cm deposition in the lower reach during low rainfall situations. The profiles (Fig.4-7) showed significant changes during higher rainfalls of 2007 and 2008.

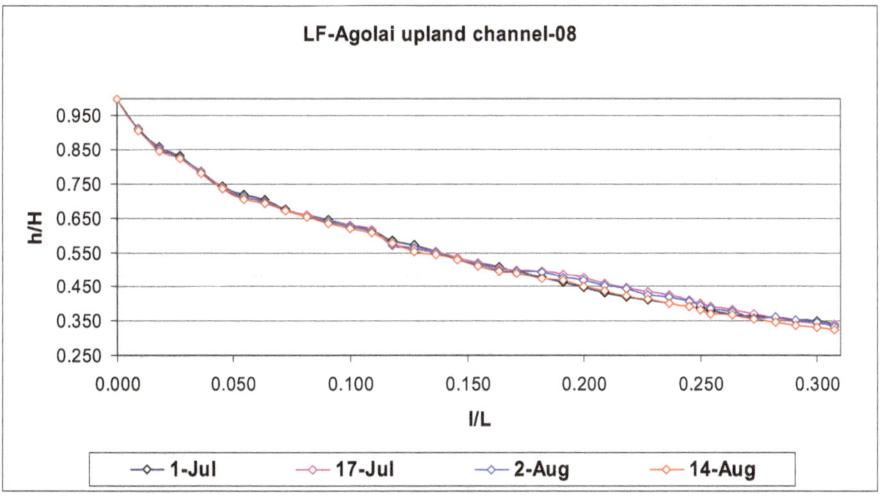


**Figure 1 :** L.F. Agolai upland channel - 08

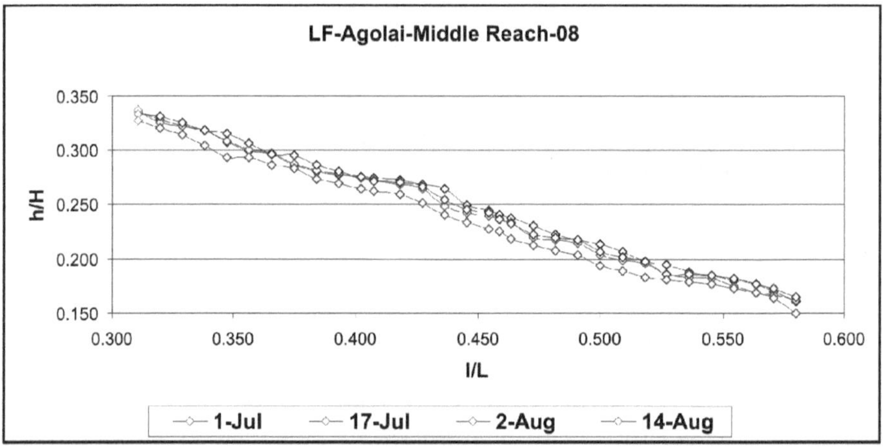


**Figure 2 :** L.F. Agolai-Middle Reach - 08

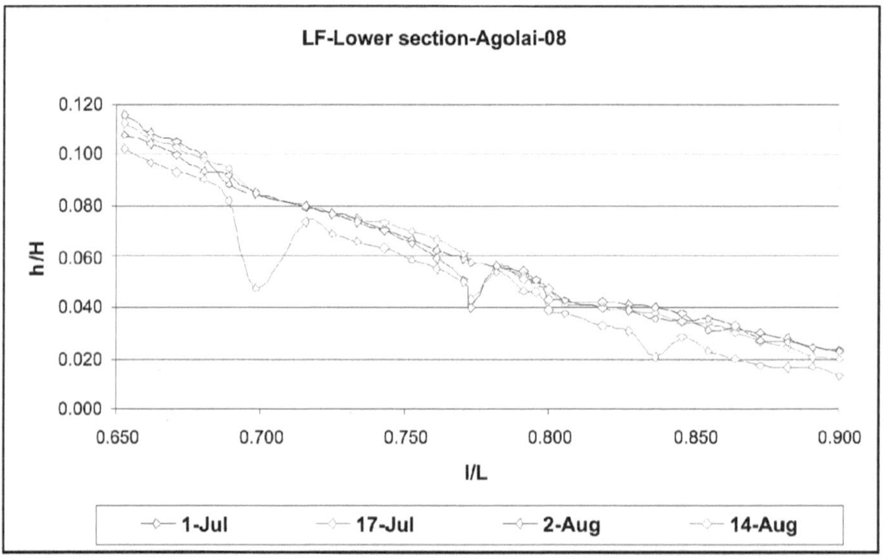


**Figure 3 :** L.F. Lower section-Agolai - 08

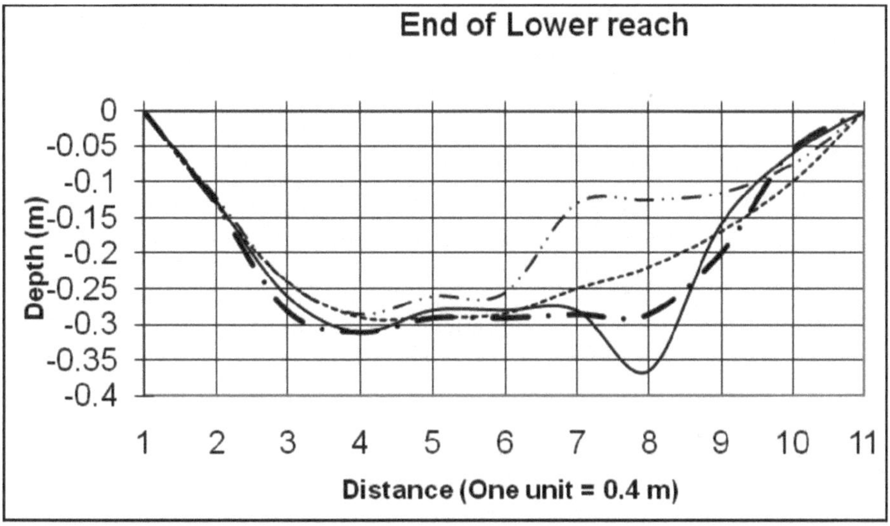


**Figure 4 :** End of Lower reach

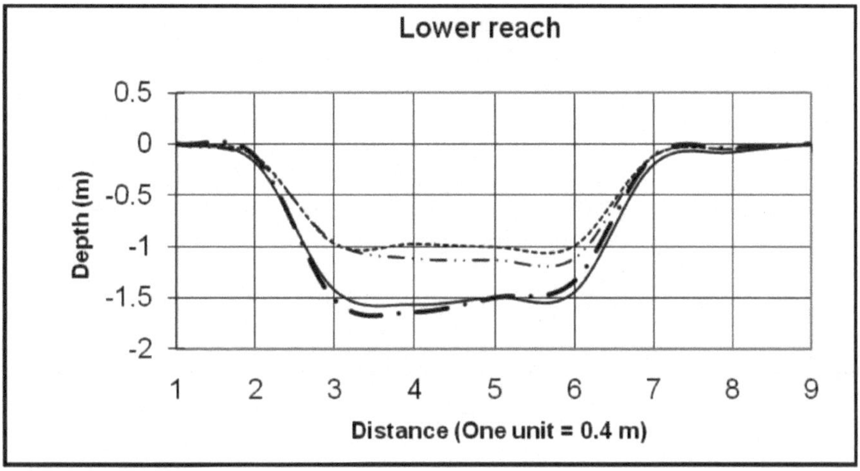


**Figure 5 :** Lower reach

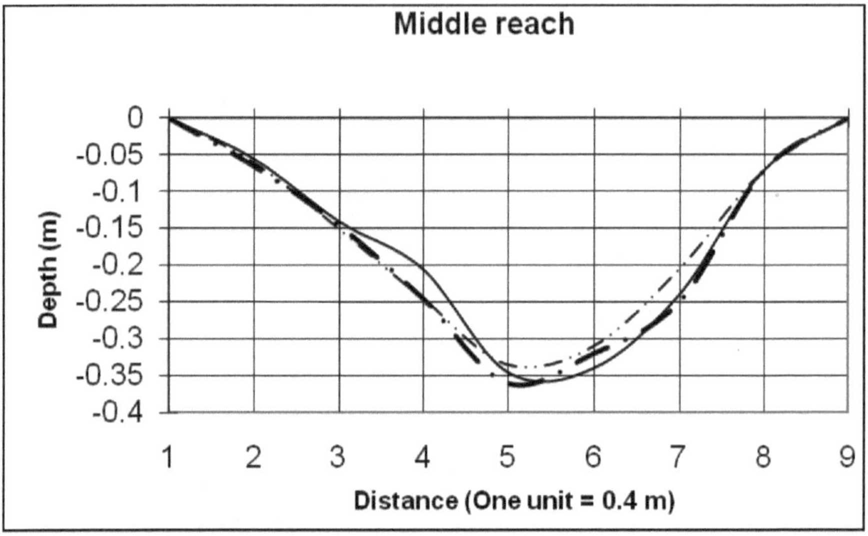


**Figure 6 :** Middle reach

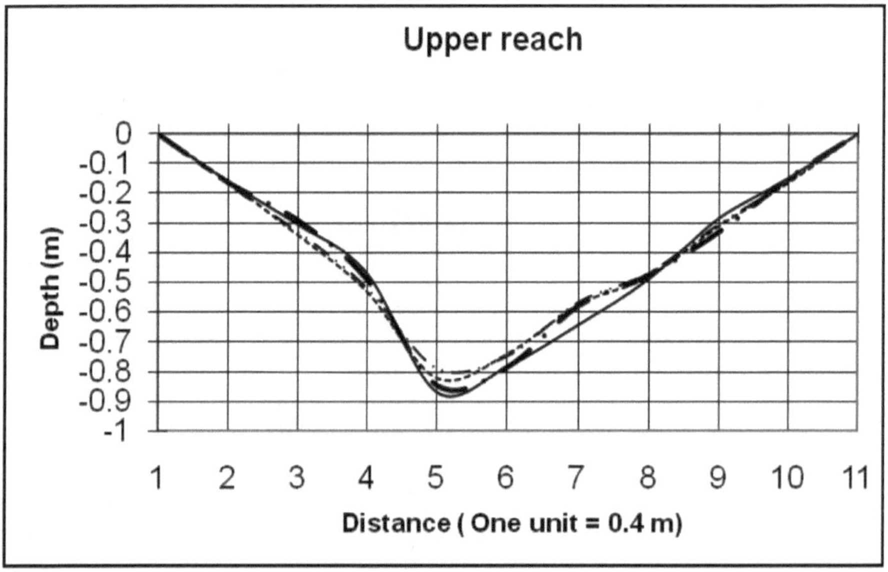


**Figure 7 :** Upper reach

Discharge was calculated at 6 cross section locations using the Manning equation $Q = 1/n\ AR^{2/3}\ S^{1/2}$, Where Q = discharge ($m^2\ s^{-1}$), A = cross-sectional area ($m^2$), R (A/P)= hydraulic radius (m), P=wetted perimeter (m), S = average gradient (m/m) and n = Manning's roughness coefficient. A rough estimate of velocity was made by measuring the time required for a float to travel a fixed distance along the stream at selected cross-sections.

*Measurement of Sediment movement*: To monitor bed load movement along the main channel, we colored groups of key particles >2 mm size and they were placed at different locations along the long profile of the channel. Length, width, breadth and weight of the particles were measured before and after a rainfall event. All the particles were grouped under 3 weight groups of >100 g, 51 - 100 g, and <50 g. In order to assess the impact of shape factor in the sediment movement, relationship between flatness index and distance moved by the sediment was studied. The degree of flattening (flatness index) of each particle was calculated using the equation: FI = E/L X 100 (where FI = Flatness index, L = major axis (length) of pebble, l = minor axis (width); E = breadth of pebble; r = smallest radius of curvature. The particles were grouped into three categories of FI : < 25, 26 – 50 and > 50. It was found that flatness had less impact in the upper section of the channel. But in the lower section, particles with lower FI, had more displacement. In comparison to lime particles (less flat), rhyolites (more flat) moved more distances in the lower reach.

*Discussion* : Two significant run-off generating events of 42 mm and 52 mm during 2007 generated peak discharges of 20 cum $s^{-1}$ (upstream) and 13 cum $s^{-1}$ (downstream) that moved sediments (bed load) to measurable distances of 43 m – 141 m in the upstream, 6-28 m in the middle section and 63-95 m in the lower section. Similarly during 2008, the higher rainfalls (113 mm on 1-2 August 2008) accumulated >6000 $m^3$ run-off which increased the water level in the pond 3-3.5 m causing about 24 cm incision and 13 cm deposits in the lower section of the channel. There was an increase in the peak discharges (7 – 22 cum/s) from upper to lower reach and moved bed loads (particle size >2mm) to a maximum distance of 400 m. Among the <50 g weight group the highest displacement (124.56 m) was recorded for the U2 particles, followed by L1 (95.48 m). Lowest movement was for M1 particles (28.80 m). Among the 50-100 g group, highest displacement was for U2 (141.8m) and lowest for L1 (07.7m). In case of 100-900g size particles, highest displacement was for the L2 particles (63.78 m) and the lowest for the M1 (6.33 m).

During 2008, the channel bed in lower reach was elevated by a fresh sand deposit of 15 cm during the rains on 17 July which was washed away by storm rains during 2$^{nd}$ Aug and a scour of 22 cm occurred in the same place by another storm during 13-14$^{th}$ August. The discharge showed a good correlation with rainfall and channel conditions in 2008 while during 2007; the discharge fluctuated with alternate peaks and lows. It was found that, in low intensity rainfall situations, channel morphometry remained less disturbed to flows, but higher rainfalls caused changes in morphometry and increased the flow and sediment. Erosion remained the dominant process till the beginning of lower reach, beyond which both erosion and deposition took place. Since, there is an increasing trend in the annual rainfall in the region; the lower catchment area is likely to receive more water and sediments from the donor area - a trend which will eventually have impact on the water and sedimentation of channel, the fan area and pond. This would certainly influence the cropping activities in the region. The preliminary results provide some field level information on process-form interactions along gravel - bed upland ephemeral channels in the monsoon - driven Thar Desert that will help to further understand the process and monitor the situation.

## Drainage Network Simulation in Sandy Terrain of Thar Desert

This study was an attempt to understand the different situations in drainage morphometry in different situations of rainfall / storm events in western Rajasthan. It is well known that in arid regions as in western part of Rajasthan, drainage basins are typically characterized by discontinuous trunk streams fed by higher number of third or lower order streams, infrequent channel processes and high magnitude floods. Understanding the drainage basin processes under such setup requires establishment of a network of monitoring instruments at a considerable cost and risk, or simulation of the network integration with increasing discharge and channel processes. Some modeling work has been carried out to understand the hydrological characteristics of ephemeral streams in Luni basin which contains several large channels originating from the Aravallies ranges in the <450 mm average annual rainfall zone (Sharma and Murthy, 1996). Unfortunately the basins of these large stream network do not fit into the definition of watersheds (50 -500 $km^2$ mean size) and micro-watersheds (<2.5 $km^2$ mean size) that are considered for developmental activities by state governments departments. Secondly, large scale topographical sheets of the sandy terrain in the Thar Desert in Rajasthan contain very little information on height, including contours and spot heights. This creates problems for delineation of watershed boundaries. In order to find a simple solution to this problem, we carried out a simulation study using ARC/INFO software.

The sparse topographic information (Fig. 8) from Survey of India Toposheets at 1:50,000 scales was used for reconstruction of a Digital Elevation model (DEM) under ARC/INFO GIS. The first phase involved the filling up of sinks or depressions in the simulated topography of the original DEM (Fig.9), modeling of other fluvial parameters such as flow direction, flow accumulation, stream network and stream ordering (Strahler) was carried out. The last step involved overlaying of the simulated channel information (Fig. 10) on the existing resource information and finding out of the variation between the two, if any. Superimposition of the present stream network (from topographical sheet), the location of village ponds and the network of partly sand covered former streams (from satellite images) on the simulated stream network revealed a near-perfect match (Fig.11). The simulation helped to understand how even the small drainage networks in the region get integrated only after sustained high-intensity rainfall.

## Conclusions

Arid regions in Western Rajasthan have divergent morphogeny and are unique to provide ample scope for fluvial geomorphological studies during storm periods despite dominance of a sand dominated terrain. Of late, this region has witnessed more frequent incidences of floods and high rainfall events, thereby; researchers have paid little more attention to fluvial process studies. In the same time, there is shift in attention away from traditional studies to remote sensing and model based analysis of fluvial process dynamics. But, field based studies still remain dominant and provides clue to dynamism of fluvial activities in arid region. The above studies depict how field based measurements, data and monitoring of events hold keys to understanding of flow of sediment and water in arid ephemeral channels and how such data are influenced by rainfall events, their magnitude and the channel morphometry. While the first study showed how measurements of bedloads can help us quantifying the sediment load that might be created during different flow events, the second study on drainage simulation helped us to understand the flow path and flow direction of such sediments created because of processes of erosion. The study also suggest that it is possible to generate tolerably good drainage networks and watersheds in a sandy terrain, provided a sufficiently accurate landform database and map as well as other supporting databases are available. In general, any departure from their coherent relationship would require an in-depth analysis of the event in response to possible climate change scenario.

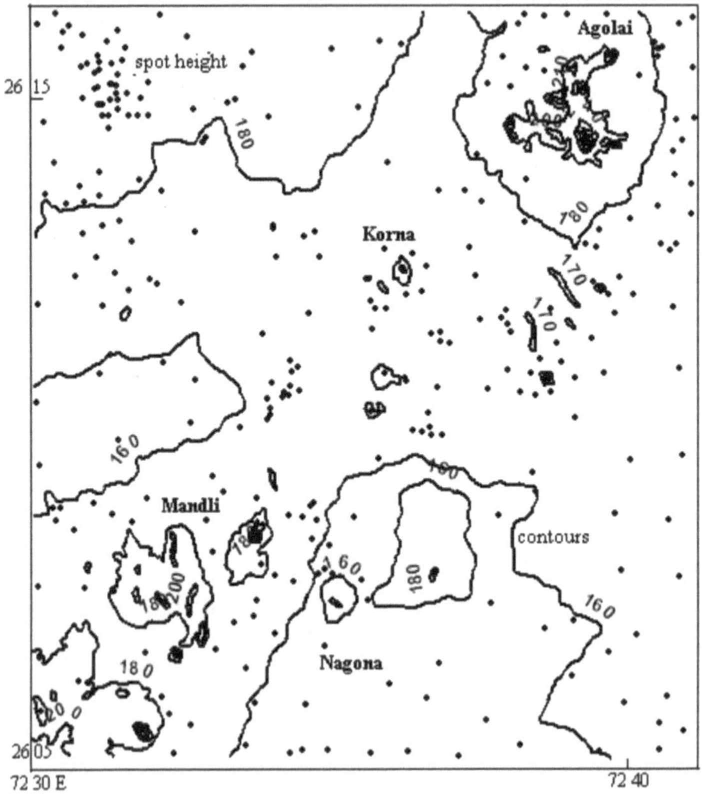


**Figure 8 :** Contours, spot heights / relative heights (Survey of India Toposheets, 1:50,000 scale)

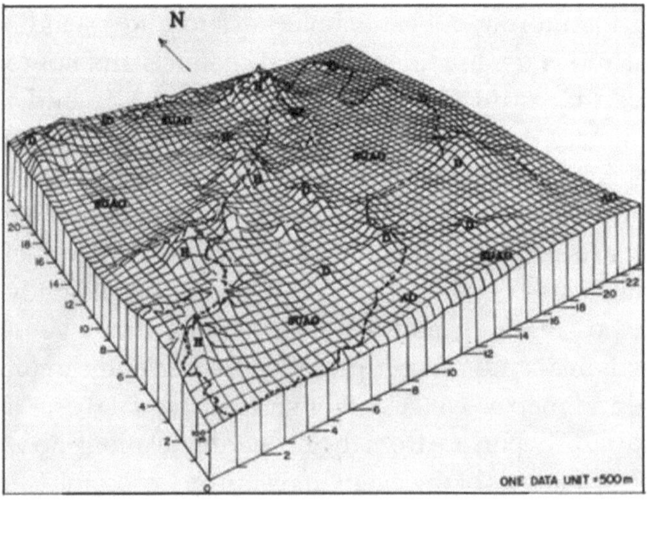


**Figure 9 :** Topography, landforms and watershed on the DEM of Korna area

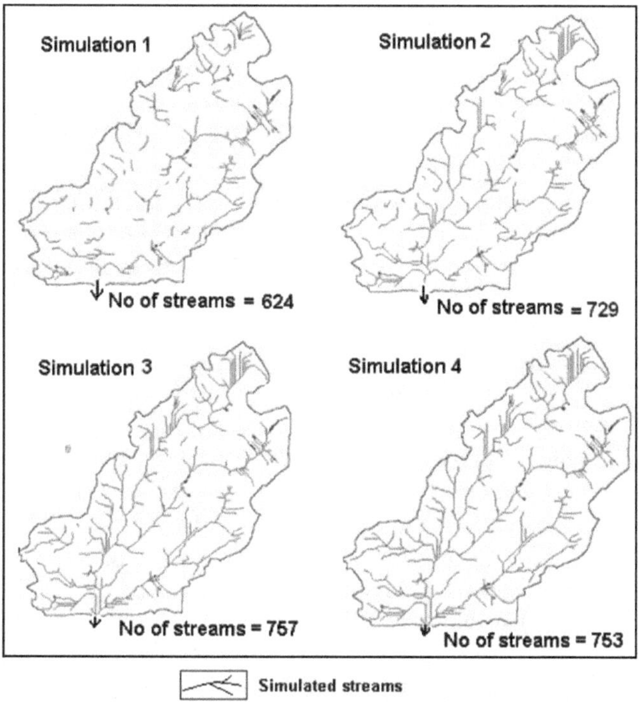


**Figure 10 :** Stream network simulation in the study area

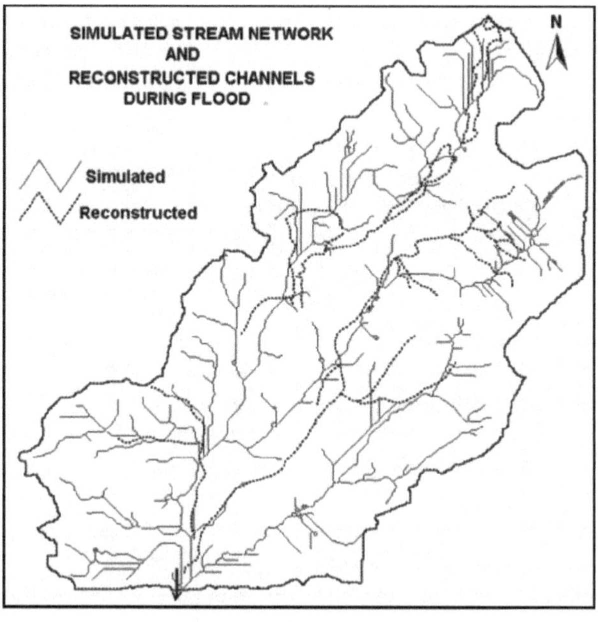


**Figure 11 :** Simulated networks and reconstructed channels during flood events in the study area

## References

Abbot, M.B., Bathurst, J.C., Cinge, A., O'Connel, P.E & Hasmussen, J. (1986). An introduction to the European hydrological system-Systeme Hydrologique Europeen, SHE.2: structure of a physical based, distributed modeling system, *Journal of Hydrology*, 87:1-77.

Anonymous (2002). Geology of Rajasthan (Northwest India), Precambrian to Recent (Eds. A.B.Roy and S.R.Jakhar). Scintific Publishers, Jodhpur. 421p

Bajpai, V.N.2004: Hydrological evolution of the Luni river basin,Rajasthan, western India: A review, proceedings, Indian Academy of Sciences, *Earth & Planetary Science* 113(3): 427-451.

Beasley,D.B. & Huggins,L.F.(1982).answers (Areal Nonpoint Source Watershed Environmental Response Simulation): User Manual.US EPA-905/9-82-001.Chicago.54p.

Graf.W.L.(1980) Fluvial processes in dryland rivers. Berlin:Springer.Verlag.346p.

Ghose, B., Pandey,S and Lal, G.1967. Quantitative geomorphology of drainage basins I central Luni basin, western Rajasthan. *Zeitschrift fur Geomorphologie*,11: 146-160.

Ghose, B.,Kar, A., and Hussain, Z.(1979): The lost courses of Sraswati river in the Great Indian Desert : New Evidences from Landsat imagery. *The Geographical Journal*,145(3): 446-451

Gupta,V.K., Waymire, E & Andwang, C.Y.(1980). A representation of an instantaneous unit hydrograh from geomorphology, *Water Resources Research*, 16:855-862.

Horton, R. E. 1945. Erosional development of streams and their drainage basins: hydrophysical approach to quantitative morphology. *Geological Society of America Bulletin* 56:275-370.

Jenson, S. K. 1991. Applications of hydrologic information automatically extracted from digital elevation models. in: Terrain Analysis and Distributed Modelling in Hydrology. Beven, K.J. and I.D. Moore (Eds). 1992. John Wiley & Sons, New York. Pp. 35-48.

Kar, A., Moharana, PC, Singh, S.K., Goyal, R.K. and Rao, A.S. (2007): Barmer flood, 2006, causes and consequences. IN, Flood of August 2006 in arid Rajasthan, Magnitude and Strategies (Eds. A.S.Faroda and D.C.Joshi), pp.26-39. State of Art Report,Scientific Contribution No. INCOH/SAR-29/2007;INCOH,Roorkee.

Kar, 1994. Lineament control on channel behavior during 1990 flood in the southeastern Thar desert. *International Journal of Remote Sensing*, 15:2521-2530.

Kar 1999a. Quaternary period in Rajasthan: Geomorphic process response to forcing and implications for terrain evolution. In, P.Kataria , Geology of Rajasthan: Status and Perspective.M.L.Sukhadia University, Udaipur;pp.175-212.

Kar, A. 1993b. Prsent day geomorphic processes as key to the reconstruction of Quaternary landform history in the Thar desrt, *Journal of Geological Society of India*, 41(6): 513-517.

Kar, A. and Ghose, B. 1984. The Drishadvati river system of India : an assessment and new findings. *The Geographical Journal*,150(2): 221-229.

Kar, A.1999c: Neotectonic and climate control on drainage evolution in the Thar Desert, In. B.S.Paliwal(ed), Geological Evolution of North western India. Scientific Publishers, Jodhpur;pp.246-259.

L.B.Leopold, L.B.,Wolman, M.G. and Miller, J.P. (1964). Fluvial processes in geomorphology. San Fransisco: Freeman,522p.

Laflen, J. M., Elliot, W. J., Flanagan, D. C., Meyer, C. R., and Nearing, M. A. (1997) WEPP-predicting water erosion using a process-based model. *Journal of Soil and Water Conservation*, 52 : 96–102.

Moharana, P.C and Kar, A. 2002. Watershed simulation in a sandy terrain of Thar desert using GIS. *Journal of Arid Environments*, 51: 487-498.

Miller, J.P.and Leopold, L.B.(1970) Simple measurements of morphologica changes in river channels and hill slopes. *Changes of Climate*, pp.421-427.

Moussa (1997). Geomorphological transfer function calculated from Digital Elevation Models for Distributed Hydrological modeling, *Hydrological Processes*, 11 : 429-449.

Rao, A.S., Purohit, R.S. Mertia, R.S., Surendra Poonia and Sharma, M.D(2006). Meterolgical aspects of floods in Arid Rajasthan, Maru Krishi Chaynika (special issue on flood-January to December, 10 (1-3) : 32-39

Sharma, K.D & Murthy, L.S.R (1996): Ephemeral flow modeling in arid regions, *Journal of Arid Environments*, 33:161-178.

Strahler, A.N.(1952): Hypsometric (area altitude) analysis of erosional topography. *Geological Society of America Bulletin*, 63: 1117-1142.

Strahler, A.N.(1964) : Quantitative analysis of watershed geomorphology. *Transactions American Geophysical Union*, 38:913-920.

Smart, J. S., and A. J. Surkan. 1967. The relation between mainstream length and area in drainage basins. *Water Resources Research* 3(4):963-974.

Schum,S.A.1960: The shape of alluvial channels in relation to sediment type.US.Geological Survey of Professional paper, 352-B.

SCHICK, A.P. LEKACHI and M.A. Hassan (1987) Vertical exchange of coarse bedload in desert streams. In Desert Sediments: Ancient and Modern. spec. pub, *Geol. Soc. London.*, 35 : 7-15

Sharma, K.D & Murthy, L.S.R (1994). Estimating transmission losses in an arid region - a realistic approach. *Journal of Arid Environments*, 27:107-112.

Young, R.A.,Onstad,C.A.,Bosch,D.D. & Anderson, W.P.(1989). AGNPS, non-point source pollution model for evaluating agricultural watersheds. *Journal of Soil and Water Conservation*, 44: 168-173.

Chapter - 4

# Saraswati River - The Source of Quenching Thirst of Thar Desert

**B.K. Bhadra and J.R. Sharma**

## Abstract

*Mention about the Great Himalayan river of Vedic Period (10000-8000 BC) is found in ancient Indian literature. The river called as 'Saraswati' in India and 'Hakra' in Pakistan, was flowing with full majesty through Thar Desert around 6000 BC and dried up during 3000 BC due to climatic and tectonic changes in Himalayan region. The relict of this lost river is still found as palaeochannels in this region. Large numbers of archaeological sites of Harappan civilization occur along the dry river bed.*

*Satellite images have unearthed the hidden course of a major lost river below the sands of Thar Desert in the Indo-Pakistan region. In the present study, remote sensing techniques have been applied using digitally processed multi-resolution and multi-temporal satellite images. The mapped course of the river is 4-10 km wide and is passing through Haryana, Punjab, Rajasthan and Gujarat. Discovering these palaeochannels is a new find and is validated through drilling of tube wells on the channels and through archaeological, hydrogeological, litholog and sediment data which conform to the magnanimity of the river as described in 'Rigveda'. Enormous quantity of potable water has been explored along these channels which may quench the thirst of the water-strived people of Thar region.*

***Keywords:*** *Remote Sensing, Palaeochannel, Saraswati, Archaeology, Ground Water.*

## 1.0 Introduction

In the western part of India, the existence of a mighty river (known as *Saraswati* in India and *Hakra* in Pakistan) is found in most of the ancient Indian literatures (Fig.1). The river Saraswati is described in the Rigveda literature as the Ambitame, Nad*itame* and *Devitame* that is best of mother, best of river and best of goddess. Existence of this river in the past and the reasons of its disappearance in Thar Desert remains a debated topic among the researchers due to lack of proper scientific database. Varying number of courses of extinct river have been reported by the earlier workers viz. 5 channels by Ghose et al. (1979); 7 channels by the Bakliwal and Grover (1988) and 1 channel by Yashpal et al. (1980). Many of these authors believed that river shifted several times westward from its initial position near Aravalli hills to occupy final position about 500 km west near Indo-Pak border.

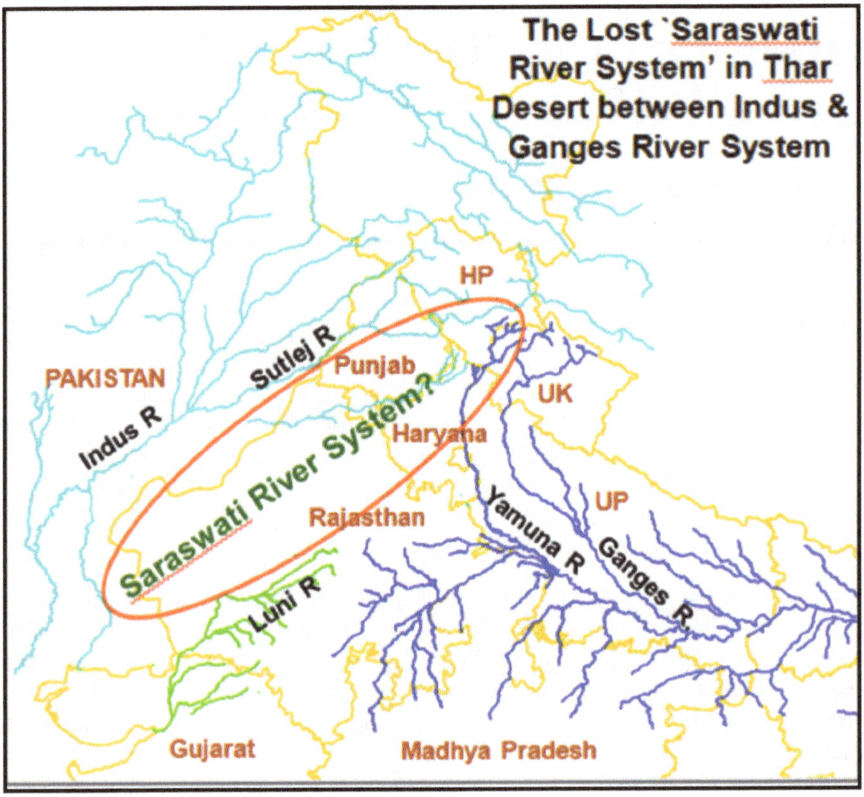


**Figure 1 :** Existence of Saraswati River System along the vast tract of alluvial/aeolian plain between Indus & Ganges River System in NW India.

The main objective of this endeavor is to carry out an unbiased and afresh study to delineate the true course of river 'Saraswati' buried below the sands of Thar Desert using satellite images. This study is carried out taking advantage of the availability of multi-spectral, multi-temporal and multi-resolution data from new generation satellites / sensors and digital image processing technics. Further, the mapped course has been validated through a variety of ground data, carried out by various survey agencies. An attempt has been made to link Vedic Saraswati with the perennial source of existing Himalayan Rivers.

## 1.1 Vedic Saraswati river in the past

The River Saraswati is said to be the oldest and holiest of the holy rivers in India. This river is described in most of the scriptures of India viz. Vedas, Brahmana and Srautasutra literature, Mahabharata, Ramayana, Bhagvat Purana, Vamana Purana and Upnishads, etc. It is believed that the river Saraswati was a life line of the people of ancient India mostly in the Vedic and Puranic ages. Subsequently for reasons not fully known, the Saraswati River vanished altogether from the Indian subcontinent. It is therefore only the ancient mythological literature that serves as a primary source of information about the lost Saraswati River. The ashrams of many famous rishish (saints) like Vagyavalka, Dadhichi, Parasurama and Parasara etc located along the banks of the river Saraswati.

Vedic Saraswati River is believed to be mightier than the existing Ganges River in the past which is corroborated by frequent mention of the name 'Saraswati' (75 times) as compared to river Ganges (only once) in Rigveda (Chauhan, 1999). The discovery of sites of Harappan civilization on the banks of Saraswati-Hakra River also indicates towards its magnanimity more than 8000 years ago. The Vedic Saraswati, a mighty and holy river of northwest India during 6000 BC, originated from Bandarpunch glacier in Garhwal Himalayas and finally discharged into the Gulf of Khambat in Gujarat coast and disappeared around 3000 BC (Radhakrishna, 1999). Presently, relict evidences of this river are found in the form of isolated dry channels, sweet water wells along these channels with good yield, archaeological artifacts etc. Several remnants of this river still exist as palaeochannels within Thar Desert region (Ghose et al, 1979; Gupta, 1996; Mishra, 1995; Radhakrishna, 1999; Valdiya, 2002; Yashpal et al, 1980). Today the Vedic Saraswati River is represented by the Ghaggar River (Gupta et al., 2004) which flows on palaeochannel of Vedic Saraswati, located in the western part of the Haryana state (Bhadra et al, 2009; Bhadra and Sharma, 2011). The Vedic Saraswati River has been flowing sub-parallel to the Indus River in NW India (Krishnan, 1953; Oldham, 1893; Stein, 1942; Wilhelmy, 1999). Plenty of archeological remains of the Harappan civilization are found along the palaeochannels of the extinct river Saraswati / Hakra.

## 1.2 Present day drainage systems of NW India

The physiography and drainage pattern of NW India can be viewed through SRTM Digital Elevation Model (Fig.2). The flood plains of the Saraswati River slope towards SW direction from Siwalik foothills (elevation > 270m) to the low lying marshy stretch of the Great Rann of Kachchh. This stretch of land is manifested by the 'Lost River Saraswati' between Indus and Ganges River Systems, dating back to middle Pleistocene. The northern plains of Haryana and Punjab are drained by three major independent river systems namely Yamuna, Sutlej and Ghaggar (Saraswati). Yamuna and Sutlej are perennial river systems while the Ghaggar is mainly ephemeral. The Sutlej River originates in the Himalayas from the holy Lake of Mansarovar in Tibet and flows towards northwest across the Himalayan ranges. Finally, it enters the plains near Ropar (Punjab) where it takes a sharp right-angled turn and flows westward.

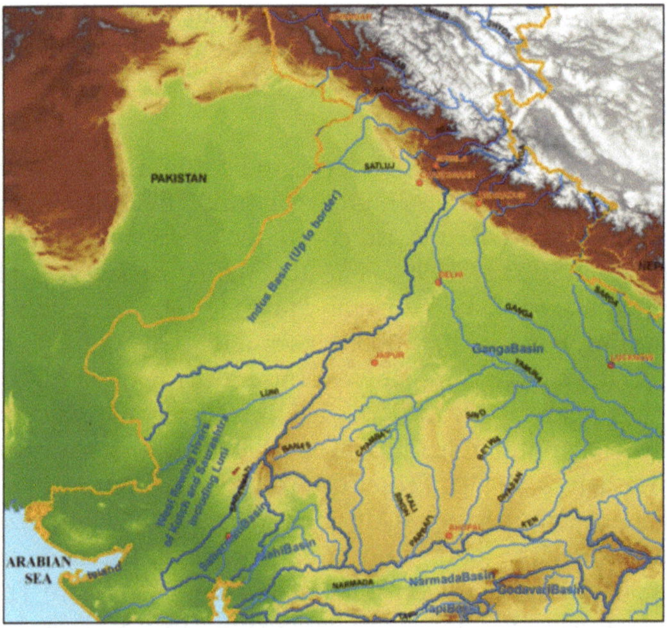


**Figure 2 :** SRTM derived DEM of NW India, showing major drainage systems and the delineated basins.

The River Yamuna, supposed to be a major tributary of Saraswati, was earlier flowing through Bata river course and joining the Markanda River and later on diverted through the Yamuna Tear Fault. The Ghaggar River raises in the Siwaliks from the Morni hills and enters the plains near Ambala (Haryana). After covering a distance of 175 Km it joins at Rasula (Patiala district) with the Saraswati which originates from the Adi Badri area in Yanmuna Nagar district. The combined river now known as Ghaggar which flows through Sirsa (Haryana), Hanumangarh and Ganganagar districts (Rajasthan) and entered into Thar Desert. The river Saraswati was probably stretched for 1600 km long and 3 to 12 km wide, significantly broad and perennial. Saraswati River used to receive water from Yamuna and Satluj and ceases its flow due to eastward shifting of Yamuna and westward deflection of Satluj. As a result of neo-tectonic activity and climatic changes, river Saraswati dwindled finally and got lost in the sands of the Thar Desert.

## 2.0 Data & Methodology

With the advent of Remote Sensing technology and the available satellite images, it is possible to trace the drainage course in the form of buried palaeochannels. Due to having synoptic viewing capability with different spatial, spectral and multi-temporal resolution, optical satellite images from Landsat ETM (30m) of 2000, IRS P3 WiFS (188m) of 2002, IRS P6 AWiFS (56m), LISS-III (23.5m) and LISS-IV (5.8m) of 2004 have been used for delineation of palaeochannel and other thematic interpretation. Further, microwave data from Radarsat SAR data (50m) of December, 2002 has also been processed to delineate palaeochannels in parts of Punjab. Digital image processing techniques like histogram equalization, linear stretching, contrast and brightness enhancement etc. have been applied on a small area of the satellite images. SRTM (90m) data has been used for generation of Digital Elevation Model (DEM) of northwest India which represents the spatial representation of the features in terms of height. GIS database have been prepared with details of mapped palaeochannels, locations of archaeological sites, existing tube wells / drilled sites, villages, roads, railway, drainage lines and administrative boundaries under ERDAS Imagine and ARC Info environment. Apart from the satellite images, a large number of collateral data in the form of maps and ground data have been collected and correlated with the delineated palaeochannels.

## 3.0 Results and Discussion

## 3.1 Palaeochannel mapping using satellite data

Palaeochannels are basically the old course of river channels which appears on the satellite image as serpentine drainage course with high moisture content (dark tone). Remote Sensing techniques helped in locating these buried channels as they are clearly seen in the satellite images (Ghose et al. 1979, 1980; Yashpal et al. 1980; Sood and Sahai, 1983; Bakliwal and Grover, 1988; Ramasamy et al. 1991; Kar, 1995; Sahai, 1999). Pioneering works in Remote Sensing has been carried out by Yashpal et al. (1980), Sahai (1993), Rajawat and Narain (1996) who have compiled various studies pertaining to the palaeo drainage network of Northwestern India. Recently the entire course of River Saraswati has been traced from satellite images by the authors like Gupta et al. (1984), Bhadra et al. (2009), Bhadra and Sharma (2011), Bhadra et al. (2005), Sharma and Bhadra (2009a and b), Sharma et al. (2006) and others.

### 3.1.1 Delineation of Palaeochannels in Rajasthan

A pioneering work has been done by Gupta et al. in 2004 to map the course of Saraswati, buried below sands of Thar Desert. They have showed the presence of Saraswati palaeochannels which are self-evident on satellite images of IRS P3 WiFS and LISS-3 satellite data (Fig.3). *"Piece-wise Histogram Stretching"* technique has been used to enhance the palaeochannel signatures on the image. This technique has been found unique in enhancing palaeochannel details in the sandy as well as alluvial and vegetated areas. The feature enhancement is carried out by way of loading sub-scenes of 1k x 1k size on computer terminal in full resolution and improving the feature contrast by histogram stretching interactively.

**Figure 3 :** Palaeo-drainage courses along indo-Pak, interpreted from IRS P3 WiFS satallite images.

Thus, a palaeochannels network of Saraswati River in Rajasthan and parts of northern Gujarat has been established across the Thar Desert (Fig.4).

**Figure 4 :** Simplified palaeochannel (PC) map in parts of Rajasthan and Gujarat. Channel-I represents the Saraswati PC, linking to Ghaggar R.

## 3.1.2 Delineation of Palaeochannels in Haryana and Punjab

The Sarasvati river system (present day drainages along with palaeochannels) in Haryana and Punjab (Fig.5) has also been studied in detail by using IRS P6 AWiFS, LISS-3 and LISS-4 satellite data (Bhadra et al., 2009; Kar and Ghose, 1984; Sharma and Bhadra, 2009a and b; Bhadra and Sharma, 2011). For delineation of palaeochannels in Haryana, digital image processing techniques like histogram equalization, linear stretching, contrast and brightness enhancement etc. have been applied on a small area of the satellite images. Drainage features are highlighted on applying local stretching on ~10X10 sq. km area out of the full LISS-3 scene (141X141 sq. km). Using this process, the possible course of Saraswati & Drishadvati palaeochannels in Haryana are delineated with proper care by avoiding the canals, existing ephemeral drainages and water logged areas.

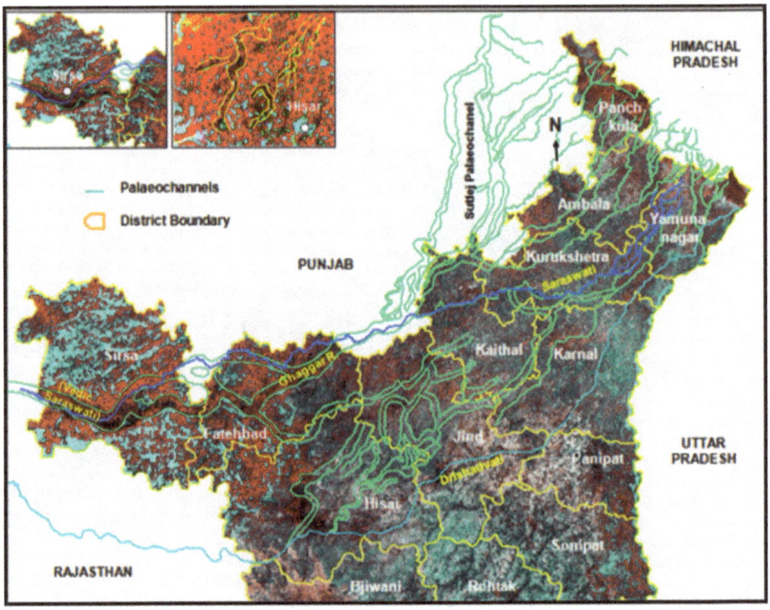


**Figure 5 :** IRS LISS-3 image showing the delineated palaeochannels in northern Haryana and Punjab. Inset shows the signature of palaeochannels (dark tone) in Sirsa and Hisar districts.

An attempt has been made to identify the continuity of Saraswati palaeochannels in eastern Punjab also. For this purpose, satellite images of IRS P6 LISS-3 data (23.5m resolution) of February, 2004 and Radarsat SAR data (50m resolution) of December, 2002 have been processed to delineate palaeochannels. In this area, Radarsat SAR image could pick up a wide moist zone (torquous channel with dark tone), possibly indicating a subsurface palaeochannels that matches with the signatures of IRS LISS-3 image. The delineated palaeochannel between Ropar and Patiala is named as *'Sutlej palaeochannel'* which is represented by N-S trending tortuous palaeochannel extending for a length of about 75 km with having width from ~1 to 6 km. The Sutlej palaeochannel is connecting the present day Sutlej River near Ropar and the Ghaggar River in the south of Patiala. Thus, the Sutlej, Saraswati and Drishadvati palaeochannels possibly contributed to the main Vedic Saraswati to form a mighty flowing River in the past.

### 3.1.3 Saraswati Delta Structure at Rann of Kachchh (Gujarat)

The Saraswati River is supposed to be the mightiest river in the proto-historic times which might have flowed through Rann of Kachchh and finally discharged into the Gulf of the Arabian Sea. The Rann is today a salt impregnated marshy tract, which gets flooded during the rainy season. A number of references have been made about the Saraswati River meeting the Sea along the existing Kori Creek (Oldham, 1893). Further, it has been mentioned that major independent rivers like Sindhu and Saraswati met the Rann of Kachchh close to one another. Recent investigations have revealed that more than one river were involved in building up the deltaic deposits in the Great Rann. According to Roy and Merh (1977) the channels represent relicts of a delta of Nara River which is a tributary of Indus River. The Saraswati delta is characterized by strikingly uneven topography, marked by areas of higher grounds (bet of river borne sandy and silty sediments) and depression with distributary channels which form palaeo-delta structure (Malik et al., 1999).

Digital image processing of IRS-P6 AWiFS and Radarsat SAR images reveals deltaic drainage pattern (Bird's Foot type) which is made up of complex intertwined channels. This structure was formed in the past by huge sediment discharge of Saraswati River within marshy land of Great Rann of Kachchh. The delta structures are restricted by the E-W fault in the south, known as Allah Band. However, some of these palaeochannels can be traced upto the Gulf of Kachchh which might have a link to the submerged Dwarka of Mahabharata times.

## 3.2 Ground validation as supporting evidence

Mapped courses of Saraswati River within Indian Territory have been validated through data from a variety of ground investigations viz. historic maps, geomorphic anomaly, archeological sites, core drilling / litholog data, ground water quality-yield-depth, age of ground water etc.

### 3.2.1 Analysis of Historical Maps

A number of historical maps of Indo-Pakistan region of Mughal period, prepared by the Italians, Dutch and British authors during 14$^{th}$ -17$^{th}$ Century A.D. have been studied. Presence of a stream parallel to Indus River in 17$^{th}$ century map indicates towards the presence of the dry channels of Saraswati. Water flows occasionally along this river during heavy monsoon or flood events that continued till 13$^{th}$ century and minor flows up to end of 16$^{th}$ century (Wilhelmy, 1969). The mapped courses of Saraswati main course runs 80 to 100 km east, parallel to the Indus river course.

### 3.2.2 Micro-geomorphic anomalies

The digital merging of PAN and LISS-III data has helped in delineation of narrower anomalies (Corn-cob structures) within and outside of the broad palaeochannels. Corn-cob structures are formed due to deposition of sand over a flowing as well as stagnant water bodies. As a result, flowing water tries to remove the deposited sand load to the best of its capacity and forms a different kind of dune pattern of normally much smaller dune size in larger dune fields. On satellite imagery, these appeared like scattered corn grains resembling a corn-cob feature. In Jaisalmer area, corn-cob structures (lake shaped as well as channel shaped forms) have been delineated on the satellite images, indicating the presence of fluvial activity in this region in the past.

### 3.2.3 Archaeological Studies

Discovery of the great Harappan site at Mahenjodaro (Indus Valley) and a large number of contemporary sites at Kalibangan (Rajasthan), Banawali and Rakhigarhi (Haryana), Dholavira and Lothal (Gujarat) along Saraswati River gave a tremendous impetus to the archaeologists to study the growth of civilization in this subcontinent (Lal, 1979, 2009). More than 1200 archaeological settlements, including many prosperous towns of Mature Harappan culture (4600-3900 BC) have been discovered in the Saraswati River Basin (Mughal, 1981; Valdiya, 2002). Compilation of all the available archaeological sites of Early Harappan, Mature Harappan and Late Harappan sites in northern states of India indicates 436 sites in Haryana, 157 sites in Punjab, 100 sites in Gujarat and only 36 sites in Rajasthan (Joshi et al., 1984; Kalyanraman, 1999; Possel, 2000).

Plotting of 36 sites of Early-Harappan and Mature Harappan period (4600-3900 BP) in Rajasthan shows that most of the archaeological sites in Ganganagar and Hanumangarh districts fall along the Ghaggar River, indicating Ghaggar to be the palaeo Sarasvati course (Gupta et al., 2004). Similarly, 436 archaeological sites of Haryana have been plotted and overlaid on the palaeochannels map (Bhadra and Sharma, 2011). It has been observed that most of the Mature Harappan sites (Kunal, Birdana, Banawali, Agroha and Rakhigarhi) lie on the Ghaggar (Saraswati) palaeochannels. Clustered locations of Harappan sites have also been observed along Sutlej palaeochannels in Punjab and northern Gujarat. Mostly, Late Harappan sites have been found to lie in eastern Haryana and Punjab. The change in the pattern of settlements of the Late Harappan time (3900-3300 B.P) also clearly brings out the large-scale migration from the middle reaches to the upper catchment area of Sarasvati River. Thus, the association of Harappan sites and the delineated palaeochannels points towards a close link with the Vedic Saraswati civilization, beyond doubt.

### 3.2.4 Drilling/ Litholog Data Analysis

Using the delineated palaeochannels maps as a guide, 14 tube wells have been drilled by Ground Water Department (GWD), Rajasthan jointly with the Central Ground Water Board (CGWB) along Tanot-Kishangarh-Longewala-Ghotaru section Jaisalmer district of Rajasthan (Fig.6). Observation of core drilled samples (Table-1) suggests that the palaeochannels most often consist of alternating layers of fine to medium and coarse grained sand and sometimes have gravel columns, indicating presence of fluvial regime. The column thickness of palaeochannels as inferred from drilling data in this area ranges from 35 to 80 m. Coarser sediments are noticed at a depth ranging from 40-125m (general depth of 55-85m) and have been encountered in 9 of the 14 tube wells drilled. Water levels in most of the bore wells drilled ranged in between 35-60 m. The geophysical resistivity surveys between Tanot and Ghantiyali in Jaisalmer region also indicate presence of palaeochannels in this region (Singh et al, 1994).

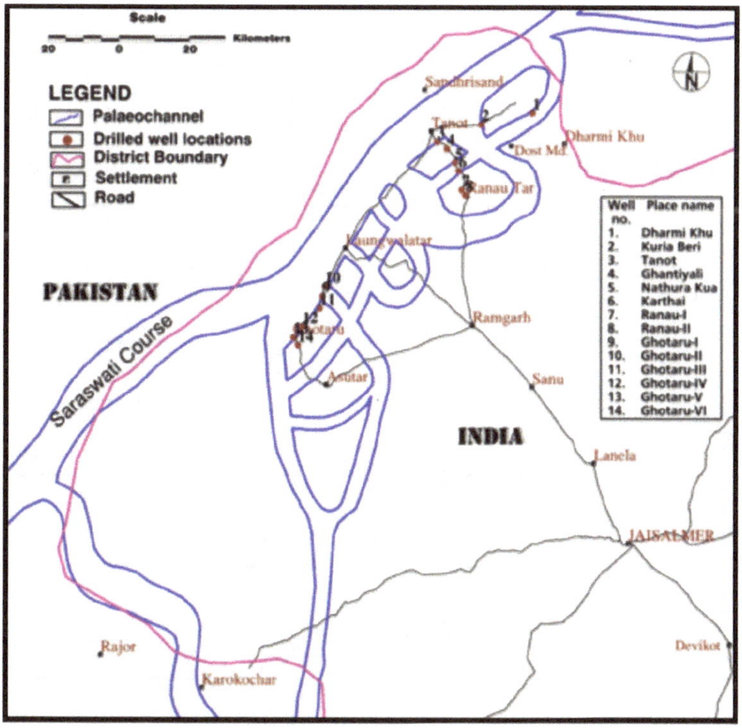


**Figure 6 :** Locations of tube wells drilled along the palaeochannels in Jaisalmer district along Tanot-Ghotaru-Longewala section.

Similarly, lithology data analysis of 304 drilling sites (Source: Ground Water Cell, Govt. of Haryana) in 9 districts of northern Haryana indicates that most of the lithologs falling near the palaeochennels have medium to coarse grained sands and associated gravel and pebbles at a depth between 10 to 100m. Mostly, the coarse grained sands/gravels have been encountered close to the Ghaggar palaeochannels. Based on the collection of subsurface lithofacies data from well logs, Saini et al. (2009) attempted to map the buried channel-floodplain systems of a part of the northwestern Haryana plains and provides evidence of buried major sand bodies at various depths. The Optical Stimulated Luminance (OSL) dates of sands at different levels in northern Haryana suggest the existence of much older palaeochannels (~26 to 28 Ka) than the Saraswati palaeochannels (2.9 to 5.9 Ka), as estimated by most of the previous workers.

**Table 1 : Data of the tube wells drilled along the palaeochannels in the Jaisalmer district, Rajasthan**

| Si. No. | Drilled well name/location | Yield (lph) | Quality (TDS) | Depth Drilled (m) | Static Water Level (m) | Aquifer material as observed in lithologs |
|---|---|---|---|---|---|---|
| 1. | Tanot (3.5 Km from Ghantivali to Tanot) | 11250** | 2650 | 125 | 33 | Mainly fine grained sand, medium grained at certain levels. |
| 2. | Ghotaru-I (12.5 Kms Ghotaru to Longewala) | 13500** | 6506 | 151 | 43 | Medium to coarse sand and gravel **(Out of main channel)** |
| 3. | Ghotaru-II (14.5 Kms Ghotaru to Longewala) | Not Developed | - | 151 | - | Fine grained sand/S.St-Kankar **(Out of main channel)** |
| 4. | Ghotaru-III (10 Kms from Ghotaru to Longewala) | 2250** | 4337 | 151 | 48 | Fine grained sad and very coarse grained sands |
| 5. | Ghotaru-IV (3 Kms from Ghotaru to Longewala) | 32400* | 3554 | 151 | 45 | Medium to fine and coarse grained sands |
| 6. | Ghotaru-V 150 m NE of Fort | 33750* | 1536 | 148 | 33 | Coarse gravelly and fine to medium grained sands, occasional clayey. |
| 7. | Ghotaru-VI (1.5 Kms from Ghotaru to Asutar | 22500* | 934 | 125 | 46 | Dominantly medium to coarse sands, fine grained and clayey sands at certain |
| 8. | Dharmi Khu (3 Kms from Kishengarh to Dharmi Khu) | 35100* | 1024 | 153 | 40 | Fine and medium grained sands |
| 9. | Ranau-I*(Ranau-Tanot Road) | 9120** | 1010 | 102 | 42 | Fine grained sand and silt with kankar; fined to medium sand |
| 10. | Ranau-II* (Close to Ranau village) | 18240** | 1000 | 120 | 58 | N.A. |
| 11. | Karthai* (9.5 km from Ranau on Tanot road) | 12312* | 1800 | 125 | 42 | Mostrly Find sand medium at certain levels |
| 12. | Nathura Kua* (4.5 Km from Tanot) | 12768** | 2656 | 120 | 36 | Fine grained sand and silt with kankar |
| 13. | Kuria Beri* | 12768** | 1295 | 131 | 32 | Mostly Fine sand |
| 14. | Ghantiyali I* (500m from Ghantiyali Mandir to Tanot) | 11400** | 2200 | 130 | 62 | Fine grained sand |

*Note : Yield by pumps normally higher than the compressor yield by a fector of 1.5 to 2.0 * Wells are drilled by the CGWB ** Compressor yield * Pump test yield*

### 3.2.5 Quality and Quantity of Groundwater in the Palaeochannels

Analysis of ground water samples indicates that the water quality is quite good (potable water) for most of the drilled tube wells, as compared to tube/dug wells away from the channel. For example, the well drilled at Tanot shows TDS in the order of 2650 ppm in comparison to the existing well at Tanot (TDS~9000 ppm) which is about 1.5 km away from the channel. However, water quality improves (TDS<1500 ppm) around the villages like Ghotaru (V & VI), Dharmi Khu, Ranau, Karthai and Kuria Beri. As per the Local Government norms, in the desert region groundwater upto EC~4000 µS/cm are supplied to villages for drinking water purposes. The ground water away from palaeochannels is saline (non-potable), EC >8,000 µs/cm at 25°C (CGWB Report, 2002 & GWD Report, 2003). Even though slight higher TDS water is encountered in some of the bore-wells drilled along the palaeochannels, these channels most often form the source of potable quality ground water in water-scared Thar Desert region.

Similarly, hydrogeological data analysis in northern Haryana shows wide variation in EC zoning pattern. Lower EC (2000-4000 µS/cm) values are observed in Kurukshetra, Kaithal and Fatehbad districts, indicating good quality water close to the demarcated palaeochannels. Poor quality water (brackish to saline) is found in the central Haryana (Rohtak district) and northwest Haryana (Sirsa district) where EC values are very high (>4000 µS/cm). The discharge contour map shows the contour line with higher values in the northern part of Kaithal & western part of Karnal & Panipat (21000-39000 cusec) where palaeochannels network exist.

### 3.2.6 Age of Groundwater in the Palaeochannels

The isotopic ages ($^{3}H$, $^{18}O$ and $C^{14}$) of 17 ground water samples from the existing wells along the palaeochannels in Jaisalmer district of Rajasthan were analyzed by the Bhabha Atomic Research Center (BARC), Mumbai on the basis of Pearson model (Table-2). The analysis indicates the variation in ground water ages from 1340 to 18880 BP at different localities from NE to SW (Rao and Kulkarni, 1997; Nair et al., 1999). These areas lie either on the palaeo channel or very close to it viz. Kuriaberi (1340 BP), Ghantiyali (550 BP), Ranau (1930 BP), Sadewala (18800 BP), Longewala (12400 BP), Ghotaru (8910 BP), Dost Mohd. (2000 BP) etc. Hence, age analysis of water samples indicate towards a palaeo source of water along the channels that may be linked to Ghaggar (Saraswati) palaeochannels. The shallow (>30m) and deep (>60m) groundwater along the palaeochannels have similar chemical and isotopic characteristics. The ground water do not have isotopic signature of present day Himalayan Rivers. However, the deep groundwater indicates towards origin from Himalayan source. Based on the relative radiocarbon ages, ground water movement (velocity) along the palaeochannels has been estimated as 5m/yr. The isotopic analysis of ground water samples from the wells drilled on palaeochannels in Jaisalmer indicated that water stored in palaeochannels is palaeo-water with no connection with Himalayan source.

**Table 2 :** Age of Groundwater along the palaeochannels in Jaisamer district, Rajasthan

| ID. No. | Location | Well type | $^{3}H$ (TR±0.5) | $\delta^{13}C$ | 14 ± 1σ (pMC) | Age (BP) (a), UC | Model Age, a (Pearson) |
|---|---|---|---|---|---|---|---|
| D1 | Dharmikua | DW | 2.1 | -9.6 | 79.5 (2.2) | 1900 | m |
| T1 | Kishengarh | TW | 0.3 | -5.7 | 47.3 (1.4) | 6190 | m |
| D3 | Kuriaberi | DW | 0.5 | -8.3 | 58.8 (1.6) | 4390 | 1340 |
| D4 | Nathurakua | DW | 0.3 | 7.9 | 69.3 (1.8) | 3000 | m |
| T2 | Ghantiyali | TW | 0.5 | -4.0 | 31.2 (1.2) | 9630 | 550 |
| D5 | Ghantiyali | DW | 0.6 | - | 54.9 (1.5) | 4960 | - |
| D7 | Gajesing ka tar | DW | 2.1 | -7.7 | 64.9 (1.9) | 3570 | m |
| T3 | Ranau | TW | 0.6 | -7.4 | 48.8 (1.5) | 5930 | 1930 |
| T4 | Sadewala | TW | 0.4 | -7.7 | 6.6 (0.7) | 22450 | 18800 |
| T5 | Loungewala | TW | 0.4 | -5.6 | 10.4 (0.9) | 18700 | 12400 |
| T7 | Ghotaru | TW | 0.4 | -7.3 | 20.7 (1.0) | 13000 | 8910 |
| D12 | Ghotaru | DW | 1.1 | - | 62.7 (1.7) | 3860 | - |
| T8 | Asutar | TW | 0.4 | -7.5 | 36.1 (1.3) | 8420 | 4540 |
| D14 | Langtala | HP | 0.3 | -6.2 | 68.6 (2.0) | 3120 | m |
| D15 | Langtala | DW | 1.0 | - | 64.8 (1.7) | 3590 | - |
| D17 | Dostmoh, Kua | DW | 1.0 | -7.6 | 49.7 (1.5) | 5780 | 2000 |
| D18 | Mituwala | DW | 0.6 | -11.0 | 57.9 (1.7) | 4520 | 3800 |

*DW : Duagwell; TW : Tubewell; DCB: Dug cum Borewell; TR : Tritium ratio; PMC; Percent Modern Carbon, UC : uncorrected; m : Modern*

*(Source : Ground Water Department (GWD), Govt. of Rajasthan. Data analyzed by BARC, Mumbai)*

### 3.2.7 Static Reserve of Groundwater in the Palaeochannels

Based on the hydrogeological investigations in Jaisalmer district, three major aquifer zones viz. A1, A2 and A3 have been identified along palaeochannels of Saraswati River (Fig.7). The zonation has been done based on the quality of ground water, yield of wells and effective aquifer thickness (Table-3). The aquifer is made up of the unconsolidated and semi-consolidated sediments, comprising of gravel and sands with thin intercalation of clay and kankar. Flow of ground water in the area is from NE to SW direction. From electrical conductivity of ground water it is inferred that on either side of palaeochannels, quality of ground water is saline to highly saline. This study shows that potable ground water reserves occur along sections running through Kishangarh, Kuriaberi, Ghantiyali Ka Mandir, Ranau, Ghotaru and Asutar.

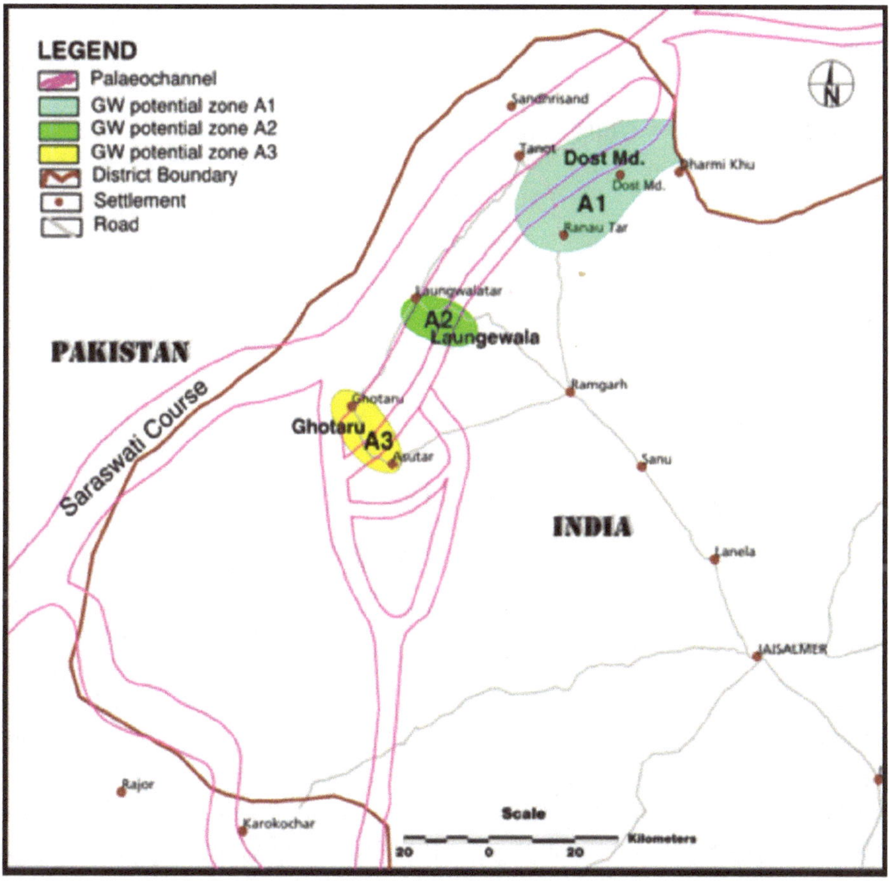


**Figure 7 :** Groundwater potential zones (A1, A2 and A3) identified in Jaisalmer district, Rajasthan (Source: GWD Report)

The isotope study conducted in the area shows that most of the tube wells and dug wells have water with negligible tritium and carbon-14 values, indicating trapped palaeo groundwater pockets. There is no monsoonal recharge in the area mainly due to low rainfall, very high rate of evaporation exceeding rainfall values and deep ground water levels. Since, the above estimation is based on limited number of drilled wells along the palaeochannels in Jaisalmer district, it is suggested that static ground water reserves must be used cautiously for drinking water supply.

**Table 3** : Hydrogeological Characteristics of the Aquifer Zones delineated along the Palaeochannels in Jaisalmer District, Rajasthan

| Potential Zones/Parameters | Zone-A1 | Zone-A2 | Zone-A3 |
|---|---|---|---|
| Formation Material | Alluvial | Alluvial | Alluvial |
| Area (sq.km) | 547 | 1100 | 150 |
| Direction of Locations | NE part | SW part | SW part |
| Water Level (m) | 31-62 | 45-63 | 39-60 |
| Water Level Fluctuation (m) | 014 | 0.02 | 0.69 |
| Average Yield ($m^3$ph) | | 136 | 22.6 |
| EC (μS/cm) | 1100-4000 | 2960-4000 | 1100-4000 |
| Aquifer Thickness (m) | 18 | 14 | 9 |
| Static GW Reserves (Mcu.m) | 590 | 92 | 81 |
| Recharge due to RF | Negligible | Negligible | Negligible |

*(Source : GWD Report, 2003)*

## 4.0 Conclusion

Based on the analysis of several satellite images coupled with the Remote Sensing techniques, the entire course of Saraswati palaeochannels have been delineated from the Himalayan foothills to the Rann of Kutchch, passing through the Thar Desert in NW India (Fig.8). To be a vibrant mighty river, the Saraswati in Vedic Period must have been contributed by any major river system of the Himalaya. Presently, Sutlej and Yamuna are the two perennial rivers which are likely to be the feeder channels of Vedic Saraswati River in the past. Beyond the range of Siwalik and Lesser Himalaya, these two rivers are fed with the permanent glaciers in the Higher Himalaya. It has been observed from the satellite images that the size of the glacier of Sutlej River is much larger than the size of the Yamunotri and Bandarpunch glaciers. But, due to tectonic changes in the past, these two perennial rivers shifted their courses viz. Satluj to the west to join river Indus and Yamuna to east to join River Ganges near Allahabad.

**Figure 8 :** Satallite Image showing the entire network of Saraswati palaeochannels (dark blue lines) in NW India.

The mapped course is validated with a variety of ground data such as archaeological sites, drilling and hydrogeological data. Radiometric ages of river sediments suggests that the age of Vedic Saraswati River (older palaeochannels) may be as old as 28,000 years and flourished during 5900 to 2900 years ago which may be represented by younger palaeochannels in Haryana (Saini et al., 2009). However, the age of trapped ground water in the palaeochannels in Rajasthan shows contemporary age of Saraswati (1340-8910 BP) and as old as 18800 BP (Nair et al., 1999).

## Acknowledgements

The authors are extremely grateful to Dr. V. K. Dadhwal, Director and Dr. Y. V. N. Krishnamurthy, Dy. Director, NRSC, Hyderabad for their valuable guidance to carry out this research work. We are equally grateful to Dr. S. Kalyanraman of Saraswati Nadi Sodh Prakalp, Chennai; Shri Darshan Lal Jain of Saraswati Nadi Sodh Sansthan, Yamunanagar; Dr. Amal Kar, CAZRI, Jodhpur, Dr. A. K. Gupta, Ex. Scientist of RRSC-W and others for their active association, encouragement and fruitful discussion at various stages of the project work.

## References

Bakliwal P.C. and Grover A.K. (1988) "Signature and migration of Sarasvati river in Thar desert western India, *Records Geological Survey of India*, 116 (3-4) : 7-86

Bhadra, B.K., Gupta, A.K. and Sharma, J.R. (2009) 'Saraswati Nadi in Haryana and its linkage with the Vedic Saraswati River – Integrated study based on satellite images and ground based information'. *Journal of Geological Society of India*, Bangalore (Springer Co-Publ.), 73 : 273-288

Bhadra, B.K. and Sharma J.R. (2011) *Satellite Images as Scientific Tool for Saraswati Palaeochannel and its Archaeological Affinity in NW India*. In: Proceedings of International Seminar '*How Deep are the Roots of Indian Civilization*' by Draupadi Trust, New Delhi, pp.104-125 (in press).

Bhadra, B.K., Sharma, J.R. and Bharadwaj, Trilok (2005) '*Delineation of Sutlej Palaeo Channel using IRS LISS-III and Radarsat SAR data in parts of Punjab: Its possible linkage with "Lost River Saraswati" in Northwest India*'. Published Abstract in Indian Geological Congress (IGC) National Conference at Delhi University during 2-4 December, 2005, pp.8-9.

Chauhan, D.S. (1999) Mythological observations and scientific evaluation of the Lost Saraswati River. In: Vedic Saraswati (Eds. B.P. Radhakrishna and S.S.Merh). *Mem. Geol. Soc. India*, 42 : 35-45

Ghose, B., Kar, A. and Hussain, Z. (1979) The lost courses of the Saraswati River in the Great Indian Desert: new evidence from LANDSAT imagery. *Geograph. Jour., London*, 42 : 35-45

Ghose B, Kar A and Hussain Z (1980). "Comparative role of Aravalli and Himalayan river systems in the fluvial sedimentation of Rajasthan desert", *Man and Environment*, 4 : 8-12

Gupta S.P. (1996). Indus - Saraswati Civilization: Origins, problems and Issues. Pratibha Prakashan, Delhi

Gupta, A.K., Sreenivasan, G. and Sharma, J.R. (2001). An attempt towards establishing inter-relationship between palaeo-drainage courses and the archaeological sites found in the Saraswati River Basin, Western Rajasthan. *NNRMS Bull.*, 26: 14-23.

Gupta. A.K., Sharma, J.R. Sreenivasan, G. and Srivastava, K.S. (2004) New findings on the course of River Saraswati. *Jour. Ind. Soc. Remote Sensing*, 32 (1) : 1-24

Joshi, J.P., Madhu Bala and Jas Ram (1984) The Indus Civilisation: A reconsideration on the basis of distribution maps: In: Frontiers of the Indus Civilisation, B.B.Lal and S.P.Gupta (Eds.), Books and Books, New Delhi, pp.511-539/

Kalyanraman, S. (1999) Saraswati River, Goddess and Civilization. In: Vedic Saraswati (Eds. B.P. Radhakrishna and S.S.Merh). *Mem. Geol. Soc. India*, 42 : 25 - 34

Kar A. (1995). Geomorphology of arid western India. Memoirs Geological Society of India, 32 : 168 - 190

Kar, A. and Ghose, B. (1984) The Drishadvati River System of India: An Assessment and New Findings. 150 (2) : 221 - 229

Krishnan, M.S. (1953) The structure and tectonics of India. *Mem. Geol. Surv. Ind.*, 81 : 109

Lal B. B. (1979) Kalibangan and the Indus Civilization. In: Agrawal, D. P. and Chakraborty, D. K. (eds) Essays in Indian Protohistory, B. R. Publ. Co., Delhi, pp.65-97.

Lal B. B. (2009) How deep are the Roots of Indian Civilisation? Archaeology Answers. Aryan Books International, New Delhi.

Malik , J.N., Merh, S.S. and Sridhar, V. (1999) Palaeo-Delta Complex of vedic saraswati and other Ancient Rivers of Northwestern India. *Mem. Geol. Soc. India*, 42 : 163 - 174

Misra, V.N. (1995). Geoarchaeology of Thar Desert, North-West India. In : *Quaternary Environments and Geoarchaeology of India* (Eds.: S. Wadia, et. al.). *Geological Society of India*, Bangalore, pp. 210-250.

Mughal, Md. Rafique (1981) New Archaeological evidence from Bahawalpur, in Indus civilization. In: New Perspectives, A.H. Dave (ed.) pp.33-42. Islamabad: Quaid-i-Azam University.

Nair, A.S., Navada, S.V. and Rao, S.M. (1999) Isotope study to investigate the origin and the age of ground water along the palaeo channels in Jaisalmer and Ganganagar districts of Rajasthan. In : *Vedic Saraswati* (Eds.: B.P. Radhakrishna and S.S. Merh). *Memoir 42, Geol. Soc. India,* pp. 315-321.

Oldham, C.F. (1893) The Saraswati and the lost river of the Indian desert. *Jour. Royal Asiatic Soc. London,* 34 : 49-76.

Possehl, G.L. (2000) Indus Age – The Beginnings. IBH & Oxford Publication Co., Delhi.

Puri V.M.K. (2001) Origin and course of Vedic Saraswati River in Himalaya - Its secular desiccation episodes as deciphered from palaeo-glaciation and geomorphological signatures. *Geol. Surv. Ind. Sp. Publ.* No,53, pp.175-191.

Radhakrishna (1999) Vedic Saraswati and the dawn of Indian Civilization. In: Vedic Saraswati (Eds. B.P. Radhakrishna and S.S.Merh). *Mem. Geol. Soc. India,* 42 : 5-13.

Rajawat A.S. and Narain A. (1996). "Reconstruction of palaeodrainage network of Northwest India" – A review of remote sensing based studies. SAC Scientific note No. SAC/RSA/RSAG/MWRD/SN/01/96.

Ramasamy S.M., Bakliwal P.C. and Verma R.P. (1991). Remote Sensing and river migration in western India. *IJRS* 12 : 2597 - 2609

Rao, S.M. and Kulkarni, K.M. (1997) Isotope hydrology studies on water resources in western Rajasthan. *Curr. Sci.,* 72 (1) : 55-61

Roy, B. and Merh, S.S. (1977) The Great Rann of Kutch: An intriguing Quaternary terrain. Rec. Res. Geol., (Hindustan Publ., Delhi), 9 : 100-108

Sahai Baldev (1993) Remote sensing of deserts: Indian Experience. *Jour. Arid Environments,* v.25, pp.17-18.

Sahai Baldev (1999) Unraveling the 'Lost' Vedic Sarasvati. In: Vedic Sarasvati (Eds.: B.P. Radhakrishna and S.S. Mer). *Memoir Geol. Soc. of India,* 42: 121-142.

Saini, H.S., Tandon, S.K., Mujtaba, A.I., Pant, N.C. and Khorana, R.K. (2009) Reconstruction of Buried channel-floodplain systems of the northwestern Haryana Plains and their relation to the 'Vedic Saraswati'. *Current Science,* 97 (11) : 1634-1643

Sharma J.R. and Bhadra B.K. (2009a) 'Satellite Imagery and Saraswati: Tracing the Lost River'. Article in Geospatial Today, April, 2009, pp.18-20.

Sharma J.R. and Bhadra B.K. (2009b) 'Vedic Saraswati course and its likely connectivity with the Himalayan River Sources: A study based on satellite imagery, DEM, Archaeological and Drilling data. In: Proceedings of International Conference at Kurukshetra, Nov., 2009, pp.55-57.

Sharma, J.R., Gupta A.K. and Bhadra B.K. (2006) Course of Vedic River Saraswati as deciphered from latest Satellite data. 'Puratatwa', Bulletin of Indian Archaeological Society, New Delhi, 2006, No.36, pp.187-195.

Singh, A.P., Babel, M.K. and Soni, V. (1994). A report on geophysical resistivity exploration to trace the path of buried Saraswati river and its tributaries, district Jaisalmer. Ground Water Department-Jodhpur, May 1994, 41p.

Singhvi, A.K. and Kar Amal (2004) The Aeolian Sedimentation Record of the Thar Desert. *Proc. Indian Acd. Sci.* (Earth Planetary Sci.), 13 (3) : 371-401.

Sood, R. K. and Sahai, B. (1983) Hydrographic changes in northwestern India. *Man and Environment,* 7 : 166-169.

Stein, A. (1942) A survey of ancient sites along the lost Saraswati river. *The Geographical Journal.*, 99 : 173-182

Valdiya K.S. (1996) "River Piracy - Saraswati that Disappeared" "Resonance" Jr. of *Academy of Sciences*, Bangalore, 1(5) : 19-28.

Valdiya, K.S. (2002) Saraswati - The River That Disappeared. University Press (India), Hyderabad, 116p.

Wilhelmy, H. (1999) The ancient river valley on the eastern border of the Indus plain and the Saraswati problem. (Extracted from Zeit. Geomorph., NF Suppl., Band-8, 1969, pp.76-93). In: Vedic Saraswati (Eds. B.P. Radhakrishna and S.S.Merh). Mem. *Geol. Soc. India*, 42 : 95-112

Yashpal, Sahai, B., Sood, R.K. and Agarwal, D.P. (1980) Remote Sensing of the 'Lost' Saraswati River. *Proc. Ind. Acad. Science* (Earth and Planetary Sciences), 89 : 317-331

# Chapter - 5

# Land Degradation Situation in Thar Region, India : Type, Extent and Associated Factors

**A.K. Bera, S. Rama Subramoniam and J.R. Sharma**

## Abstract

*Land degradation is defined as human induced or natural process that negatively affects the lands to function effectively. It is temporary or permanent lowering of the productive capacity of land (UNEP, 1992). Several definitions of land degradation have been suggested by different authorities to express the degree of deterioration in potential of land in terms of physical, chemical and biological aspects. It has both environmental and economic consequences. The information on land degradation is needed for a variety of purposes such as to improve productivity levels in degraded lands through reclamation programmes, rational land use planning and also to bring more areas under cultivation.*

*Thar desert comprising the western part of Rajasthan, northern part of Gujarat, and some parts of Punjab and Haryana (Santra et al., 2004). The area is characterized by not only in high variability in average annual rainfall but also its erratic distribution and intensity. Rainfall ranges from 100-500 mm per year and almost entire amount is received during July to September. The temperature fluctuate form near freezing to as high as 50°C. The desert terrain is dominated by gently sloping plains, broken by some dunes and low barren hills. Major soil orders found in the Thar desert region includes Entisols, Inceptisols, Aridisols, Alfisols and Vertisols. The vegetation is controlled by edaphic conditions and communities vary distinctively among various soil types.*

*Soil erosion due to wind and water is the main cause of land degradation. The process of erosion occurs when rain and wind strike the soil aggregates and separate the soil particles. These detached particles are then transported through the action of runoff water and wind. Eroded deep soil with weathered parent material as substrata have flexibility to be altered again through application of fertiliser, organic matter and other management practices. On the other hand, eroded shallow soils with hard rock as substrata cannot be renewed by above means, and so they degrade due to erosion resulting into loss of productivity. Wind erosion is predominant and it gradually removes silt, clay and organic matter from the surface soil leaving the coarse material behind. In severe conditions, the sand begins to drift and form unstable dunes which can encroach on surrounding productive lands. Wind erosion also causes serious problems for infrastructural facilities such as roads, railway tracts, canal and settlements which may get buried during dust storms. Sporadic high intensity rainfall in Thar desert region results in water erosion in the form of sheet, rill and gullies. Sheet erosion takes place on slopes because of overland flow. Rills form in areas where overland flow accumulates. Excess accumulation of flow results in gully erosion. Rainfall erosivity, soil erodibility, topography, vegetation cover, management and conservation practices are the major factors affecting the soil erosion.*

*Database on soil loss due to water and wind erosion is important for advocating the conservation measures properly (Shyampura and Sehgal, 1995). Qualitative assessment of erosion in western Rajasthan on 1:1 million scale was carried out by Narain et al. (2000) using field characteristics and satellite image interpretation. Map of degraded lands on 1: 50000 scale for Thar desert region is being prepared by Department of Space, Govt. of India using kharif, rabi and summer season Resourcesat-1 LISS III satellite data for the period 2005-06.*

## Introduction

Land, a natural component of an ecosystem, serves many purposes of man. It is composed of several elements namely, soil, water, flora and fauna. Land degradation is defined as human induced or natural process that negatively affects the lands to function effectively. It is temporary or permanent lowering of the productive capacity of land (UNEP, 1992). Numerous definitions of land degradation have been suggested by different authors to express the degree of deterioration in potential of land (physical/chemical/ biological aspects). Besides economic uses, land performs environmental regulatory (ecological) functions also. Land degradation may be referred as loss in certain intrinsic qualities or a decline in the land's capability to perform vital functions (both economic and ecological). It has both environmental and economic consequences.

From the viewpoint of agriculture, land degradation is said to happen when the productivity of a land use or production system declines (both in terms of quantity and kind) continually in relation to normal weather condition and non-changing management practices. It can be described as a sort of weakening in the capability of land to produce benefits when put to a particular land use under a specific set of management options (Blaikie and Brookfield, 1987). Degradation is most likely to take place where land use and land attributes are mismatched (FAO, 1976). The information on land degradation is useful for a variety of purposes such as to improve productivity levels in degraded lands through reclamation programmes, rational land use planning and also to bring more areas under cultivation.

## Stages of Land Degradation

There are various stages of land degradation. Land degradation leading to loss in total productivity is not a sudden phenomenon. Rather, it tends to occur along a retrogressive pathway. In the initial stages, the reduction in potential productivity is low. 'Slight degradation' represents a persistent productivity loss of 10% (Dregne and Chou, 1992) to 15% (Sehgal and Abrol, 1994). It can be overcome by adopting appropriate management practices. If land degradation is allowed to proceed, productivity declines further, a stage is reached when conventional agronomic practices are found inadequate to regain this lost capacity. This stage of degradation is called 'moderate degradation' and corresponds to a potential productivity loss of 10 to 25% (Dregne and Chou, 1992) or 33% (Sehgal and Abrol, 1994). Only ameliorative management can restore these degraded lands. Land degradation is considered severe when the productivity loss reaches 50% to 66% (Dregne and Chou, 1992; Oldeman, 1988; and Sehgal and Abrol, 1994). Upto this point, land degradation is reversible because restoration of land is still possible, although at high costs and with major engineering works. Further degradation makes a land unreclaimable economically. This stage of land degradation is pracically irreversible. Reversible degradation process should be monitored constantly and serviced timely. Land might be degraded for one purpose and not necessarily for another. Apparently, it is usespecific, management-sensitive, and thus not necessarily permanent (Johnston and Lewis, 1995).

## Impact of Climatic Variation and Anthropogenic Factors

The fluctuating rainfall pattern is the most important concern in semi-arid and arid regions. More than fifty percent of the rainfall often falls in less than 10% of the rainy days associated with intensive storm events. It is this high variability that makes these regions climatically unstable and particularly prone to drought and subsequent land degradation. When coefficients of variation of annual precipitation frequently exceeds 30%, these regions become susceptible to land degradation and desertification (UNEP-UNCOD, 1978; Ahmad and Kassas, 1987). Although evidence is not enough to link drought to the onset of land degradation, the consequences of land degradation are most pronounced under drought conditions (Dregne, 1978, Hare, 1985, and Mainguet, 1994).

Theoretically, land degradation starts when either the potential productivity associated with a land use system becomes non-sustainable or the land is not able to perform its environmental regulatory function. This indicates that land, normally a renewable natural resource, has lost its ability to recover (partially or totally) from a disturbance. Degradation is mainly due to interaction of the land with its users/ organisms. Of all the users/organisms dependent on land for their survival, humans assert predominant influence on it. Land has a finite capacity to support diverse organisms called "carrying capacity". If the carrying capacity is exceeded persistently, land gets progressively degraded and loses the ability to renew itself. Further, mismatches of land use and land attributes aggravate degradation. Since humans determine land use, definition land degradation should be considered a social problem (Blaikie and Brookfield, 1987). However, restorative management with proper inputs and technologies, can reverse the negative effects of exploitation by human. Lack of capability to invest in restoring inputs, small and marginal farmers are forced to exploit their limited resources often leading to land degradation. In addition, land use policies (typically ownership rights and rules governing tenancy) do not encourage the adoption of relatively permanent land conservation technologies (Syers et al., 1996). Therefore, the facts link land degradation directly to human actions and indirectly to climatic variations. While decline in productivity is the primary measure of land degradation, its effects on land are visible through the loss in quality of soil, water and vegetation - the three ecologically and economically important attributes of land.

Along with anthropogenic factors, a range of natural factors endemic to territories susceptible to drought and desertification are supposed to influence the progress of land degradation (Blaikie and Brookfield, 1987; Barrow, 1991 and Lal, 1997). Year-round aridity limits bio-productivity and slows down the processes of soil development, resulting in poor quality soils (Stewart et al., 1991). These environmental disadvantages make dryland ecosystems more fragile (readily susceptible to damage).

## Geographical Setting of the Thar Desert

The Thar, sometimes called the Great Indian Desert, includes the arid portions of western India and eastern Pakistan. Arid portion of western India covers the western part of Rajasthan, northern part of Gujarat, and some parts of Punjab and Haryana (Santra et al., 2004). However, there is ambiguity to demarcate the area which is naturally desert and the area with arid appearance resulted from the activities of man. Facts at a glance for western part of Rajasthan under Thar desert region are given in Table 1. This region is consists of 12 administrative districts namely, Barmer, Bikaner, Churu, Ganganagar, Hanumangarh,

Jaisalmer, Jalor, Jhunjhunu, Jodhpur, Nagaur, Pali and Sikar (Figure 1). The false colour composite (FCC) of the study area is depicted in figure 2.

**Table 1 :** Facts at a glance for western part of Rajasthan under Thar desert region

| Extent | N 24° 37' 00"- N30°10' 48"<br>E69°29' 00"- E76°05' 33" |
|---|---|
| Area | 208184 sq km |
| Range of mean annual rainfall | 100-500 mm |
| Administrative districts | 12 |
| Human population (2001) | 22496790 |
| Rural | 17571142 |
| Urban | 4925648 |
| Human population density | 108 per sq. km |
| Total villages | 15323 |
| Total sub districts | 86 |
| Total blocks | 85 |
| Total towns | 88 |
| Village density | 1 per 13.62 sq km |
| Livestock population (1997) | 28220000 |
| Net sown area | 52.45 % (109201 sq km) |
| Major crops | Kharif: Bajra, Guar, Moth, Jawar, Cotton & Sesamum<br>Rabi: Rapeseed, Mustard, Wheat & Gram |
| Forest area | 4436.28 sq km |
| Roads | 36112 Km |
| Railways | 3240 Km |

*Source: Census of India, 2001*

**Figure 1 :** Various districts of western Rajasthan under Thar desert region

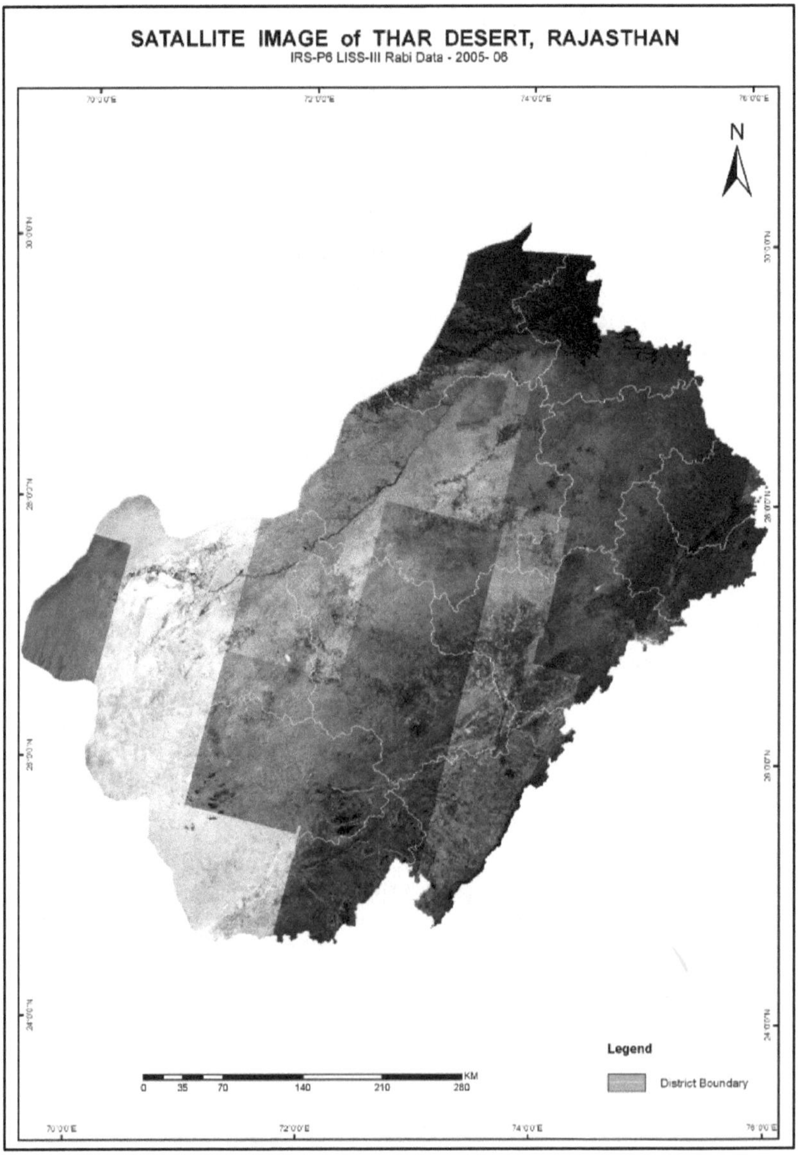


**Figure 2 :** FCC Image of Thar region of Rajasthan

The region is one of the most populated deserts of the world. According to 2001 population census, the region has a population of 21496000, the average density being 108 persons per sq. km. As per 1997 livestock census, there are 28220000 livestock which has increased significantly over the decades. The Jaisalmer district has highest livestock to population ratio of 4.88 (Table 2).

**Table 2 :** Human and livestock population in various districts of western Rajasthan

| District | Population census 2001 (in ' 000) | Livestock census 1997 (in ' 000) | Ratio |
|---|---|---|---|
| Barmer | 1964 | 4177 | 2.13 |
| Bikaner | 1674 | 2531 | 1.51 |
| Churu | 1923 | 2100 | 1.09 |
| Ganganagar & Hanumangarh | 2305 | 2321 | 1.01 |
| Jaisalmer | 508 | 2477 | 4.88 |
| Jalor | 1448 | 1814 | 1.25 |
| Jhunjhunu | 1913 | 1248 | 0.65 |
| Jodhpur | 2881 | 3789 | 1.32 |
| Nagaur | 2774 | 3228 | 1.16 |
| Pali | 1819 | 2738 | 1.51 |
| Sikar | 2287 | 1797 | 0.79 |
| **Total** | **21496** | **28220** | **1.31** |

*Source: Census of India, 2001*

The main characteristic feature of this region is low annual rainfall which varies from 500 mm in the eastern part of the region to less than 100 mm in the westernmost part. Almost the entire amount is received during July to September. This scanty rainfall is often erratic in intensity and distribution with about 40-60% coefficient of variation in mean annual rainfall (Table 3), leading to prolonged droughts conditions frequently. Droughts are a recurring feature in many parts of this region with variation only in its magnitude from year to year. The region is gifted with plenty of sunshine but poor water availability for crop production. The temperature varies from near freezing to as high as 50°C. Strong to very strong winds with storms are experienced during May to July when south west monsoon sets in. The strongest winds with a mean speed of 27.2 kmph in the month of June are experienced in Jaisalmer district. The mean monthly wind speeds at different stations are given in Table 4.

**Table 3 :** Mean annual rainfall and its coefficient of variation in various districts of western Rajasthan

| Meteorological station | Mean annual rainfall (mm) | Coefficient of variation of annual rainfall (%) |
|---|---|---|
| Barmer | 267.7 | 58 |
| Bikaner | 293.7 | 55 |
| Churu | 366.3 | 40 |
| Ganganagar & Hanumangarh | 213.6 | 43 |
| Jaisalmer | 185.0 | 59 |
| Jalor | 369.1 | 53 |
| Jhunjhunun | 404.4 | 37 |
| Jodhpur | 361.4 | 53 |
| Nagaur | 315.3 | 50 |
| Pali | 419.5 | 45 |
| Sikar | 429.9 | 40 |

**Table 4 :** Mean monthly wind speed at meteorological stations in various districts of western Rajasthan

| Meteorological station | Wind speed (kmph) in month | | | | | | | | | | | |
|---|---|---|---|---|---|---|---|---|---|---|---|---|
| | Jan. | Feb. | Mar. | Apr. | May | Jun. | Jul. | Aug | Sept. | Oct. | Nov. | Dec. |
| Barmer | 7.6 | 7.5 | 9.1 | 10.6 | 12.9 | 14.2 | 12.4 | 10.6 | 9.6 | 7.2 | 5.3 | 6.4 |
| Bikaner | 4.6 | 5.1 | 6.5 | 7.3 | 10.1 | 13.3 | 12.8 | 11.0 | 9.4 | 5.3 | 3.5 | 3.7 |
| Ganganagar & Hanumangarh | 4.0 | 5.0 | 6.4 | 6.8 | 8.0 | 10.7 | 13.3 | 8.0 | 6.2 | 4.7 | 3.3 | 3.4 |
| Jaisalmer | 8.6 | 8.2 | 10.9 | 12.7 | 18.3 | 27.2 | 24.8 | 21.7 | 16.1 | 8.5 | 5.5 | 6.5 |
| Jodhpur | 8.9 | 8.8 | 9.8 | 10.2 | 15.0 | 18.5 | 16.6 | 12.9 | 10.6 | 6.6 | 5.8 | 7.3 |
| Sikar | 6.1 | 6.2 | 7.9 | 8.0 | 9.4 | 11.6 | 9.2 | 7.8 | 7.3 | 5.1 | 4.3 | 4.9 |

*Source: Climatological Tables of Observatories in India (1931 - 1967), India Meteorological Department, 1967*

As a whole the region is dominated with interspersed sandy and medium to fine textured surface materials. Gravelly and skeletal soils are found in mountains and hilly areas. Footsteps and water courses are not widespread. High soil salinity is observed in the uncultivated fine textured soils. Major soil orders found in the region includes Entisols, Inceptisols, Aridisols, Alfisols and Vertisols. Gently sloping plains broken by some dunes and low barren hills dominate the desert terrain.

Various soil types, gravely and rocky areas have different vegetation communities which are controlled by edaphic conditions. Trees and shrubs are common on open soils. Five distinct plant communities namely, salt desert, clay desert, stony desert, sand desert and riverside thickets are found. The vegetation of the salt desert is extremely scarce, dominated by bushes and a few summer annuals. The clay deserts are very dry with grasses dominating along with some bulb and tuberous plants. The stone desert areas are unfavourable to plants and support only a sparse growth of small woody plants. The sand deserts support grasses, a variety of shrubs and trees. The riverside thickets include several species of trees and shrubs.

Agriculture with rainfed cropping is dominant in western part of Rajasthan. Although net sown area is about 52% of the total area, but fallow lands and culturable wastelands are also put to kharif cropping during favourable monsoon years. Major land utilisation pattern is given Table 5.

**Table 5 :** Land utilization pattern (2002-2003) in various districts of western Rajasthan

| Districts | Land excluding fallow lands | Fallow lands (ha) | | Total cropped area (ha) | Forest (ha) |
|---|---|---|---|---|---|
| | Culturable wasteland (ha) | Fallow lands other than current fallow | Current fallow | | |
| Barmer | 266514 | 431260 | 624577 | 1149126 | 31026 |
| Bikaner | 823813 | 332707 | 1193411 | 311747 | 82543 |
| Churu | 12154 | 106300 | 613949 | 576573 | 6442 |
| Ganganagar | 38482 | 152011 | 230177 | 722633 | 59340 |
| Hanumangarh | 9252 | 52648 | 155162 | 897786 | 15048 |
| Jaisalmer | 2594091 | 168473 | 384431 | 91125 | 23291 |
| Jalore | 25496 | 131920 | 171204 | 651217 | 20096 |
| Jhunjhunun | 6473 | 26872 | 117002 | 431516 | 39613 |
| Jodhpur | 14415 | 625119 | 1072304 | 229284 | 6996 |
| Nagaur | 14821 | 185813 | 591513 | 836900 | 17989 |
| Pali | 49317 | 173849 | 280787 | 379136 | 80245 |
| Sikar | 10619 | 63993 | 124061 | 561977 | 60999 |

*Source: www.indiastat.com*

## Process of Land Degradation in the Thar Desert

Based upon the nature of damage to the land, processes of degradation are classified as physical (erosion, soil organic carbon loss, compaction, waterlogging), chemical (salinization, acidification, nutrient imbalances), and biological (rangeland degradation, deforestation, loss in biodiversity) (Lal, 1997). Once the balance is upset among the basic elements of land, degradation sets in and manifested in several different ways such as (a) soil loses topsoil (by erosion) and some essential nutrients (resulting in nutrient imbalances), accumulates harmful chemicals (by salinization/alkalinization), or develops physical deformities (compaction or textural discontinuity in the profile including hard-setting and pan formation); (b) water accumulates close to or above the soil surface (waterlogging) or becomes scanty or salty; (c) vegetation loses productivity of useful plants due to deforestation, overgrazing by livestock, and invasion by less useful species (resulting in loss of biodiversity). Their combined results are disastrous. Poor vegetation cover associated with loss of organic matter is responsible for accelerated erosion.

Soil erosion due to wind and water is the main cause of land degradation in Thar desert region. The process of erosion occurs when wind or rain strike the soil aggregates and separate the soil particles. These detached particles are then transported through the action of wind and runoff water. Wind erosion starts when there is minimum wind shear stress that can detach and force the movement of naturally held soil particles from their place. Soils with smooth dry and loose surfaces offer the least resistance to blowing wind and so, arid regions are most vulnerable to wind erosion. Decreasing infiltration and increasing runoff provide the basic conditions for water erosion. In the natural condition, soils are generally protected from accelerated erosion by the aboveground and below ground parts of plants. Aboveground plants cover, provided by stems and leaves, diminishes the potential of wind and water to erode soils by acting as barriers to their destructive forces. Roots reduce erosion by binding and anchoring soil particles. In addition, when accelerated by human actions, erosion increases in intensity and soil displacement greatly exceeds soil formation, resulting in soil degradation. A reduced ground cover due to overgrazing of rangelands, deforestation, or destruction of vegetation results in a greater intensity of erosion (Singh et al., 1992). Soils having stable aggregates resist erosion because of more water infiltration and reduced runoff. With vegetation removed, soils lose organic matter and the mineral soil particles tend to disperse. Dispersed soil particles also are blown away more readily by wind. Aridisols and Entisols, two dominating soil orders in arid and semiarid region, are prone to wind erosion due to the mostly dry surface and lack of stable structure. In severe conditions, the sand begins to drift and form unstable dunes which can encroach on surrounding productive lands. Wind erosion also causes serious problems for infrastructure facilities such as roads, railway tracts, canal and settlements which may get buried during dust storms. The self crusting tendency of soils in arid and semiarid region makes them vulnerable to water erosion as well (Dregne, 1976). Sporadic high intensity rainfall in Thar desert region results in water erosion in the form of sheet, rill and gullies. Sheet erosion takes place on slopes because of overland flow. Rills form in areas where overland flow accumulates. Excess accumulation of flow results in gully erosion.

Irrespective of soil type, slope factor determines the erosive power of running water. The greater will be the erosive power when the slope is longer and steeper. In general, rainfall erosivity, soil erodibility, topography, vegetation cover, management and conservation practices are the major factors affecting the soil erosion.

Application of fertiliser, organic matter and other management practices can alter eroded deep soils if it has weathered parent material as substrata. However, eroded shallow soils with hard rock as substrata cannot be transformed by above means. Wind erosion gradually removes silt, clay and organic matter from the surface soil leaving the coarse material behind.

## Indicators of Land Degradation

Land degradation is not an abrupt phenomenon but a gradual process. The costs of stopping land degradation are not high when action is taken early. Once severe, the land must be abandoned because its reclamation becomes economically non-viable. Indicators need to be developed that can predict the onset of land degradation. Potential indicators should meet the requirements such as quantitative, sensitive enough to give early warning of the impending change, widely applicable, able to assess the present status and trend, able to distinguish change due to natural cycles as opposed to anthropogenic interventions, and relevant to ecologically significant phenomena. Although there are many potential indicators but critical values to highlight the inception of degradation are not available for Thar desert region. Efforts are, therefore, required to develop threshold limits or symptoms of change that can predict well in advance (before the effect becomes visible). Measures to amend the ongoing land use practices so as to arrest the process before it intensifies and causes its damaging effect. This approach will help the least costly way of arresting the further spread of land degradation.

## Population-Poverty-Land Degradation Nexus

Western Rajasthan region is one of the most populated deserts of the world. Poverty is a multi-dimensional phenomenon. It involves not only lack of financial means but also various kind of non-monetary deprivation such as lack of ability to participate in political, social and cultural institutions and decision making. As the farms become smaller due to increasing population pressure on more or less fixed land resources, production per person declines and landlessness increases, which aggravates poverty (FAO undated document available on-line). Micro-level study through interviewing farmers whose plots have been degraded showed that poor farmers didn't adopt conservation measures, overgrazed the grazing land because of excessive animal population and used unbalanced fertilizers - all indirectly led to land degradation and desertification (Sharma et al., 2007). The lack of means for investment in basic restorative management as well as lack of knowledge prevents farmers from halting land degradation.

## Assessment of Land Degradation in the Thar Desert Region

Database on soil loss due to water and wind erosion is important for advocating the proper conservation measures in Thar desert region (Shyampura and Sehgal, 1995). Conventional methods for measuring land degradation are either not precise or expensive. Whereas satellite remote sensing data has shown potential to assess land degradation in addition to efficient vegetation cover mapping. A reliable, cost effective and timely method is needed to determine the extent of these problems. Remote sensing data acquired in the visible and near infrared regions have shown encouraging results in providing information on spatial extent of degraded lands. By virtue of its synoptic viewing capability, repitivity and spectral sensitivity, it is a valuable tool for obtaining relevant data on degraded areas rapidly. Additional information such as topographic data, soils and geology maps, ground observations, and imagery from different dates may improve the accuracy of mapping. Field indicators and relevant ancillary information for different kind of erosion are given in Table 6. Visual interpretation of false colour composite (FCC) from IRS P6 LISS III satellite imagery with 23.5 m spatial resolution and subsequent field verification have led to the identification of different land degradation features. Interpretation elements namely, colour / tone, texture, pattern, size, shape and association is given in Table 7.

Flash floods are rare phenomena in arid region of western Rajasthan. But as and when the event happens due to heavy intense rains of long duration, it causes devastating damage to the fragile ecosystem. Damages to agricultural lands include sheet erosion, formation and widening of rills and gullies. Major flash floods occurred in arid region of western Rajasthan during 1979, 1990 and 2006. During 1979 flash flood, nearly 108799 ha of agricultural land was under severe impairment by the formation of new rills and gullies (0.5 to 3 m deep) and river sand casting (0.20 to 2.80 m thick deposition). Crop standing over an area of 87186 ha was also completely damaged (Sharma and Vangani, 1982). Detailed studies were conducted by CAZRI, Jodhpur on the nature and extent of damage caused by flash flood of 1990 through field surveys and remote sensing data. The excess runoff water from many tributaries of the Luni river system spilled over the adjoining alluvial plains and caused widespread damages (Sharma and Vangani, 1992). Flash flood of 2006 caused extensive damage in Barmer district and about 244.51 sq. km area was inundated. Many depressional areas like Kawas and Malwa villages were under prolonged inundation (Bothale et al., 2007).

Qualitative assessment of erosion in western Rajasthan on 1:1 million scale was carried out by Narain et al. (2000) using field characteristics and satellite image interpretation.

**Table 6 :** Field indicators for assessing various types of land degradation (water and wind erosion) in western Rajasthan

| Land degradation process | Land degradation type | Field indicators | Physiography | Land cover | Soil type | Climate |
|---|---|---|---|---|---|---|
| Water erosion (W) | Sheet erosion | Muddy runoff dirng rainy season<br>Soil color is lighter than surrounding soils.<br>Concretions / coarse fragments on surface | Plains / valleys / pediments with >1-3% slope class | Crop lands/ fallows/ land with / without scrub/ degraded forests. Grass cover and thick canopy forests reduce erosion rates. Poor vegetal cover enhances erosion rate. | Predominantly in soils with fine texture, low organic matter and weak structure. | Arid and semi-arid climates. Erosion rate is more with high intensity rainfalls. |
| | Rills | Conspicuous tiny rivelets or finger-shaped channels. | Plains / valleys / pediments with >1-3% slope class | Predominant in crop lands/ fallows followed by land without scrub/ scrub land / degraded forests. Poor vegetal cover enhances erosion rate. | Predominantly in soils with fine texture, low organic matter and weak structure | Arid and semi-arid climates. Erosion rate is more with high intensity rainfalls. |
| | Gullies | Well defined and permanent incised land neither cultivable nor traversable. | Occurs on >5% slope lands.<br>lower element of slope and gradually creeps to upper slopes. | Mostly land with / without scrub. | Predominant in loams and associated textures. | Arid and semi-arid climates. Erosion rate is more with high intensity rainfalls. |
| | Ravines | Well defined and permanent incised land neither cultivable nor traversable.<br>Network of deepened gullies. | Associated with major streams / river network. | Mostly land with / without scrub. | Predominant in loams and associated textures. | Arid and semi-arid climates. Erosion rate is more with high intensity rainfalls. |
| Wind erosion (E) | Sheet erosion | | Aeolian plains | Barren land / land without scrub associated with no / very poor vegetal cover | Soils with weak structure like sands and loamy sands. | Deserted regions associated with scanty rain and strong winds. |
| | Partially stabilized dunes | Sand dunes covered with sparse vegetal cover. | Aeolian plains | Grass / scrubs. | Soils with weak structure like sands and loamy sands. | Deserted regions associated with scanty rain and strong winds. |
| | Stabilized dunes | Sand dunes covered with moderately dense vegetal cover. | Aeolian plains | Grass / scrubs/ trees | Soils with weak structure like sands and loamy sands. | Deserted regions associated with scanty rain and strong winds. |
| | Un-stabilized dunes | Sand dunes devoid of any vegetation. | Aeolian plains | Barren | Soils with weak structure like sands and loamy sands. | Deserted regions associated with scanty rain and strong winds. |

*Source: Nationwide Mapping of Land degradation Project Manual. 2007. NRSC, Hyderabad, ISRO, Department of Space, Govt. of India*

**Table 7 :** Satellite image (FCC) interpretation elements for various types of land degradation in western Rajasthan

| Land degradation process | Land degradation type | Colour / Tone (On standard FCC) | Texture (on LISS-III data) | Pattern | Size | Shape | Association |
|---|---|---|---|---|---|---|---|
| Water erosion (W) | Sheet erosion (sh) | Slightly brighter than surrounding land of its class | Smooth to medium | Contiguous patches | Small to large | Irregular | Sloping cultivated / lands with poor vegetal cover during rainy season. |
| | Rills (ri) | Brighter than surrounding land of its class | Medium | Discrete to contiguous patches | Small to medium | Irregular | Sloping cultivated lands. |
| | Gullies (gu) | Brighter than surrounding land / gray in color depending on soil colour. | Medium to slightly coarse | Discrete to contiguous patches | Small to medium | Irregular | First order streams. |
| | Ravines(rs / rm) | Medium gray to dark gray | Slightly coarse for shallow ravines and coarse for deep ravines | Contiguous patches | Large to very large | Irregular | Stream / river banks |
| Wind erosion (E) | Sheet erosion (sh) | Various shades of yellow and light gray combination. | Smooth to medium | Contiguous / mottling (in cultivated areas) | Large to very large | Regular / Irregular | Desertic plain areas with active sand movement |
| | Partially stabilized dunes (dp) | Light grey to medium grey with light yellowish tones. | Medium | Contiguous / discrete patches | Small to medium | Regular / Irregular | Desert sandy dunal area |
| | Stabilized dunes (ds) | Medium grey with light yellowish tones during dry season. Pink mottles during rainy season. | Medium to coarse | Discrete patches | Small to medium | Regular / Irregular | Desert sandy dunal area |
| | Un-stabilized dunes (du) | Various shades of yellow and very light gray combination. | Smooth to medium | Contiguous / discrete | Medium to large | Irregular | Desert sandy dunal area |

*Source: Nationwide Mapping of Land degradation Project Manual. 2007. NRSC, Hyderabad, ISRO, Department of Space, Govt. of India*

Map of degraded lands on 1: 50000 scale for Thar desert region has been prepared by Department of Space, Govt. of India using kharif, rabi and summer season Resourcesat-1 LISS III satellite data for the period 2005-06 (Table 8). Spatial distribution of degraded lands is shown in figure 3. Various classes of land degradation are easily noticeable from field photographs (Figure 4). Status of wastelands (i.e. degraded land which can be brought under vegetative cover with reasonable effort) was also assessed using the same dataset (Table 9).

**Table 8 :** Area (ha) and % of total district area (in bracket) under different types of land degradation in various districts of western

| Class | Barmer | Bikaner | Churu | Ganganagar | Hanumangarh | Jaisalmer | Jalor | Jhunjhunun | Jodhpur | Nagaur | Pali | Sikar |
|---|---|---|---|---|---|---|---|---|---|---|---|---|
| Wsh1 | 0.00 (0.00) | 3946.48 (0.14) | 986.62 (0.06) | 0.00 (0.00) | 0.00 (0.00) | 23678.90 (0.62) | 986.62 (0.16) | 12826.07 (2.16) | 0.00 (0.00) | 0.00 (0.00) | 6906.35 (0.56) | 5919.73 (0.77) |
| Wri2 | 8879.59 (0.31) | 2959.86 (0.11) | 0.00 (0.00) | 0.00 (0.00) | 0.00 (0.00) | 12826.07 (0.33) | 4933.11 (0.79) | 1973.24 (0.33) | 11839.45 (0.52) | 10852.83 (0.61) | 29598.63 (2.39) | 3946.48 (0.51) |
| Wgu3 | 10852.83 (0.38) | 3946.48 (0.14) | 0.00 (0.00) | 0.00 (0.00) | 0.00 (0.00) | 3946.48 (0.10) | 13812.69 (2.22) | 8879.59 (1.50) | 4933.11 (0.22) | 1973.24 (0.11) | 986.62 (0.08) | 8879.59 (1.15) |
| Wrs4 | 0.00 (0.00) | 0.00 (0.00) | 0.00 (0.00) | 0.00 (0.00) | 000 (0.00) | 0.00 (0.00) | 0.00 (0.00) | 4933.11 (0.83) | 0.00 (0.00) | 0.00 (0.00) | 0.00 (0.00) | 10852.83 (1.40) |
| Esh1 | 109514.93 (3.86) | 612691.65 (22.49) | 147006.53 (8.73) | 155886.12 (14.20) | 241722.15 (25.03) | 368996.26 (9.61) | 15785.94 (2.54) | 54264.16 (9.15) | 88795.89 (3.89) | 27625.39 (1.56) | 986.62 (0.08) | 68076.85 (8.80) |
| Eds1 | 334464.52 (11.78) | 626504.34 (23.00) | 155886.12 (9.26) | 29598.63 (2.70) | 103595.21 (10.73) | 470618.22 (12.26) | 2959.86 (0.48) | 18745.80 (3.16) | 76956.44 (3.37) | 33545.11 (1.89) | 0.00 (0.00) | 43411.32 (5.61) |
| Edp2 | 442992.83 (15.61) | 1224396.67 (44.94) | 678795.25 (40.33) | 191404.48 (17.44) | 163779.09 (16.96) | 1098109.18 (28.60) | 4933.11 (0.79) | 28612.01 (4.83) | 201270.69 (8.81) | 19732.42 (1.11) | 986.62 (0.08) | 70050.09 (9.06) |
| Edu3 | 1973.24 (0.07) | 63143.74 (2.32) | 10852.83 (0)64 | 0.00 (0.00) | 0.00 (0.00) | 147993.15 (3.85) | 0.00 0.00 | 6906.35 (1.17) | 1973.24 (0.09) | 0.00 (0.00) | 0.00 (0.00) | 1973.24 (0.26) |
| Ssa1 | 4933.11 (0.17) | 15785.94 (0.58) | 1973.24 (0.12) | 986.62 (0.09) | 986.62 (0.10) | 1973.24 (0.05) | 4933.11 (0.79) | 0.00 (0.00) | 10852.83 (0.47) | 13812.69 (0.78) | 4933.11 (0.40) | 986.62 (0.13) |
| Ssa2 | 2959.86 (0.10) | 0.00 0.00 | 986.62 (0.06) | 0.00 (0.00) | 0.00 (0.00) | 1973.24 (0.05) | 1973.24 (0.32) | 986.62 (0.17) | 4933.11 (0.22) | 1973.24 (0.11) | 3946.48 (0.32) | 0.00 (0.00) |
| Ssa3 | 7892.97 (0.28) | 986.62 (0.04) | 7324 (0.12) | 0.00 (0.00) | 0.00 (0.00) | 10852.83 (0.28) | 1973.24 (0.32) | 0.00 (0.00) | 3946.48 (0.17) | 9866.21 (0.56) | 0.00 (0.00) | 0.00 (0.00) |
| Sso1 | 986.62 (0.03) | 0.00 (0.00) | 0.00 (0.00) | 0.00 (0.00) | 0.00 (0.00) | 0.00 (0.00) | 986.62 (0.16) | 0.00 (0.00) | 986.62 (0.04) | 0.00 (0.00) | 20719.04 (1.67) | 0.00 (0.00) |
| Sso2 | 0.00 (0.00) | 0.00 (0.00) | 0.00 (0.00) | 0.00 (0.00) | 0.00 (0.00) | 0.00 (0.00) | 1973.24 (0.32) | 0.00 (0.00) | 0.00 (0.00) | 0.00 (0.00) | 0.00 (0.00) | 0.00 (0.00) |
| Sss1 | 0.00 (0.00) | 0.00 (0.00) | 0.00 (0.00) | 0.00 (0.00) | 0.00 (0.00) | 0.00 (0.00) | 15785.94 (2.54) | 986.62 (0.17) | 7892.97 (0.35) | 0.00 (0.00) | 5919.73 (0.48) | 0.00 (0.00) |
| Sss2 | 5919.73 (0.21) | 0.00 (0.00) | 0.00 (0.00) | 0.00 (0.00) | 0.00 (0.00) | 0.00 (0.00) | 18745.80 3.01 | 0.00 (0.00) | 0.00 (0.00) | 1973.24 (0.11) | 0.00 (0.00) | 0.00 (0.00) |
| Sss3 | 0.00 (0.00) | 0.00 (0.00) | 0.00 (0.00) | 0.00 (0.00) | 0.00 (0.00) | 0.00 (0.00) | 13812.69 (2.22) | 0.00 (0.00) | 0.00 (0.00) | 986.62 (0.06) | 0.00 (0.00) | 0.00 (0.00) |
| Hmd | 3946.48 (0.14) | 11839.45 (0.43) | 986.62 (0.06) | 0.00 (0.00) | 0.00 (0.00) | 41438.08 (1.08) | 0.00 (0.00) | 0.00 (0.00) | 3946.48 (0.17) | 5919.73 (0.33) | 1973.24 (0.16) | 0.00 (0.00) |
| Tbs | 79916.30 (2.82) | 5919.73 (0.22) | 0.00 (0.00) | 0.00 (0.00) | 0.00 (0.00) | 571253.56 (14.88) | 18745.80 (3.01) | 12826.07 (2.16) | 121354.38 (5.31) | 53277.53 (3.01) | 30585.25 (2.47) | 8879.59 (1.15) |
| Tms | 0.00 (0.00) | 0.00 (0.00) | 0.00 (0.00) | 0.00 (0.00) | 0.00 (0.00) | 0.00 (0.00) | 0.00 (0.00) | 1973.24 (0.33) | 0.00 0.00 | 0.00 (0.00) | 0.00 0.00 | 986.62 (0.13) |
| Lsp1 | 0.00 (0.00) | 10852.83 (0.40) | 0.00 (0.00) | 986.62 (0.09) | 0.00 (0.00) | 147993.15 (3.85) | 0.00 (0.00) | 0.00 (0.00) | 0.00 (0.00) | 986.62 (0.06) | 0.00 (0.00) | 0.00 (0.00) |
| Lsp2 | 0.00 (0.00) | 986.62 (0.04) | 0.00 (0.00) | 0.00 (0.00) | 1973.24 (0.20) | 0.00 (0.00) | 986.62 (0.16) | 0.00 (0.00) | 0.00 (0.00) | 0.00 (0.00) | 0.00 (0.00) | 0.00 (0.00) |

*Source: Unpublished results of the project "Natiowide Mapping of Land Degradation using Multi-temporal Satellite Data" (2010). NRSC, Hyderbad, ISRO, Department of Space, Govt. of India*

**Symbols used for description of the class**

| Symbol | Description |
|---|---|
| Wsh 1 | Sheet erosion - Skight |
| Wri 2 | Rills erosion - Moderate |
| Wgu 3 | Gullies - Severe |
| Wrs 4 | Ravines - shallow – Very severe |
| Esh 1 | Sheet erosion – slight |
| Eds 1 | Stabilized dunes - slight |
| Edp 2 | Partially-stabilize dunes – Moderate. |
| Edu 3 | Un-stabilized dunes - Severe |
| Ssa 1 | Saline - Slight |
| Ssa 2 | Saline - Moderate |
| Sso 1 | Sodic – Slight |
| Sso 2 | Sodic – Moderate |
| Sss 3 | Saline Sodic - Severe |
| Hmd | Mining. Surface/open cast mines including mine dumps |
| Tbs | Barren rocky/Stony waste. Rock outcrops/ sheet rock |
| Tms | Miscellaneous – Riverine sands/sea ingress areas etc |
| Lsp 1 | Surface ponding – Slight. Waterlogging with 4-6 moths of ponding affecting one crop (excluding paddy) |
| Lsp 2 | Surface ponding – Moderate. Permanent waterlogging with more than 6 months ponding affecting more than one crop (excluding paddy) |

**Table 9 :** Area (sq. km) under different categories of wastelands in various districts of western Rajasthan

| Wasteland Category | Barmer | Bikaner | Churu | Ganganagar | Hanumangarh | Jaisalmer | Jalor | Jhunjhunu | Jodhpur | Nagaur | Pali | Sikar |
|---|---|---|---|---|---|---|---|---|---|---|---|---|
| Gullied And/ Or Ravenous Land(Medium) | 0.00 | 59.16 | 0.00 | 0.00 | 0.00 | 0.00 | 51.89 | 1.42 | 11.72 | 0.00 | 14.35 | 21.43 |
| Gullied And/ Or Ravenous Land(Deep) | 20.22 | 37.25 | 0.00 | 0.00 | 0.00 | 0.00 | 20.42 | 27.46 | 0.00 | 0.00 | 0.00 | 2.64 |
| Land Wth Dense Scrub | 1032.84 | 1430.39 | 52.82 | 21.95 | 13.66 | 8352.40 | 233.51 | 30.23 | 439.60 | 291.82 | 1188.03 | 50.38 |
| Land With Open Scrub | 384.68 | 265.73 | 343.38 | 89.39 | 33.61 | 3870.31 | 80.84 | 324.37 | 945.42 | 289.53 | 361.50 | 392.10 |
| Land Affected By Salinity/Alkalinity(Medium) | 123.12 | .00 | 14.71 | 0.00 | 0.00 | 11.36 | 11.91 | 0.00 | 48.97 | 18.09 | 77.78 | 6.93 |
| Land Affected By Salinity/Alkalinity(Strong) | 9.93 | 0.00 | 0.39 | 0.00 | 0.00 | 40.67 | 36.66 | 0.00 | 6.18 | 102.56 | 7.72 | 4.48 |
| Under-Utilised/Degraded Forest(Scrub Domin) | 124.69 | 70.87 | 0.00 | 0.00 | 23.12 | 145.87 | 161.06 | 169.99 | 57.53 | 127.64 | 197.16 | 309.74 |
| Under-Utilised/Degraded Forest(Agriculture) | 0.79 | 0.00 | 0.00 | 0.00 | 0.00 | 0.04 | 0.00 | 0.00 | 0.00 | 0.00 | 4.16 | 0.00 |
| Degraded Pastures/ Grazing Land | 279.09 | 485.71 | 128.12 | 0.00 | 21.46 | 85.98 | 286.88 | 56.71 | 661.95 | 528.26 | 395.64 | 68.08 |
| Degraded Land Under Plantation Crop | 0.00 | 0.00 | 0.00 | 0.00 | 0.00 | 0.00 | 0.00 | 0.00 | 0.00 | 0.00 | 0.00 | 0.00 |
| Sands-Riverine | 0.00 | 0.00 | 0.00 | 0.00 | 0.00 | 0.00 | 0.03 | 15.48 | 4.10 | 0.00 | 69.59 | 19.85 |
| Sand-Desertic | 402.27 | 1111.56 | 58.42 | 154.59 | 330.54 | 528.99 | 285.73 | 19.21 | 1084.12 | 327.84 | 8.87 | 178.00 |
| Sand-Semi Stab.-Stub>40m | 1728.70 | 1278.94 | 0.00 | 1088.21 | 20.44 | 6905.69 | 20.75 | 0.00 | 145.66 | 0.00 | 0.00 | 0.00 |
| Sand-Semi Stab.-Stub> 15-40m | 2598.27 | 5907.50 | 109.99 | 171.83 | 0.00 | 5058.99 | 132.30 | 0.00 | 1349.97 | 248.36 | 0.00 | 0.00 |
| Mining Wastelands | 1.95 | 0.00 | 0.04 | 0.00 | 0.00 | 0.84 | 0.91 | 0.00 | 20.59 | 4.27 | 5.85 | 0.12 |
| Industrial Wastelands | 0.00 | 0.00 | 0.00 | 2.57 | 0.00 | 0.00 | 0.61 | 0.00 | 0.99 | 0.00 | 1.37 | 0.00 |
| Barren Rocky/ Stony Waste | 299.06 | 19.63 | 4.94 | 0.00 | 0.00 | 1228.13 | 173.30 | 32.69 | 500.21 | 92.67 | 289.24 | 28.33 |

*Source: Wastelands Atlas of India (2010). Published by National Remote sensing Centre (NRSC), Hyderabad, ISRO, Department of Space, Govt. of India*

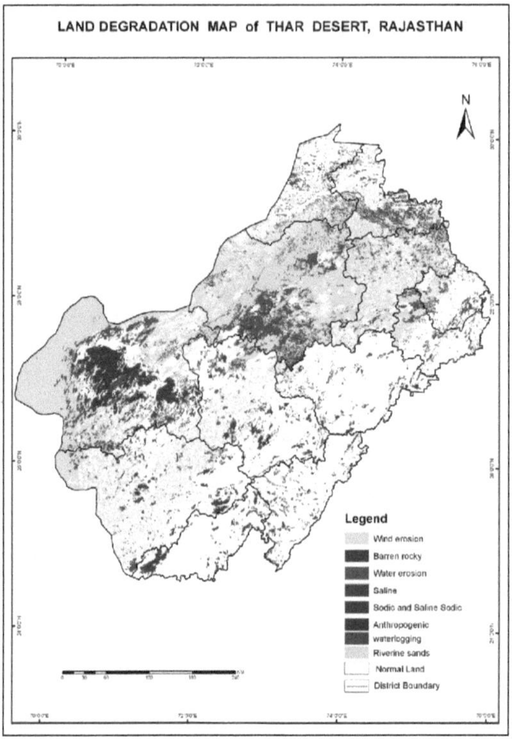


**Figure 3 :** Distribution of land degradation in Thar desert of Rajasthan

**Figure 4 :** Field Photograph showing various land degradation classes

## Conclusions

Accelerated wind erosion as well as water erosion due to flash flood are primarily responsible for land degradation in Thar desert region. Land is a precious resources and its conservation is life. We should fight the land degradation menace by the use of available conservation technologies. Satellite remote sensing has the capability of providing adequate information for assessment, monitoring, management and control of land degradation processes which are dynamic and temporal in nature. People's participation and co-operation is must in organising soil and water conservation, especially in dealing with fragmented and scattered land holdings. There is a great need to educate the farmers on the benefits of soil and water. In spite of the fact that erosion control procedures developed at experimental stations in different countries have proved to be quite effective, their wide application is hampered because of a number of financial and socio-economic reasons. However, the role of prevailing climatic variations and influence of global warming in intensifying land degradation need to be considered when designing strategies for combating it.

## References

Ahmed, Y. J. and Kassas, M. 1987. Desertification: Financial Support for the Biosphere. Hodder and Stoughton, London, UK.

Barrow, C. J. 1991. Land Degradation : Development and Breakdown of Terrestrial Environments. Cambridge University Press, Cambridge, UK.

Blaikie, P. and Brookfield, H. 1987. Land Degradation and Society. Methuen, London, UK.

Bothale, R. V., Paliwal, R., Dutta, D. and Sharma, J. R. 2007. Torrential rains and flash floods in Rajasthan desert - role of geoinformatics for rapid action plan and disaster management. *ISG Newsletter,* 13 (2&3):36-41.

Dregne, H. E. 1978. Desertification : man's abuse of the land. *Journal of Soil and Water Conservation,* 33:11-14.

Dregne, H. E. and Chou, Nan-Ting. 1992. Global desertification dimensions. In: Degradation and Restoration of Arid Lands (Dregne, H. E. ed.). International Center for Arid and Semiarid Studies, Texas Tech University, Lubbock, Texas, USA. Pp. 249-282.

Dregne, H.E. 1976. Soils of the Arid Regions. Elsevier, Amsterdam, The Netherlands.

FAO (undated on-line). The problem of land degradation. Available on-line www.fao.org/docrep/v9909e/v9909e02.htm.

FAO. 1976. A Framework for Land Evaluation. FAO Soils Bulletin 32. Rome, Italy.

Hare F. K. 1985. Climatic variability and change. In: Climatic Impact Assessment (Kates, R. W., Ausubel, J. H. and Bererian, M. eds.). John Wiley and Sons Ltd., Chichester, West Sussex, UK. Pp. 7-68.

Johnson, D. L. and Lewis, L. A. 1995. Land Degradation : Creation and Destruction. Cambridge, Blackwell, UK.

Lal, R. 1997. Soil quality and sustainability. In: Methods for Assessment of Soil Degradation (Lal, R. Blum, W. H. Valentine and Stewart, B. A. eds.). CRC Press, Boca Raton, Florida, USA. Pp. 17-30.

Mainguet, M. 1994. Desertification-Natural Background and Human Mismanagement. Springer-Verlag, Berlin, Germany.

Narain, P., Kar, A., Balak Ram, Joshi, D. C. and Singh, R. S. 2000. Wind Erosion in western Rajasthan, CAZRI, Jodhpur.

Oldeman, L. R. 1988. Global Assessment of Soil Degradation (GLASOD). Guidelines for General Assessment of Status of Human Induced Soil Degradation. International Soil Reference and Information Center, Wageningen, The Netherlands.

Santra, P., Mertia, R. S. and Narain, P. 2004. Spatial pattern of wind erosion in western Rajasthan. In: Geoinformatics Application for Sustainable Development (eds Singh, A. K. et al.), Water Technology Centre, IARI, New Delhi, 2004, pp. 239-243.

Sehgal, J. L. and Abrol, I. P.1994. Soil Degradation in India: Status and Impact. Oxford and IBH Publishing Co. Pvt. Ltd., New Delhi, India.

Sharma, J. R., Jayaraman, V. and Bera, A. K. 2007. EO based study on desertification and poverty nexus. 58th International Astronautical Congress, 24 - 28 September, 2007, Hyderabad.

Sharma, K. D. and Vangani, N. S. 1982. Flash flood of July 1979 in the Luni basin - a rare event in the Indian desert. *Hydrological Sciences,* 27:385-397.

Sharma, K. D. and Vangani, N. S. 1992. Characteristics of a flash flood in the Luni basin : north west India. *Irrigation and Power Journal,* 49:31-41.

Shyampura, R. L. and Sehgal, J. L. 1995. Soils of Rajasthan for optimizing land use. NBSS Publ. No. 51, NBSS & LUP, Nagpur.

Singh, N. T. 1992. Dryland salinity in the Indo-Pakistan subcontinent. In: Degradation & Restoration of Arid Lands (Dregne, H. E. ed.). International Centre for Arid and Semi-arid Land Studies, Texas Tech University, Lubbock, Texas, USA, pp. 179-248.

Stewart, B. A., Lal, R. and El-Swaify, S. A. 1991. Sustaining the resource base of an expanding world agriculture. In: Soil Management for Sustainability (Lal, R. and Pierce, F. J. eds.). Soil and Water Conservation Society, Ankey, Iowa, USA, pp. 125-145.

Syers, J. K., Lingard, J., Pieri, C., Ezcurra, E. and Faure, G. 1996. Sustainable land management for the semiarid and sub-humid tropics. *Ambio* 25: 484-491

UNEP. 1992. World Atlas of Desertification. Edward Arnold, Sevenoaks, UK

UNEP-UNCOD. 1978. United Nations Conference on Desertification, 29 August - 9 September, 1977.

## Chapter - 6

# Sustainable Utilization of Wastelands and their Economic Significance in Western Rajasthan, India

**Mahesh Kumar Gaur and J.R. Sharma**


## Abstract

*India shares 16% of the world population, while its land is only 2% of the total geographical area of the world. Naturally, the pressure on the land is often beyond its carrying capacity. Therefore, the productive lands, especially the farmlands in the India are in the constant process of various degrees of degradation and are fast turning into wastelands. At present, approximately 68.35 million hectare area of the land is lying as wastelands in India. Out of these lands, approximately 50% lands are such non-forest lands, which can be made fertile again if treated properly. It was unprotected non-forestlands, which suffered the maximum degradation mainly due to the tremendous biotic pressure on it. In the last 50 years it is India's lush green village forests and woodlots have been deforested to the maximum. Voluminous increase in population has created an excessive demand for productive land. At the same time land degradation because of desertification, soil salinity, waterlogging, floods/droughts, excessive soil erosion and unscientific agricultural practices has resulted in the creation of vast stretches of wastelands. This has necessitated adoption of scientific measures for increasing land productivity and bringing more areas under cultivation/forests. In the present study, the multi-temporal satellite images of IRS LISS-III were used to map wastelands dynamics over different seasons. An attempt has also been made to evaluate the potential of moderate spatial resolution LISS-III data (23.5 m) from the Indian Remote Sensing Satellite for delineation of wastelands in a portion of the western Great Indian desert of Rajasthan state. Visual interpretation based on image characteristics and a prior knowledge of the study area was used to delineate wasteland classes. Wasteland classes such as seasonal and permanent waterlogged could be further separated into wasteland classes such as land with dense scrub, land with open scrub, degraded pastures/grazing lands and degraded land under plantation using higher resolution satellite data.*

## 1.0 Introduction

Land is a non-renewable resource. It is central to all primary production systems of the human activities. Since independence, degradation has received an impetus causing country's landmass to suffer from different types of degradations. Degradation of land is caused by biotic and abiotic pressures. An ever increasing population places enormous demands on land resources for various activities. This is particularly acute in India, which has only 2.4 per cent of the world's geographical area but supports over 16 per cent of the world's population. It has 0.5 per cent of the world's grazing area but has over 18 per cent of world's cattle population. These pressures have led to drastic changes in the proportion of land utilised for agricultural activities, urbanisation and industrial development. Western Rajasthan is facing even worst situation where more than 19.06 per cent area was classified under various categories of wastelands during 1991-92 whereas it is 26.16 per cent mapped in 2008-09 using multi-date IRS LISS-III data. But, there is a positive sign also. Area under various wasteland categories has decreased by 0.01 per cent over 2006-06. Such a vast area under wastelands lying unutilised and unproductive is alarming and serious threat to the livelihood and food security of the Thar Desert. It needs a serious grassroots-level micro-planning for the better utilisation of this large chunk of the land. Authors of this paper endeavours to present such a solution for this densely populated desert of the world. Furthermore, this study examines how various variables could affect the dynamics of poverty and natural resources degradation relationship in the fragile desert environment.

## 2.0 Objectives

Objectives of the study are:

1. To generate land use / land cover change detection data base for the period 2008-09 over 2005-2006 using three seasons (Kharif, Rabi and Zaid) LISS-III satellite data.
2. To provide a state of land use report with details that can be used as input for planning exercise at district and regional level.
3. To generate tables of the land cover parameters and the changes-over the previous census cycle.
4. To prepare LULC Information System for easy query and retrieval of geo-database.
5. To report the areas of major land use change with appropriate scale maps and assessment report on causative factors and remedial measures

## 3.0 Data used

Land use / land cover change detection mapping is done based on the digital classification of multi-spectral and multi-temporal satellite data. Preparation of Land use / land cover change detection map requires information on different thematic layers viz. settlements and transport network, drainage pattern and water bodies. These layers have been prepared from the information available in satellite data and from Survey of India topomaps and meteorological data.

### a. Satellite data

For delineation and mapping of land use / land cover, multi-temporal orthorectified IRS LISS-III data has been used for kharif, rabi and zaid seasons of 2005-2006 and 2008 - 2009.

### b. Collateral data

(i) Topographic Maps: Following 360 Survey of India digital topographic maps on 1:50000 scale were used for identification of base features and for planning ground truth data collection.

| | | | | | | | | | | | | | | |
|---|---|---|---|---|---|---|---|---|---|---|---|---|---|---|
| 39 L | 8 | 12 | | | | | | | | | | | | |
| 39 P | 16 | | | | | | | | | | | | | |
| 40 E | 11 | 12 | 14 | 15 | 16 | | | | | | | | | |
| 40 F | 5 | 9 | 10 | 13 | 14 | | | | | | | | | |
| 40 I | 1 | 2 | 3 | 4 | 5 | 6 | 7 | 8 | 9 | 10 | 11 | 12 | 14 | 1516 |
| 40 J | 1 | 2 | 3 | 4 | 5 | 6 | 7 | 8 | 9 | 10 | 11 | 12 | 13 | 1415 | 16 |
| 40 K | 1 | 5 | 6 | 9 | 10 | 11 | 13 | 14 | 15 | 16 | | | | |
| 40 L | 13 | | | | | | | | | | | | | |
| 40 M | 1 | 2 | 3 | 4 | 5 | 6 | 7 | 8 | 9 | 10 | 11 | 12 | 13 | 1415 | 16 |
| 40 N | 1 | 2 | 3 | 4 | 5 | 6 | 7 | 8 | 9 | 10 | 11 | 12 | 13 | 1415 | 16 |
| 40 O | 1 | 2 | 3 | 4 | 5 | 6 | 7 | 8 | 9 | 10 | 11 | 12 | 13 | 1415 | 16 |
| 40 P | 1 | 2 | 5 | 6 | 9 | 10 | 13 | 14 | | | | | | |
| 44 D | 2 | 3 | 4 | 5 | 6 | 7 | 8 | 9 | 10 | 11 | 12 | 13 | 14 | 1516 |
| 44 F | 12 | 16 | | | | | | | | | | | | |
| 44 G | 2 | 3 | 4 | 5 | 6 | 7 | 8 | 9 | 10 | 11 | 12 | 13 | 14 | 1516 |
| 44 H | 1 | 2 | 3 | 4 | 5 | 6 | 7 | 8 | 9 | 10 | 11 | 12 | 13 | 1415 | 16 |
| 44 K | 1 | 2 | 3 | 4 | 5 | 6 | 7 | 8 | 9 | 10 | 11 | 12 | 15 | 16 |
| 44 L | 1 | 2 | 3 | 4 | 5 | 6 | 7 | 8 | 9 | 10 | 11 | 12 | 13 | 1415 | 16 |
| 44 O | 3 | 4 | 7 | 8 | 12 | | | | | | | | | |
| 44 P | 1 | 2 | 3 | 4 | 5 | 6 | 7 | 8 | 9 | 10 | 11 | 12 | 15 | 16 |
| 45 A | 1 | 2 | 3 | 4 | 5 | 6 | 7 | 8 | 9 | 10 | 11 | 12 | 13 | 1415 | 16 |

| | | | | | | | | | | | | | | | |
|---|---|---|---|---|---|---|---|---|---|---|---|---|---|---|---|
| 45 B | 1 | 2 | 3 | 4 | 5 | 6 | 7 | 8 | 9 | 10 | 11 | 12 | 13 | 1415 | 16 |
| 45 C | 1 | 2 | 3 | 4 | 5 | 6 | 7 | 8 | 9 | 10 | 11 | 12 | 13 | 1415 | 16 |
| 45 D | 1 | 2 | 5 | 6 | 9 | 13 | | | | | | | | | |
| 45 E | 1 | 2 | 3 | 4 | 5 | 6 | 7 | 8 | 9 | 10 | 11 | 12 | 13 | 1415 | 16 |
| 45 F | 1 | 2 | 3 | 4 | 5 | 6 | 7 | 8 | 9 | 10 | 11 | 12 | 13 | 1415 | 16 |
| 45 G | 1 | 2 | 3 | 4 | 5 | 6 | 7 | 8 | 9 | 10 | 11 | 12 | 13 | 1415 | 16 |
| 45 H | 1 | 2 | 5 | | | | | | | | | | | | |
| 45 I | 1 | 2 | 3 | 4 | 5 | 6 | 7 | 8 | 9 | 10 | 11 | 12 | 13 | 1415 | 16 |
| 45 J | 1 | 2 | 3 | 4 | 5 | 6 | 7 | 8 | 9 | 10 | 13 | | | | |
| 45 K | 1 | | | | | | | | | | | | | | |
| 45 M | 1 | 2 | 3 | 4 | 5 | 6 | 7 | 8 | 9 | 10 | 11 | 13 | 14 | 15 | |
| 45 N | 1 | | | | | | | | | | | | | | |
| 53 D | 3 | 4 | | | | | | | | | | | | | |
| 54 A | 1 | 2 | | | | | | | | | | | | | |

(ii) Legacy Data: Legacy data on land use/land cover (AWiFS), Wastelands data generated on 1:50000 scale, NRDB database, and previously generated Base details have been used (wherever required) as a reference during delineation of various LULC categories.

## 4.0 Methodology and approach

The broad steps involved in the generation of the wasteland database are as follows.

## a. Data and base map preparation

Preparing the data is always a primary requirement before undertaking interpretation and subsequent analysis. Preparation of datasets for the use in the current project is described here under. (Fig 1)

- Multi-temporal LISS-III data covering Kharif (August - November), Rabi (January - March), Zaid (April - May) seasons to address spatial and temporal variability in land cover classes were used. In the event of non availability of cloud free and good quality of LISS-III data, the use of multi-sensor data was also contemplated.
- Multi-temporal datasets were ortho-rectified to correct the effect of relief and georeferencing with reference framework using LCC projection and WGS 84 datum.
- The base map details are digitized from SOI topomaps and further updated using satellite images. The base details are road, rail, canals, settlements etc.

## b. Development of interpretation key

It essentially consists of a set of image elements or characteristics, which help in the recognition or identification of various land use / land cover classes systematically on the satellite imagery during visual interpretation and digital classification. Ideally an interpretation key consists of two components viz.;

(i) Collection of annotated / captioned images: complete scene needs to be studied thoroughly for distinction of features in different possible band combinations across different seasons to understand the spectral response patterns. Image subsets for each land cover need to be prepared and annotations / captions for each are provided.

(ii) Description of each land cover features needs to be provided in the key in specific reference to all interpretation elements viz.; Tone. Texture, Size, Shape, Association, Pattern, etc.

## c. Preliminary interpretation

The land cover polygons as seen in the satellite data were delineated on-screen using standard classification system and a preliminary interpretation map was prepared.

## d. Classification

A hybrid classification approach (digital and visual classification) is used in the study. Digital classification for crop was performed for all images for three seasons. The classified categories were than further segregated between pure crop and other vegetation through a constrained NDVI threshold value, selected interactively in image processing. The other land use / land cover classes are delineated based on visual interpretation technique.

## e. Smoothening

The output of the digitally classified cropped area is jagged and there is staircase affect of polygon boundary due to raster data. Which appears in different patch within onscreen digitized other smooth categories in the vector. Therefore, customized software was written which converts the raster polygon into smooth shapes and the same was applied on aggregated crop outputs. During this phase the classes with required accuracies were aggregated to the nearest wasteland class.

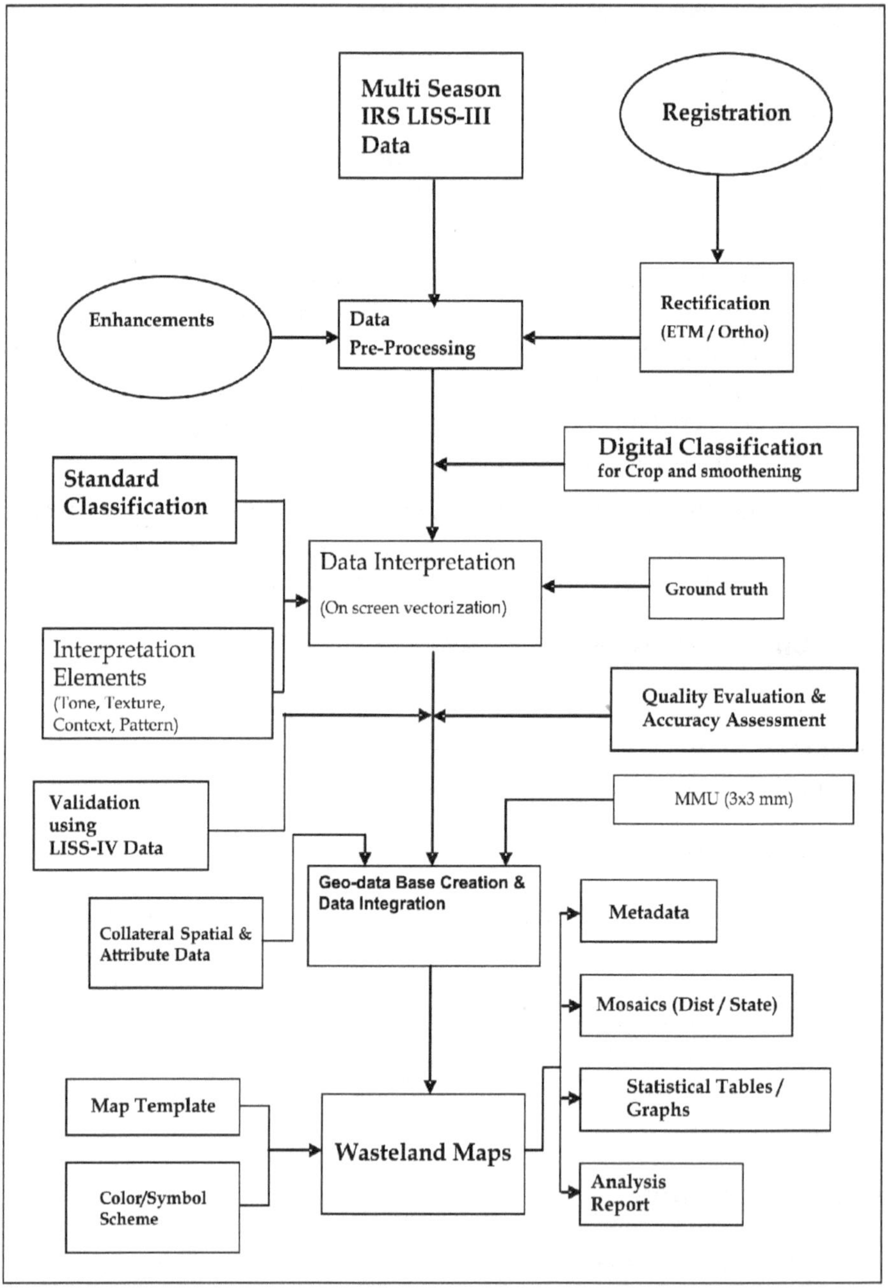


**Figure 1 :** Flow chart of methodology

## f. Preparation for fieldwork

This step involved identification of statistically sound sample grids to verify on the ground. On the basis of interpretative uncertainty highlighted during the preliminary interpretation, a portion of map units were selected for field checking. This first set of point was selected for integration, if necessary, by an extra set of checks to assure a good statistical representation of the land cover classes.

## g. Ground data collection and verification

Ground truth / field verification is an important component in mapping and its validation exercise. Utmost care and planning is required for collecting ground data and verification. To facilitate a good ground truthing exercise the following steps were followed:

(i) Identification and listing of all the doubtful areas for the ground verification. Referring all such doubtful areas with respect to the toposheet to know their geographical location and accessibility on the ground.

(ii) Preparation of field traverse plan to cover maximum doubtful areas in the field thereby ensuring that each traverse covers, as many land use / land cover classes as possible, apart from the doubtful areas, and

(iii) The number of points to be covered for each category was based on the formula provided in the accuracy assessment. These observations were required both for quality checking as well as accuracy estimation, in addition to use in interpretation.

Once the planning was complete, a pre-field interpreted map and satellite FCC with proper contrast enhancement was generated on photo glossy paper. Field details were noted directly on the map for post-ground truth corrections and modifications for compilation in the laboratory.

Information collected through ground data forms, enables improvements in classification accuracy, confidence level and estimates. The data was collected based on a field verification form that gave an exhaustive description of sample site visited. A proforma for ground data as given in the manual was used.

(i) Supportive details like the status of land utilization, cropping pattern, irrigation facilities, reclamation of wasteland, were entered in the ground truth field verification form for facilitating. Ground truth proforma gives the detailed description of various terrain parameters collected for proper image analysis.

(ii) Field information collected, was later archived as softcopy information information collected, was later archived as softcopy information

(iii) Geographic coordinates for the land cover visited were recorded in Degree-Minute-Second (dd/mm/ss) format. For the purpose of recording the coordinates, GPS was used. Survey of India toposheets on 1:50000 were also used to translate the position in terms of latitude and longitude.

(iv) A clear picture of the panoramic position of the land cover has been obtained so that adjacency issues could be understood. In order to have better visibility of land surface feature at each ground survey point, photographs were taken and recorded them with directions

### h. Final interpretation based on correction / modification and transfer of post-field details

Correction and modifications of misclassified and doubtful land use / land cover and forest cover classes is carried out and further final classification is performed. The delineation of wasteland categories made during the preliminary interpretation phase were modified based on information collected during the fieldwork, existing maps such as wasteland, biodiversity, wasteland data, etc.

### i. Minimal mapping area

It is a concept applied by cartographers when addressing the smallest area that can be shown on a map. The minimum mapping has been considered as 2.25 ha. on 1:50K scale. However, for few important classes variable mapping area was adopted to preserve the local variations e.g. MMU for scattered rain fed crop area is kept as 1.5 ha for the state.

## 5.0 Study Area

The desertic tract of Rajasthan is characterised by a low and erratic rainfall, high atmospheric aridity, abundant solar radiation, strong dust raising winds, sparseness of vegetation cover and majestic dunes. It occupies an area of about 208,228 km$^2$ or 61.0 and 6.35 per cent of Rajasthan State and country respectively. The management of natural resources has become increasingly exploitative and cause of an acceleration of land degradation processes. The region has a rich cultural heritage. It extends between 24°37'00" N to 30°10'48" N latitudes and 69°29'00" E to 76°5'33" E longitudes. The Aravalli hills runs on the eastern margin, thus separating from eastern part of the State. Thar Desert is spread over 12 districts of Rajasthan state namely, Barmer, Bikaner, Churu, Ganganagar, Hanumangarh, Jaisalmer, Jalore, Jhunjhunun, Jodhpur, Nagaur, Pali and Sikar.

As per the Census of India, 2011, more than 27,115,542 persons (39.51% of state population) inhabit here, of which 14,069,930 are male and 13,045,612 are female population. Density of population is 174 persons per sq. km. with sex ratio as 923 females.

**Climate :** The tract has a mean annual rainfall of 450 to 470 mm at its eastern boundary with the semi-arid zone. From here it declines in westerly direction to less than 100 mm per year to the west of Jaisalmer town. The mean annual rainfall for the region as a whole comes to 310 mm. As against this the mean annual potential evapotranspiration (PET) varies from 1600 mm in the east to more than 2160 mm in the west and hence a high moisture deficit. Normals of monthly rainfall potential evapotranspiration, and aridity index for district headquarters are given together with co-efficient of its variation in Table 1. It will be seen that besides low rainfall, the region has a high inter-annual variability; the co-efficient of variation being 36 per cent in the north-east to as high as 66 per cent in the west. Additionally, there are frequent situation of late onset and early withdrawals or of mid-season droughts.

**Table 1 :** Mean annual rainfall, co-efficient of variation and aridity

| Station | Mean annual rainfall (mm)* | C.V. | Mean annual P.E. (mm) | Aridity Index (%) |
|---|---|---|---|---|
| Barmer | 264.0 | 63 | 1857 | 86 |
| Bikaner | 287.5 | 48 | 1771 | 84 |
| Churu | 357.7 | 38 | 1578 | 77 |
| Ganganagar | 242.7 | 57 | 1662 | 85 |
| Jaisalmer | 183.4 | 65 | 2064 | 91 |
| Jalore | 374.3 | 52 | 1561 | 76 |
| Jodhpur | 367.3 | 52 | 1842 | 80 |
| Jhunjhunun | 399.3 | 37 | 1562 | 74 |
| Pali | 418.3 | 54 | 1650 | 75 |
| Nagaur | 338.6 | 50 | 1641 | 79 |
| Sikar | 456.8 | 43 | 1532 | 70 |

The rainy season (end of June to middle of September) is generally marked by high atmospheric humidity and moderate amplitude of diurnal temperature variation. The autumn season is brief and winter sets in by the beginning of December. The winter is mild, sunny and calm. A few days of frost do occur especially in the north and west. The winter period hardly receives any rainfall, except in the northern-most part. Here 8 - 10 per cent of annual rain occurs in winter in response to westerly disturbances.

These areas co-relate very strongly with the incidence of poverty in the country. The prevention of land degradation and the augmentation of the carrying capacity of land to provide food, fuel and fodder requirements have, therefore, been a primary concern of the Government.

**Land-Forms :** Though dunes and sandy plains are dominant, the region is endowed with a large variety of landforms that owe their origin to endogenic and exogenic processes over a long drawn and complex evolutionary history. At the eastern end of the arid zone lie the ancient Aravalli Mountains. From here after a narrow belt of foot-plains, starts the vast expanse of nearly flat aggraded alluvial plains that are traversed by a number of streams. The monotony of the plains is broken here and there by rock outcrops and hills with well formed piedmonts. Isolated hills and plateau exist also outside the alluvial zone. The parallel slope retreat has caused the formation of pediments, which are buried at some places and exposed at others. This description, however, does not fit the rocky terrains of Jaisalmer, where the Tertiary and pre-Tertiary sedimentary rocks have resulted in a large sized, high level structural plains or Hammadas. This tract is distinguished also by the presence of level to gently sloping bouldery, gravelly and pavemented plains.

**Soils :** Dunes (Torripsamment-Dunes) are a dominant formation in 40.61 per cent of the area and a sub-dominant associate in 31.55 per cent. The soils here are highly sandy with clay and silt content in range of 2.5 to 4.6 per cent and 1.4 to 3.7 per cent respectively. The sand fraction is dominated by particles in the range of 0.12 to 0.18 mm through 11 to 20 per cent of medium and coarse sands together are present also. The soils are light yellowish brown, structureless - single grain, highly friable and weakly calcareous. The soil retains 50 - 60 mm of available water per meter depth. The soil of the associated sandy plains (Torripsamments-Plains) are similar to dune soils except that they have a little more of clay and silt content in the sub-soil and a slightly higher water storage capacity. Locally, the accumulative inter-dunes and depressions are loamy sand to sandy loam and highly calcareous, occasionally also gypseous. The Light Brown Sandy soils (Psammentic Camborthids) occur in about 25 per cent of the area. The surface soils here also is mostly sand but the sub-soil is generally loamy fine sand, very weakly blocky, and friable.

Soils are very low in organic carbon or humus and low (tending towards medium) in available phosphorous and medium to high in available potassium. The micro-nutrient levels are generally adequate. Of course, available nitrogen levels are low and a good response to application of nitrogenous fertilizers is obtained on all soils in years of good rainfall. Under irrigation, not only nitrogen but also phosphorous application is essential for sustained good yields.

Nutrients, other than nitrogen, are generally adequate even in the highly sandy soils of dunes and plains. Even nitrogen is not a critical factor here in establishment and maintenance of a healthy vegetation cover such as is permitted by rainfall of the tract.

**Natural Vegetation:** Major part of the desertic tract belongs to the Thorn Forest type at the broad level of classification. However, taking into account the habitat characteristics, a number of sub-divisions have been made, amongst which Psammophytic Desert and Mixed Xeromorphic Woodlands predominate. The other sub-types present are Mixed Xeromorphic Thorn Forest on hills, Lithophytic Scrub Desert on rocky gravelly plains, Mixed Xeromorphic Riverine Thorn Forest and Halophytic Scrub Desert vegetation of saline depressions.

The region has nearly seven hundred plant species which are distributed among 351 genera and 87 families.

## 6.0 Results

About 54536.87 sq. km lands of the arid Rajasthan are under different categories of wastelands. Most of these lands are government holdings classified as barren and uncultivable lands and culturable waste under revenue system. Unrestricted uses viz., grazing, cutting of trees and shrubs, mining of building and road materials have aggravated the aeolian, fluvial and salinity problems. Thereby, further degrading these lands. District-wise wasteland statistics for year 2005-06 and 2008-09 are in given in Tables 2 & 3 and variations in the various wasteland classes is presented in Table 4.

**Table 2 :** Wasteland Status, 2005-06

| S. No. | State / District | Gullied/ Ravinous lands | Land with/ without Scrub | Waterlogged/ Marshy land | Saline/ Alkaline Area | Degraded notified forest land | Degraded pastures/ grazing land | Degraded land under plantation crops | Sands- Inland / Coastal | Mining / Industrial Wastelands | Barren Rocky Area | Steep Sloping Area | Total Wastelands | Total Geographical Area | % to total geographical area |
|---|---|---|---|---|---|---|---|---|---|---|---|---|---|---|---|
| | | 1 | 2 | 3 | 4 | 5 | 6 | 7 | 8 | 9 | 10 | 11 | 12 | 13 | 14 |
| | Rajasthan | 4952.77 | 27152.76 | 289.66 | 2722.99 | 12541.89 | 12208.44 | 21.14 | 40639.51 | 128.65 | 4799.02 | 182.28 | 105639.11 | 342239 | 30.87 |
| | | | | | | | | | | | | | | | |
| 1 | Bikaner | 73.19 | 415.86 | 35.88 | 4.05 | 78.84 | 669.23 | 0 | 2098.47 | 13.77 | 0 | 0 | 3389.29 | 27244 | 12.44 |
| 2 | Barmer | 0 | 1193.15 | 0 | 254.19 | 126.97 | 1734.71 | 0 | 5893.09 | 3.7 | 291.3 | | 9497.11 | 28387 | 33.46 |
| 3 | Churu | 0 | 3.35 | 0 | 15.17 | 32.03 | 1264.75 | 0 | 318.39 | 5.7 | 1.2 | 0 | 1640.59 | 16830 | 9.75 |
| 4 | Hanumangarh | 0 | 0 | 108.51 | 9.43 | 0 | 5.55 | 110.66 | 0 | 120.3 | 2.19 | 0 | 355.64 | 9656 | 3.69 |
| 5 | Jaisalmer | 0 | 6958 | 19 | 119 | 165 | 0 | 0 | 26432 | 0 | 905 | 0 | 34598 | 38401 | 90.1 |
| 6 | Jalore | 56.05 | 422.28 | 0 | 56.98 | 141.86 | 0 | 0 | 438.21 | 0 | 130.75 | 0 | 1246.13 | 10640 | 11.71 |
| 7 | Jhunjhunun | 299.11 | 134.65 | 0 | 0 | 288.67 | 0 | 0 | 114.27 | 0.37 | 11.2 | 0 | 848.27 | 5928 | 14.31 |
| 8 | Jodhpur | 91.46 | 2197.2 | 0 | 169.62 | 34.93 | 1267.55 | 0 | 3103.77 | 25.63 | 90.82 | 0 | 6980.98 | 22850 | 30.55 |
| 9 | Nagaur | 44.82 | 507.37 | 0 | 216.54 | 127.25 | 862.35 | 0 | 504.72 | 16.52 | 67.7 | 0 | 2347.27 | 17718 | 13.25 |
| 10 | Pali | 15.57 | 928.77 | 0 | 668.73 | 616.18 | 864.67 | 0 | 53.96 | 0 | 130.35 | 16.5 | 3294.73 | 12387 | 26.6 |
| 11 | Sikar | 123.42 | 83.79 | 0 | 28.74 | 560.14 | 353.26 | 0 | 151.8 | 0 | 23.83 | 0 | 1324.98 | 7732 | 17.14 |
| 12 | Sri Ganganagar | 0 | 912.57 | 124.98 | 0.23 | 0 | 0 | 0 | 722.95 | 0 | 0 | 0 | 1760.73 | 10978 | 16.04 |
| | Total Area | 703.62 | 13756.99 | 288.37 | 1542.68 | 2171.87 | 7022.07 | 110.66 | 39831.63 | 185.99 | 1654.34 | 16.5 | 67283.72 | 208751 | 61.00 |
| %age of Rajasthan | | 0.21 | 4.02 | 0.08 | 0.45 | 0.63 | 2.05 | 0.03 | 11.64 | 0.05 | 0.48 | 0.00 | 19.66 | | |

**Table 3 :** Wasteland Status in Western Rajasthan During 2008-09

| S.No. | DESCRIPTION | BARMER | BIKANER | CHURU | GANGANAGAR | HANUMANGARH | JAISALMER | JALORE | JHUNJHUNUN | JODHPUR | NAGAUR | PALI | SIKAR |
|---|---|---|---|---|---|---|---|---|---|---|---|---|---|
| 1 | Gullied and/or ravinous land (Medium - Deep) | 19.96 | 93.19 | 4.13 | 0.00 | 0.00 | 0.00 | 54.89 | 41.01 | 10.02 | 33.48 | 9.42 | 89.29 |
| 2 | Land with Dense Scrub | 616.42 | 1497.75 | 118.35 | 79.01 | 23.47 | 7745.99 | 275.55 | 71.42 | 378.84 | 287.78 | 1203.90 | 112.64 |
| 3 | Land with Open Scrub | 719.65 | 420.80 | 219.48 | 82.53 | 40.06 | 3559.67 | 88.61 | 289.29 | 908.38 | 482.98 | 412.43 | 316.02 |
| 4 | Waterlogged and Marshy Land (Permanent) | 0.17 | 0.37 | 1.32 | 9.63 | 6.26 | 0.00 | 0.00 | 0.00 | 0.00 | 22.61 | 0.00 | 0.09 |
| 5 | Waterlogged and Marshy Land (Seasonal) | 21.98 | 0.00 | 0.00 | 8.44 | 0.00 | 4.76 | 0.00 | 0.00 | 27.57 | 0.00 | 0.00 | 0.00 |
| 6 | Land affected by salinity/alkalinity | 108.80 | 0.00 | 14.53 | 0.00 | 0.00 | 50.53 | 45.84 | 0.00 | 55.73 | 94.03 | 80.01 | 10.54 |
| 7 | Under-utilised/degraded forest | 113.02 | 70.87 | 0.00 | 0.00 | 23.12 | 113.31 | 152.43 | 168.36 | 57.97 | 120.17 | 191.92 | 293.26 |
| 8 | Degraded pastures/grazing land | 156.85 | 482.74 | 115.89 | 0.00 | 10.95 | 85.23 | 268.74 | 50.04 | 642.11 | 419.37 | 340.13 | 46.21 |
| 9 | Sands-Riverine | 2.30 | 0.00 | 0.00 | 0.00 | 0.00 | 0.00 | 0.02 | 11.38 | 0.24 | 0.00 | 36.54 | 1.56 |
| 10 | Sands-Desertic | 161.70 | 958.55 | 8.50 | 172.20 | 100.96 | 550.11 | 239.28 | 14.21 | 994.67 | 92.60 | 3.03 | 41.78 |
| 11 | Sands-Semi Stab.-Stab>40m | 1307.92 | 1010.65 | 0.00 | 0.00 | 0.00 | 6871.00 | 15.96 | 0.00 | 126.28 | 0.24 | 0.00 | 0.00 |
| 12 | Sands-Semi Stab.-Stab 15-40 m | 1384.07 | 5359.53 | 9.51 | 948.57 | 97.77 | 4895.27 | 114.55 | 0.00 | 1336.23 | 119.98 | 0.00 | 0.00 |
| 13 | Mining Wastelands | 3.09 | 0.00 | 0.04 | 0.00 | 0.16 | 0.65 | 0.91 | 0.31 | 20.52 | 6.28 | 5.85 | 0.35 |
| 14 | Industrial Wastelands | 0.15 | 2.62 | 0.00 | 2.57 | 0.00 | 0.00 | 0.65 | 0.00 | 1.12 | 0.00 | 1.37 | 0.00 |
| 15 | Barren Rocky Area | 399.60 | 20.07 | 2.64 | 0.00 | 0.00 | 1222.42 | 172.86 | 27.00 | 501.36 | 62.04 | 275.62 | 27.11 |
| **TOTAL WASTELANDS** | | **5015.68** | **9917.14** | **494.41** | **1302.95** | **302.74** | **25098.94** | **1430.30** | **673.03** | **5061.05** | **1741.56** | **2560.24** | **938.85** |
| Non Wasteland Area | | 23449.45 | 20299.03 | 13306.50 | 9397.08 | 9621.09 | 13226.31 | 9194.50 | 5245.94 | 17581.18 | 16026.78 | 9821.91 | 6777.16 |
| **TOTAL GEOGRAPHICAL AREA** | | **28465.13** | **30216.17** | **13800.90** | **10700.03** | **9923.83** | **38325.24** | **10624.80** | **5918.97** | **22642.22** | **17768.34** | **12382.15** | **7716.00** |

## Description of Different Wasteland Classes and Their Extent

Neglect and derelict, wastelands are generally the places where social and environmental abuse is more evident. These areas are everywhere: urban landfills, post-industrial terrains, left over zones in cities and peripheries, harbours, dismantled military installations, which succumb to the ever changing urban reality.

Wastelands can result from inherent/imposed disabilities such as by location, environment, chemical and physical properties of the soil or financial or management constraints. Study area contains Gullied and/or ravinous land, Land with Scrub, Waterlogged and Marshy Land, Land affected by salinity/alkalinity, Under-utilised/degraded forest, Degraded pastures/grazing land, Sands, and Industrial Mining Wastelands, and Barren Rocky area.

### 1. Mining/ Industrial Area

### 1.1 Industrial

These are areas where the human activity is observed in the form of manufacturing along with other supporting establishments of maintenance. Heavy industry, thermal, cement, petrochemical, engineering plants etc., are included under this class. An area of 46.64 sq. km. is occupied under this class.

### 1.2 Mine/ quarry

These are the areas subjected to removal of different earth material (both surfacial and sub-surface) by manual and mechanized operations. Large scale quarrying and mechanization results in mining and mine dumps. It includes surface rocks and stone quarries, sand and gravel pits, brick kilns, etc. Mining dumps are confined to the surroundings of the mining area. An area of 38.17 sq. km. is occupied under this class in 2008-09 whereas it was 34.47 sq. km. in 2005-06.

### 1.3 Industrial/mine dump

These are areas of stockpile of storage dump of industrial raw material or slag/effluents or waste material or quarried/mixed debris from earth's surface. An area of 8.47 sq. km. is occupied under this class in 2008-09 whereas it was 5.53 sq. km. in 2005-06. It also includes industrial dump which seen conspicuously around urban areas and other areas where industrial activity is prominent.

## 2. Scrub Land

This is a land, which is generally prone to deterioration due to erosion. Such lands generally occupy topographically high locations, excluding hilly/mountainous terrain. An area of 13169.35 sq. km. was occupied under this class in 2005-06 whereas in 2008-09 it reduced to 12411.12 sq. km. They appear in light yellow to brown to greenish blue depending on the surface moisture cover and vary in size from small to large having either contiguous or dispersed pattern. Scrublands are associated with moderate slopes in plains and foot hills and are generally surrounded by agricultural lands.

### 2.1 Dense scrub

These areas possess shallow and skeletal soils, at times chemically degraded, extremes of slopes, severely eroded and lands subjected to excessive aridity with scrubs dominating the landscape. An area of 13169.35 sq. km. was occupied under this class in 2005-06 which reduced to 12411.12 sq. km. in 2008-09. Further, they have a tendency for intermixing with cropped areas.

### 2.2 Open scrub

This category possesses sparse vegetation or devoid of scrub and have a thin soil covers. An area of 7539.91 sq. km. is occupied under this class in 2008-09 whereas it was 7500.11 sq. km in 2005-06.

## 3. Salt-Affected Land

Salt affected land is generally characterized as land that has adverse effects on growth of most plants due to the action or presence of excess soluble salts (saline) or high exchangeable sodium. Alkali land has an exchangeable Sodium percentage (ESP) of about 15, which is generally considered as the limit between normal and alkali soils. The predominant salts are carbonates and bicarbonates of Sodium. These salts are essentially delineated because of their high reflectance. They appear in white to light blue color depending upon moisture content ranging in small to medium sizes. Generally they are irregular in shape and discontinuous, closely spaced or dispersed depending on their location. Saline waste (including natural saline depression or 'playas' and other salt-affected land) occur dominantly in the south-eastern part of the region covering Barmer, Jalore, Pali, and Nagaur districts.

An area of 285.27 sq. km. and 174.75 sq. km. is under moderate and strong salt-affected lands (0.22 per cent of the total area) in the arid region of Rajasthan. It was 313.15 sq. km. and 208.86 sq. km. in 2005-06. There is a reduction of 27.88 sq. km. and 34.11 sq. km. in both categories.

## 4. Gullied / Ravinous Land

Gullies are formed as a result of localized surface run-off affecting the unconsolidated material resulting in the formation of perceptible channels causing undulating terrain. Gullies develop from rills which are tiny water channels with a few centimeters deep, formed as a resultant impact of heavy rainfall and wearing action of run-off generated there from. They are commonly found in sloping lands, developed as a result of concentrated run-off. Further classification of this category is possible based on the depth, width, bed slope, frequency and morphology of bed material of the ravines. Such lands occur precisely in more than 300 mm rainfall zone covering Jalore, Pali, Jhunjhunun and Sikar districts covering 54.89, 9.42, 41.02 and 89.29 sq. km. respectively in 2008-09.

They appear in light yellow to bluish green in color depending on the surface moisture and depth of erosion. They vary in size, shape with irregular broken network pattern. They are mostly associated with stream courses and sloping grounds with good rainfall and entrenched drainage. An area of 330.23 sq. km. and 108.03 sq. km. was occupied under this class in 2005-06 but drastically reduced in 2008-09 by 269.52 sq. km. and 85.87 sq. km.

## 5. Sandy Area

These are the areas that have stabilized accumulation of sand, in coastal, Riverine or inland areas, that can be either desertic or coastal. An area of 31395.43 sq. km. was occupied under this class in 2005-06 which reduced to 26987.19 sq. km. in 2008-09. They appear as white to light yellow/ bluish depending on moisture content and at time light red when vegetation is associated with the class, vary in size, with regular to irregular shape with contiguous to linear pattern Mostly, they are located in deserts and riverbeds and predominantly occur along the Indo-Pak border in Jaisalmer, Bikaner and Barmer districts getting less than 200 mm annual rainfall.

### 5.1 Desertic sand

Desertic sands are those confined to arid environment where the rainfall is scanty. These lands are characterized by accumulation of sand 'developed in situ or transported by Aeolian processes. An area of 4490.43 sq. km. was occupied under this class in 2005-06 which reduced to 3337.60 sq. km. in 2008-09.

## 5.2 Riverine sand

Riverine sands are those that are seen as accumulations in the flood plain as sheets which are the resultant phenomena of river flooding. An area of 108.76 sq. km. was occupied under this class in 2005-06 which reduced to 52.04 sq. km. in 2008-09.

## 6. Barren/Rocky/Stony Waste

These are rock exposures of varying lithology often barren and devoid of soil and vegetation cover. They occur amidst hill-forests as openings or as isolated exposures on plateau and plains. Such lands can be easily discriminated from other categories of wastelands because of their characteristic spectral response. An area of 2664.15 sq. km. was occupied under this class in 2005-06 which increased to 2710.72 sq. km. in 2008-09. They appear in greenish blue to yellow to brownish in color depending on the rock type. They vary in size with irregular to discontinuous shape with a linear to contiguous or dispersed pattern. They are located in steep isolated hillocks/hill slopes, crests, plateau and eroded plains associated with barren and exposed rocky/stony wastes, lateritic out-crops, mining and quarrying sites. Such lands mainly occur in Jaisalmer, Pali and Jodhpur districts constituting 1222.42, 275.62 and 501.36 sq. km. area in 2008-09.

## 7.0 Discussion

Wastelands of deserts and drought-prone areas have been subjected to severe forms of degradation. The capacity of these lands is limited due to environmental factors. Pressures of human and livestock population have further compromised them.

## 8.0 Suggestions

Degraded and wasteland assessment is an essential input for planning and conservation. Scientists believe that space science and technology can help in enhancing efficiency and productivity of agricultural system. "Correct estimation of green cover and water resources will help in the management of precious assets." Advanced yield forecasting can be a good resource for planning to effectively run the public distribution system.

As rehabilitation measures are highly expensive, the study stresses on priority-based planning for land conservation.

Guidelines for Watershed Development (Revised 2001) Department of Land Resources, Ministry of Rural Development, Government of India include following parameters for the reclamation of wastelands.

- Land development including *in situ* soil and moisture conservation measures like contour and graded bunds fortified by plantation, bench terracing in hilly terrain; and nurseries for fodder, timber, fuel wood, horticulture and non-timber - forest produce
- Afforestation including block plantations, agro-forestry and horticultural development. Shelter-belt plantations, sand dune stabilisation, etc.
- Drainage line treatment with a combination of vegetative and engineering structures.
- Development of small water harvesting structures such as low-cost farm ponds, nalla bunds, check dams and percolation tanks and ground water recharge measures.
- Renovation and augmentation of water resources, desiltation of tanks for drinking water and irrigation.
- Pasture development either by itself or in conjunction with plantations.
- Repair, restoration and upgrading of existing common property assets and structures in the watershed to obtain optimum and sustained benefits from previous public investments.
- Crop demonstrations for popularising new crops and crop varieties or innovative crop management practices.
- Promotion and propagation of non-conventional energy saving devices and energy conservation measures.

## 9.0 Conclusion

Wasteland has a significant role in the overall production process. Continued neglect of the wasteland results in decrease in the fuel and fodder availability, which in turn increase the pressure on cultivated land. Hence, there is need to put a halt to further degradation of these lands and initiate corrective measures to improve the production capacity. Erosion and degradation are principal causes for the collapse of many a civilization in the past and wide spread rural poverty at present. Land degradation become more acute with the ruthless denudation of forest, uncontrolled grazing and encroachment further aggregated the situation. At present hardly any land in western Rajasthan is free from one hazard or the other environmental degradation.

Owing to increase in population and not many alternate opportunities for livelihood, the pressure on land for multivarious uses has increased. Due to well-known umpteen number of reasons, the quality and quantity of land is decreasing over the years. The situation, unless addressed and attended on priority, things may grow behind vulnerable groups. The process of rapid expansion of arable farming started by the beginning of middle of the last century. As a result, much of the agriculturally usable land in the climatically favourable tract from Sikar-Jhunjhunun in the north-east through Pali upto Jalore in the south was brought under the plough. There has been two sets of trends. The first being intensification of agriculture i.e., reduction of fallow farming in the already established agricultural tract. In the Ganganagar-Hanumangarh district, the contribution of expanding irrigation has been the significant factor. The second direction of reduction of wastelands has been bringing under cultivation, lands lying waste in Churu. There is, therefore, urgent need to give prime thrust to arrest land degradation/ wastelands. The sustainable agriculture - a need of the hour is not merely to improve productivity of cultivated lands but also to rectify wastelands for economic use.

## Acknowledgement

Authors acknowledge Regional Remote Sensing Centre-West/ Indian Space Research Organization for their unconditional support and facilitation for carrying out this study. Assistance rendered by Mr. Sumit Maheshwari and Bahul Vyas is acknowledged.

**Table 4 :** Wasteland Change Analysis (2005-06 and 2008-09), Western Rajasthan

| WL0809 | DESCRIPTION | WL0506 | WL0809 | CHANGE |
|---|---|---|---|---|
| 1 | Gullied and/or ravinous land (Medium) | 330.23 | 269.52 | -60.71 |
| 2 | Gullied and/or ravinous land (Deep) | 108.03 | 85.87 | -22.15 |
| 3 | Land with Dense Scrub | 13169.35 | 12411.12 | -758.24 |
| 4 | Land with Open Scrub | 7500.11 | 7539.91 | 39.80 |
| 5 | Waterlogged and Marshy Land (Permanent) | 44.84 | 40.46 | -4.38 |
| 6 | Waterlogged and Marshy Land (Seasonal) | 67.69 | 62.75 | -4.94 |
| 7 | Land affected by salinity/alkalinity (Moderate) | 313.15 | 285.27 | -27.88 |
| 8 | Land affected by salinity/alkalinity (Strong) | 208.86 | 174.75 | -34.11 |
| 9 | Under-utilised/segraded forest (Scrub dominated) | 1388.10 | 1299.77 | -88.33 |
| 10 | Under-utilised/degraded forest (Agriculture) | 4.16 | 4.66 | 0.50 |
| 11 | Degraded pastures/grazing land | 2934.37 | 2618.25 | -316.12 |
| 12 | Sands-Riverine | 108.76 | 52.04 | -56.72 |
| 13 | Sands-Desertic | 4490.43 | 3337.60 | -1152.82 |
| 14 | Sands-Semi Stable (Stab>40m) | 11219.66 | 9332.05 | -1887.61 |
| 15 | Sands-Semi Stable (Stab 15-40 m) | 15576.58 | 14265.49 | -1311.09 |
| 16 | Mining Wastelands | 34.47 | 38.17 | 3.70 |
| 17 | Industrial Wastelands | 5.53 | 8.47 | 2.94 |
| 18 | Barren Rocky Area | 2664.15 | 2710.72 | 46.57 |
| | TOTAL WASTELANDS AREA | 60168.47 | 54536.87 | -5631.60 |
| | Non Wasteland Area | 148315.32 | 153946.91 | 5631.59 |
| | *TOTAL GEOGRAPHICAL AREA* | *208483.79* | *208483.78* | *-0.01* |

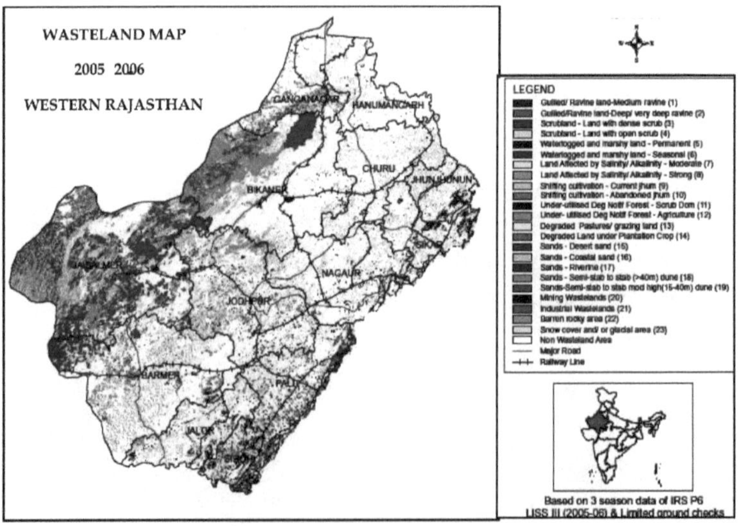


**Figure 2 :** Wasteland status in 2005-06

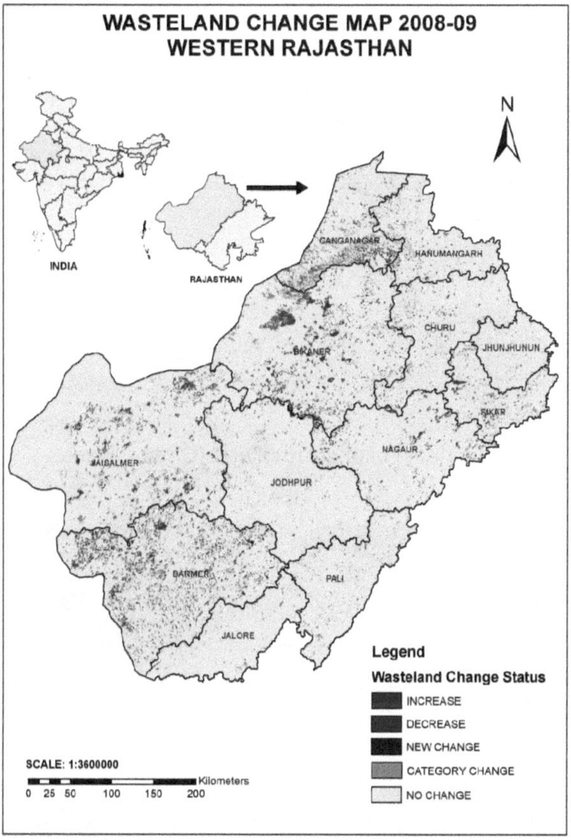


**Figure 3 :** Wasteland change detection status in 2008-09

Sand Dunes near Chouhtan Village (Barmer District)

Mining near Ratangarh Village (Churu District)

Gypsum Quarry near Mohangarh (Jaisalmer District)

Salt Affected Land near Dudhwa Village (Barmer District)

Ravinuous Land near village Jhujhu village (Bikaner District)

Waterlogged area near Ket ki dhani (Hanumangarh District)

## References

CAZRI. 2008. Meteorological Laboratory.

Dwivedi, R.S., Sreenivas, K. and Ramana, K.V., 1999, Inventory of salt-affected soils and waterlogged areas: a remote sensing approach, *International Journal of Remote Sensing*, 20(8) : 1589-1599

Gaur, Mahesh. 1992. Impact of Drought on Desertification Process in North-Western Rajasthan (A study of Jodhpur district - based on Remote Sensing Technique). Unpublished PhD Dissertation, JNV University.

Gautam, N.C. and Narayan, L.R.A., 1988, Wastelands in India, Pink Publishing House, Mathura, India,pp 96.

http://greencleanguide.com/2011/01/18/wastelands-types-and-status-in-india/

http://www.indiawaterportal.org/node/12607

http://dolr.nic.in/iwdp1.htm

Mann, H.S. and Chatterji, P.C. 1978. Impact of mining operations on the ecosystem in Rajasthan. *Ecology and Coal Resources Development.* 2: 615-625.

Jaga Novaline, R.M., Sundaram, A. and Natarajan, T., 1993, Wasteland development using Geographic Information System techniques, *International Journal of Remote Sensing*, 14(17) : 3249-3257

NRSA. 2006. Manual: National Land Use Land Cover Mapping using Multi-temporal Satellite Data. (NRC-LULC-50 K; May 2006)

NRSA. 2007. Manual: National wasteland monitoring using multi-temporal satellite Data.

Saha, S.K., Kudrat, M. and Bhan, S.K., 1990, Digital processing of Landsat TM data for wasteland mapping in parts of Aligarh District (Uttar Pradesh), India, *International Journal of Remote Sensing*, 11(3) : 485-492

Saxena, S.K. and Chatterji, P.C. 1988. Ecological imbalances caused by mining in Rajasthan and their restoration. In: Mining and Environment in India. Himalayan Research Group. Publication Series, New Delhi, pp. 223-241.

Singh, S. and Shankarnarayan, K.A. 1986. Dynamics, morphology and management of sand dunes in Great Indian Desert. *Current Practices in Geotechnical Engineering*, 3:23-47.

Tewari, J.C., Harsh, L.N. and Venkateswarlu, J. 1993. Some aspects of plantation forestry research in western Rajasthan. In: Afforestation of Arid Lands (Eds. A.P. Dwivedi and G.N. Gupta), Scientific Publishers, Jodhpur. Pp. 61-72.

# Chapter - 7

# Crop Identification Using Hyperspectral Satellite Data and Field Spectra

**Chilka Sharma, Manoj Joseph, S. Palaria and J.R. Sharma**

## Abstract

*This paper portray the generation of spectral signature library for semi arid area crops using hyperspectral data and the scope of the application of the generated digital library for the classification of the satellite imagery of similar regions.*

*Hypersprctral data have narrow bands which help to provide additional information related to observed features. In case of this agriculture study crop related information are gathered using hyperspectral sensors The variation in the observations are due to physic-chemical characteristics such as biochemical composition, physical structure, water content and plant ecophysical status. Using the hand held spectroradiometer the reflectance is captured for the bajra, moong, moth crops from the Rajasthan, India's semi arid region. The observations have been taken for different dates considering the phenology of the crops. Hyperion data from EO-1 satellite have been used to implement the classification operation based on the spectral signature library of bajra. It is observed that the image with the crops can be classified according to the phenology difference.*

*Key Words: Hyperspectral, spectral library, spectroradiometer, semi-arid region,crop*

## Introduction

Imaging Spectrometry or hyperspectral technology, as an advanced tool that provides high spectral resolution data for each pixel from a far distance significantly broadens the utility for further mapping of the soil surface, agriculture crops, mineral deposition etc from a more precise chemical and physical point of view. With the development of hyperspectral sensors, spectral features related to characteristic absorption bands of each targets can be mapped with more detail. The

Hyperion sensor onboard NASA's Earth Observing 1 (EO-1) satellite is the first space borne hyper spectral instrument to acquire both visible/ near-infrared and shortwave infrared spectral data. With its 242 potential bands and a spatial resolution of 30m, the sensor bears the potential to provide data for both a detailed land use classification and an accurate estimation of bio geophysical and chemical properties of heterogeneously vegetated areas.

All vegetation contains similar biochemical constituents, but these vary in their proportions. In the laboratory spectroscopy is successfully used to determine biochemical constituents contributing to forage quality. Remote sensing is an important tool for mapping and monitoring vegetation. The hand held spectroradiometer is the source to acquire hyperspectral data from the objects. It captures the radiance in form of reflected portion of sun incidence rays from the object.

The aim of this study is that vegetation associations can be differentiated using their hyperspectral reflectance in the visible to shortwave infrared spectral range. The reflectance from field using spectroradiometre of various crops like bajra (pearl millet), Moong (green gram), Moth (gram), Tur (Pigeon Pea) were captured on different dates. The variation in the curves of different crops and time is significant. The spectral library is generated for the different types of crops and also for difference in regards of date of sowing. For scaling up of crop phenology and productivity the spectral library is beneficial.

A part of the Luni block, Jodhpur district, and Pokaran block, Jaisalmer district of Rajasthan, India is selected for the study. The field data was collected from Jodhpur where as the satellite data was used for Jaisalmer (due to limited availability) keeping in mind the contiguity of resources and similar cropping patterns. Crop discrimination was performed using Spectral Angle Mapper (SAM) classification for hyperspectral data.

The Spectral Angle Mapper is a technique to classify hyper spectral data by determining the similarity between an end member spectrum and a pixel spectrum in an n-dimensional space. Smaller angles represent closer matches to the reference spectrum. In the present analysis, field spectra of different targets measured using HR1024 spectro-radiometer are used as end member spectrum input for Spectral Angle Mapper classification. The field spectra are very well matched with the image and an overall classification accuracy of 75 per cent has been observed. The detailed methodology of the process and classification outputs is presented for discussion. The study involve the test of the reflectance spectra of the vegetation types for differences between type classes and ensures that, although vegetation spectra consist of similar detectable absorption features making them an important source of information about the biochemical constitution of vegetation, there are significant differences between vegetation types, both in absolute reflectance and in curvature.

## Material and Methods

**Study Area -** A part of the Jodhpur district of Rajasthan, India is selected as the study area.

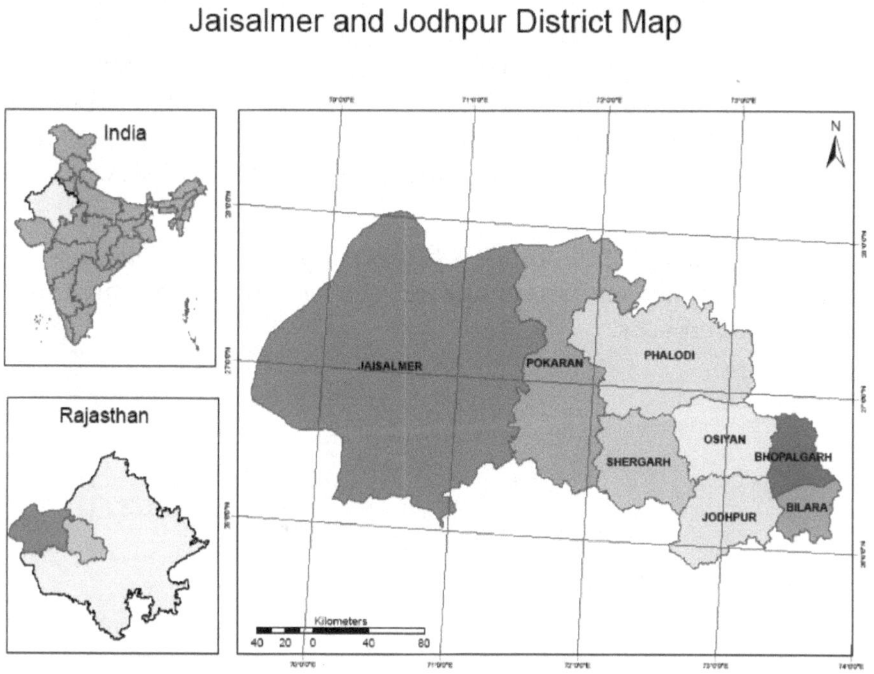


**Figure 1 :** Location of Observation

**Material -** The field data was collected from south Jodhpur whereas the satellite data was used of middle west Jodhpur (due to limited availability) keeping in mind the contiguity of resources and similar cropping patterns.

Hyperion -2$^{nd}$ September 2010; Scene lat/long :

TL - 27°33′23″N 72°6′58″E

BR - 27°15′57″N 72°7′17″E

Field spectra : The spectral signature were collected using HR1024 Spectro-radiometer on 7$^{th}$ September 2010 and 28$^{th}$ September 2010 for various crops.

## Observations

## Reflectance from crops

The field spectra were collected on two dates for kharif crops. The number of observations for single crop was observed and further used for identification and classification through satellite data.

## Methodology

The reflectances from the field were captured and analyzed for discriminating the various crops of the region such as pearl millet, green gram, bean and pigeon pea. The curves for the crops are mapped and the difference in the curves is interpreted below in figure 2. The curve below is displaying the reflectance of pearl millet on different dates and of different variety. The variation is very clearly observed in the near IR bands.

The digital spectral library is build using the ENVI software. This library is generated on basis of number of observations and can be used for future references.

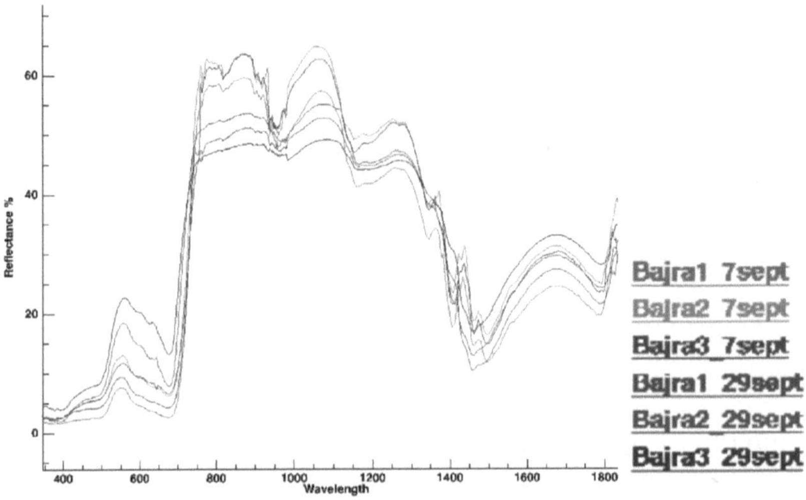


**Figure 2 :** Reflectance curve of Bajra

In the above figure 2 the curve for bajra crop is plotted and the variation of reflectance and absorption by the crop is clearly visible in visible and near infra red range.

False color composite image is generated for processing the hyperion data using the bands widths at 450 nm, 520 nm and 750 nm as blue, green and red bands respectively seen in figure 3.

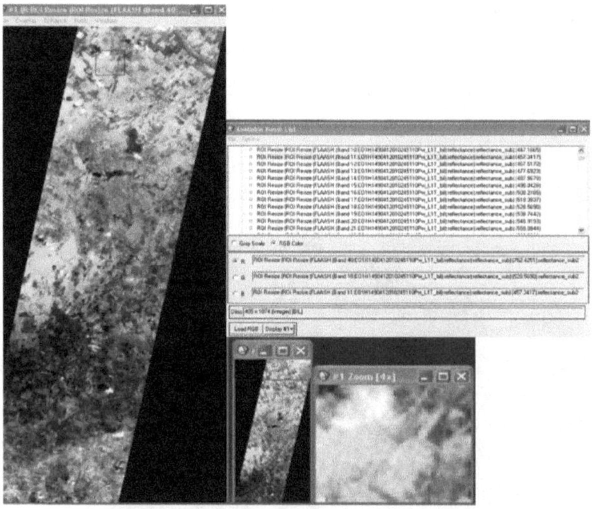

**Figure 3 :** Hyperion image – FCC (R-0.75ìm, G-0.52ìm, B-0.45ìm)

Further, the reflectance of pearl millet is used to match and classify the hyperion satellite data using spectral angle mapper algorithm and the digital spectral library.

## Result and Discussion

The digital spectral library generated for semi arid crops shows the difference in the range of near infra red majority and it is due to the chlorophyll content and structure of the plant. Further, the water content in crops also varies and is observed in the cures at 940nm, 1040nm and 1400nm.

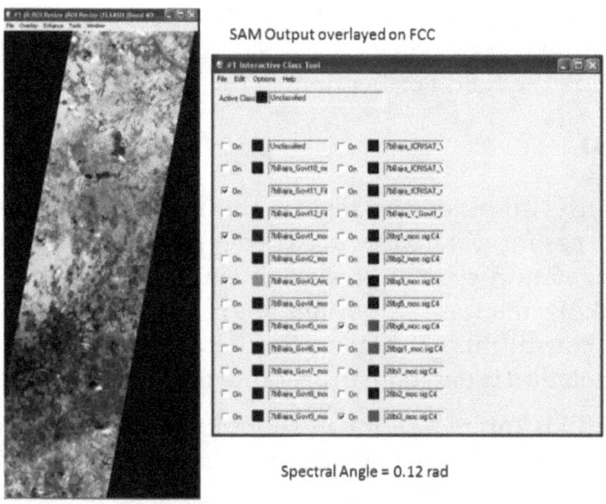


**Figure 4 :** Classification of Hyperion data

The hyperion image is classified using the pearl millet field spectra. Spectral angle mapper algorithm was used to classify the image and found that portion of image matches with first date of observation and some portion matches with second date. This could be due to difference in the sowing dates. Hence the phenology can also be mapped using the hyperspectral data. (Fig. 4)

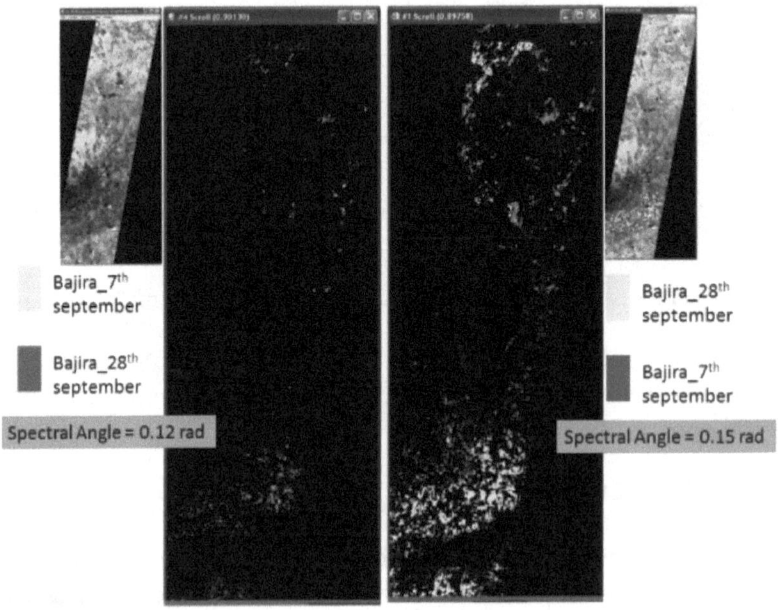


**Figure 5 :** Classification using different spectral angles

The above figure 5 shows the output of the classification performed by spectral angle mapper with 0.12 and 0.15 radians. The accuracy of the classification can be enhanced by selecting the appropriate radian for considering the field verification.

## Conclusion

The study involves the test of the reflectance spectra of the vegetation types for differences between type classes and ensures that, although vegetation spectra consist of similar detectable absorption features making them an important source of information about the biochemical constitution of vegetation, there are significant differences between vegetation types, both in absolute reflectance and in curvature.

The study is limited by the availability of hyperion data and the observations from field is not of the same location due to time constraint.

The study is still under process for refining the work of digital spectral library generation and integration of field and satellite hyperspectral data for larger area crop productivity - stress mapping and supporting the locals to achieve sustainability.

## Acknowledgement

The authors would like to thanks RRSC - W for providing the support for field observations and performing the analysis.

## References

Alexander Platonov , Prasad S. Thenkabail, Chandrashekhar M. Biradar , Xueliang Cai, Muralikrishna Gumma , Venkateswarlu Dheeravath , Yafit Cohen , Victor Alchanatis , Naftali Goldshlager , Eyal Ben-Dor , Jagath Vithanage , Herath Manthrithilake , Shavkat; Kendjabaev and Sabirjan Isaev ; Water Productivity Mapping (WPM) Using Landsat ETM+ Data for the Irrigated Croplands of the Syrdarya River Basin in Central Asia Sensors 2008, 8, 8156-8180; DOI: 10.3390/ s8128156

Chandrashekhar M. Biradar, Prasad S. Thenkabail, Alexander Platonov, Xingming Xiao, Roland Geerken, Praveen Noojipady, Hugh Turral, Jagath Vithanaged; Water productivity mapping methods using remote sensing. *Journal of Applied Remote Sensing,* Vol. 2, 023544 (6 November 2008)

Goodenough, D.G.; Pearlman, J.; Hao Chen; Dyk, A.; Tian Han; Jingyang Li; Miller, J.; Olaf Niemann, K.; , "Forest information from hyperspectral sensing,"Geoscience and Remote Sensing Symposium, 2004. IGARSS '04. Proceedings. 2004 *IEEE International* , vol.4, no., pp. 2585- 2589 vol.4, 20-24 Sept. 2004

M. Abbasi, A.A. Darvishsefat, M. E. Schaepman; Spectral reflectance of rice canopy and red edge position (rep) as indicator of high-yielding variety July 5-7, 2010, IAPRS, Vol. XXXVIII, Part 7B

M. Abbasi, A.A. Darvishsefat, M. E. Schaepman. Spectral Reflectance Of Rice Canopy And Red Edge Position (Rep) As Indicator Of High-Yielding Variety. Wagner W., Székely, B. (eds.): ISPRS TC VII Symposium – 100 Years ISPRS, Vienna, Austria, July 5-7, 2010, IAPRS, Vol. XXXVIII, Part 7B

Pearlman, J.; Carman, S.; Segal, C.; Jarecke, P.; Clancy, P.; Browne, W.; , "Overview of the Hyperion Imaging Spectrometer for the NASA EO-1 mission,"Geoscience and Remote Sensing Symposium, 2001. IGARSS '01. *IEEE 2001 International*, vol.7, no., pp.3036-3038 vol.7, 2001

Prasad S. Thenkabail1. Global Croplands And Their Water Use—Advanced Remote Sensing Methods And Approaches .ISPRS Archives XXXVIII-8/W3 Workshop Proceedings: Impact of Climate Change on Agriculture

P. Zorogastúa, R. Quiroz, M. Potts, S. Namanda, V. Mares, L. Claessens, SSAUtilization of high-resolution satellite images to improve statistics for the sweetpotato cultivated area of Kumi district, Uganda ISBN 978-92-9060-330-6 Natural Resources Management Division Working Paper No. 2007-5

Rouse, J. W., R. H. Haas, J. A. Schell, and D. W. Deering (1973). Monitoring vegetation systems in the Great Plains with ERTS, Third ERTS Symposium, NASA SP-351 I, 309-317.

Thenkabail, P.S., Biradar C.M., Noojipady, P., Dheeravath, V., Li, Y.J., Velpuri, M., Gumma, M., Reddy, G.P.O., Turral, H., Cai, X. L., Vithanage, J., Schull, M., and Dutta, R. 2009b. Global irrigated area map (GIAM), derived from remote sensing, for the end of the last millennium. *International Journal of Remote Sensing*. 30(14): 3679-3733. July, 20, 2009.

Thenkabail, P. S.; Dheeravath, V.; Biradar, C. M.; Gangalakunta, O. P.; Noojipady, P.; Gurappa, C.; Velpuri, M.; Gumma, M.; Li, Y. 2009c. Irrigated Area Maps and Statistics of India Using Remote Sensing and National Statistics. *Journal Remote Sensing*. 1:50-67. http://www.mdpi.com/2072-4292/1/2/50.

The Benefits of the 8 Spectral Bands of WorldView- 2 Wh i t e P a p e r - March2010 Digital Globe, 2010

Use of Hyperspectral Remote Sensing in Agriculture Applications: Ground based studies, Space Application Centre, ISRO, December2007.

Van Niel, T.; McVicar, T.; Campbell, S.; Liang, S.; Kaul, M.; Pearlman, J.; Clancy, P.; Segal, C.; , "The Coleambally agricultural component of Hyperion instrument validation," Geoscience and Remote Sensing Symposium, 2001. IGARSS '01. *IEEE 2001 International*, vol.7, no., pp.3230-3232 vol.7, 2001

# Chapter - 8

# Data Mining for Extracting Spatial and Temporal Pattern of Drought

**Aditi Sharma, P. Dhyani and M.P. Punia**

## Abstract

*Droughts are normal recurring climatic phenomena that vary in space, time, and intensity. The spatial and temporal variability and multiple impacts of droughts provide challenges for mapping and monitoring on regional scales. Large historical datasets are required in drought studies to understand the complex inter-relationship between the climatic and meteorological data. Extraction of valuable information from such large data archives demands an automated and efficient way. Data mining is answer to above problem as it has the potential to search for hidden pattern and identify the relationship between the datasets. For the current research, rainfall data from 1951-2006 was used to compute Standardized Precipitation Index (SPI) and NOAA-AVHRR NDVI images for the period 1981-2006 was used to calculate Vegetation Condition Index (VCI). SPI and VCI were used as input parameters in the Apriori Algorithm for generating the association rules. Interesting rules were developed identifying the relationship between user-specified target episodes and meteorological events. Therefore, spatial and temporal pattern of drought was achieved with 88% confidence.*

***Keywords:*** *Drought, Data mining, Associations rule, Standardized Precipitation Index (SPI), Vegetation Condition Index (VCI)*

## 1. Introduction

With increasing number of civilian satellite repositories holding a surfeit of data and the synergetic use of remotely sensed earth observation data for earth system studies, the need for data mining applications in describing, explaining and predicting spatial patterns is well understood by researchers (Ester et al.,1998; Ester et al.,1999; Peuquet et al.,2000). Data mining is the process of discovering meaningful patterns, profiles and trends. (Mena,1999).

Data Mining can solve problems related to drought by extracting useful information from large historical data archives. As large collections of historical data are essential in identifying relationships among different climatic and meteorological parameters, proper handling of data and quick processing becomes necessary. This highlights that efficient data management and information extraction is important in drought research. The difficulty of handling huge amount of data, understanding the complex results derived from atmospheric phenomenon and extracting the spatial and temporal pattern of drought can be resolved with data mining techniques. In this regard, the current study tries to explore one of the data mining techniques for drought application.

## 2. Study Area and Data used

The State of Karnataka is confined within 11.5 degree North to 18.5 degree North latitudes and 74 degree East and 78.5 degree East longitude. The annual rainfall in the State varies roughly from 50 to 350 cm. The rainfall increases significantly in the western part of the State and reaches its maximum over the coastal belt. The south-west monsoon is the principal rainy season during which the State receives 80% of its rainfall. Rainfall in the winter season (January to February) is less than one per cent of the annual total, in the hot weather season (March to May) about 7% and in the post-monsoon season about 12%.

For this study, Standardized Precipitation Index from 1981-2006 and Vegetation Condition Index for Karnataka State was used. Using both the drought indices SPI and VCI together adds an advantage to the study as is saves from false drought identification. Considering a case of irrigated agricultural area, SPI finds the area as drought prone area which is false. Hence combined usage of SPI and VCI is highlighted.

## 3. Methods

### 3.1 Data mining

Data mining can be defined from drought research point of view as a process that analyze periodic system variables like climatological and meteorological indices to discover and establish relationship between them thereby specifying drought characteristics like onset, severity, extent and duration which in turn would help in efficient drought monitoring. Some the of the data mining algorithms and models are decision trees, association rules, clustering, classification, regression, sequential patterns, and time series forecasting. Association rule is used for extracting the drought pattern in the current study.

## 3.2 Association rule generation

In a real-world application such as drought where time series is an important factor, it is important to study the relationships of the parameters that cause drought. In time series data mining algorithms data is organized as sequence of events and each event has a time of occurrence. Association rules are used to show the relationships between the data items. Association rule is defined as $X \Rightarrow Y$, where X is the rule antecedent and Y is its consequent and $X \cap Y = \phi$. Support and confidence are two widely used metrics in measuring the interestingness of association rules. Support {X} is a measure of statistical significance, and is defined as the ratio of the count of the events in X divided by the total number of events. This means, support indicates the percentage of episodes in the dataset that contain both X and Y (Harms et al.,2002). In association rules: if X then Y, the support of the antecedent is also called as the rule coverage. The confidence of the rule is defined as the ratio of support $\{X \cup Y\}$, which is the percentage of events that occur in both X and Y, divided by the support {X}. This is a measure of the conditional probability (Tadesse et al.,2004).

For a rule, $X \Rightarrow Y$

**Support** is $sup(X \Rightarrow Y)$ and (Equation 1)

**Confidence** is $conf(X \Rightarrow Y) = sup(X \Rightarrow Y) / sup(X)$ (Equation 2)

Thus, Confidence measures the strength of the rule, whereas support measures how often it should occur in the database. Typically, large confidence values and a smaller support are used.(Dunham, 2004)

Drought-related data is viewed as event sequences. An event sequence is a collection of time ordered events that happen within a finite time period. For example, {0.54, 2.07, 0.48, -0.48, -0.84, -1.77, 0.13, -1.45, -0.63, -0.76, -0.18, 0.65} is an event sequence which presents the twelve-month Standardized Precipitation Index (SPI) value from January to December in 1984 at Aland station. The time interval within which the events occur is called a sliding window. The width of the sliding window determines the length of time interval. An episode is a collection of events in a particular order occurring within a given window width. There are various association rule generation algorithms. Apriori algorithm was used for this work.

## 3.3 Apriori algorithm

The Apriori algorithm is the most well known association rule algorithm and is used in most commercial products. It uses the following property, which we call the large item set property:

Any subset of a large item set must be large.

An item set is a non-empty set of items. The large item set are also said to be downward closed because if an item set satisfies the minimum support requirements, so do all of its subsets (Dunham, 2004).

The problem of mining association rules is decomposed into two parts :

- Find all combinations of items that have support above minimum support. Call those combinations frequent item sets.
- Use the frequent item sets to generate the desired rules. The general idea is that if, say, ABCD and AB are frequent item sets, then we can determine if the rule AB CD holds by computing the ratio r = support(ABCD)/support(AB). The rule holds only if r >= minimum confidence. Note that the rule will have minimum support because ABCD is frequent. The Apriori algorithm used for finding all frequent item sets is given below

**Procedure** AprioriAlg (D: Dataset of transactions, min-support: Minimum Support Threshold)

**begin**

$L_1$ := frequent-1-ietmsets(D);

**for** ( k := 2; $L_{k-1}$<> Ø; k++) **do** {

$C_k$= apriori-gen($L_{k-1,}$ min-support) ; // new candidates

**for** each transaction t in the dataset **do** {

$C_t$ = subset ($C_k$, t);

**for** each candidate c $\in C_k$ contained in t **do** }

c.count++

$L_k$ = { c$\in$ $C_k$ | c.count >= min-support}

return L := $U_k L_k$ }

**end;**

The algorithm now scans the database. For each transaction, it determines which of the candidates in $C_k$ are contained in the transaction using a hash-tree data structure and increments the count of those candidates. At the end of the pass, $C_k$ is examined to determine which of the candidates frequent, yielding $L_k$ is. The algorithm terminates when $L_k$ becomes empty (Joshi, 1997).

## 3.4 Data processing

In this study, association rules are used to identify the relationship between SPI and VCI thereby finding the occurrence of drought in the study area from year 1981-2006. Preparing and processing the time series data is required to be done for applying Apriori data mining algorithm. The five steps required to process the data in order to generate association rule are:

**Step 1:** Discretizing the data:

SPI value of time scale 1 and monthly VCI values were used for generating the association rules. The sliding window width used for both the dataset is 1. SPI values show the rainfall deviation from mean for a given location at a given time and VCI quantifies the weather component of NDVI thereby reflecting the health/condition of vegetation. Both SPI and VCI was discreetized independently into 5 clusters as shown in table 1.

**Table 1 :** Discretization criteria for SPI and VCI into its five respective categories

| SPI values | Category | VCI values | Category |
|---|---|---|---|
| -3 to -1.5 | Extremely dry | 0 to 34 | Very low |
| -1.49 to -1.0 | Dry | 35 to 45 | Low |
| -0.99 to 0.99 | Normal | 46 to 65 | Normal |
| 1 to 1.5 | Wet | 66 to 85 | High |
| 1.5 to 3 | Extremely wet | 86 to 100 | Very high |

**Step 2:** Formulating Target Episodes:

As we are interested in finding the drought pattern, so the target episodes are extremely dry and dry categories for SPI, and for VCI it is very low and low.

**Step 3:** Determining the minimum confidence and support for apriori analysis :

Before running the algorithm minimum confidence and support are required to be selected. As said earlier, confidence measures the strength of the rule, and support measures its occurrence in the database. The minimum confidence is chosen to avoid the rules below the confidence values.

**Step 4:** Generating the rules:

After completing step 3, association rules were generated using the apriori algorithm. Many rules were generated.

**Step 5:** Selection of rules:

Out of the several rules generated, selection of best rule depends on the 'interesting measure'. This means comparing and selecting the better rules among the ones that are generated based on some criteria. For this study support and confidence were used to measure the goodness of the rules.

## 4. Results and Discussion

Association rule generation was used in order to identify relationship between VCI and SPI from the year 1981 to 2006. Among the set of rules, two interesting rules were picked on the basis of minimum confidence. The rules are:

If SPI is extremely dry then VCI is found to be very low, with 88% confidence.

If SPI is very dry and dry then VCI is found to be low, with 71% confidence.

Details are given in table 2.

**Table 2 :** Rules developed for identifying the drought pattern

| Rule number | Rule $(X \Rightarrow Y)$ | Confidence $(X \cup Y)/X$ | Support $(X \cup Y)$ | Coverage $\|X\|$ |
|---|---|---|---|---|
| 1 | If SPI is extremely dry then VCI is found to be very low | 0.8787 | 29 | 33 |
| 2 | If SPI is very dry and dry then VCI is found to be low | 0.7083 | 34 | 48 |

Spatial and temporal pattern of drought can be easily identified from association rule. Figure 1 shows the yearwise drought occurrence in Karnataka from 1981 to 1996 and figure 2 shows drought occurrence from 1997 -2006. It can be interpreted from first image of figure that mostly drought occurred in central part but some pockets of drought can be seen in western, northern as well as in south-eastern part of Karnataka.

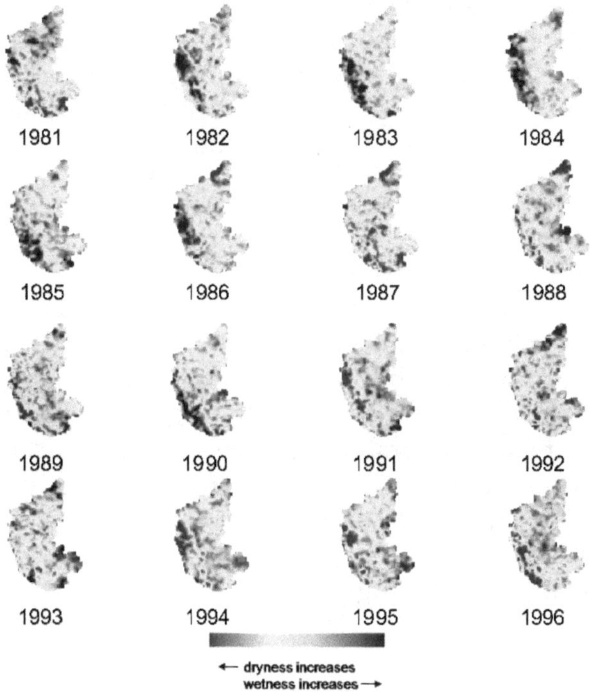


**Figure 1 :** Drought pattern for Karnataka (1981-1996)

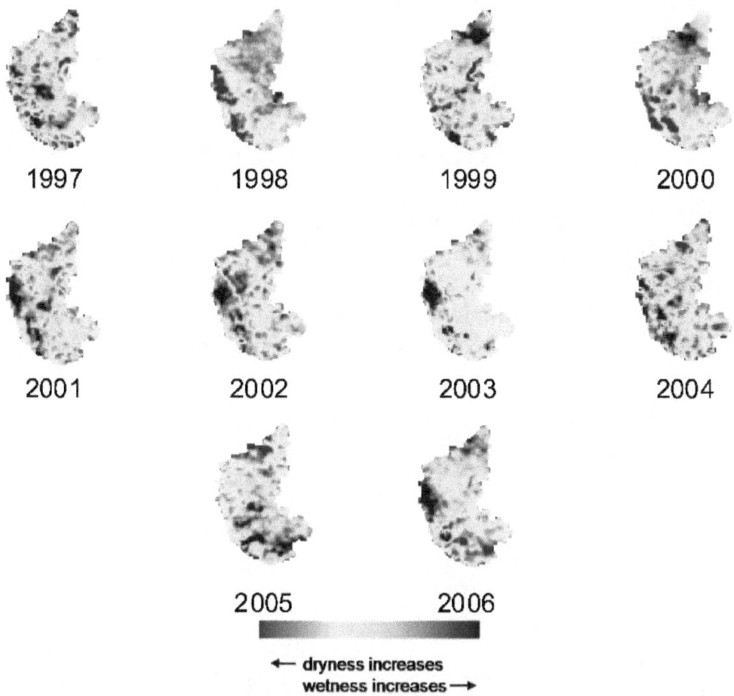


**Figure 2 :** Drought pattern for Karnataka (1997-2006)

On analysing the pattern, we find that coastal Karnataka is almost free from drought except for one or two years. It can be observed that year 1987 has been the worst year for Karnataka as many places suffered from severe drought in that year. It can also be seen that drought intensity is highest in northern most tip of Karnataka.

## 5. Conclusion

Dealing with a large volume of NDVI and rainfall dataset for a time-series of 56 years make the study not only complicated but make it also difficult to conclude but for better results and to study the spatial pattern, large number of observations are needed. Data mining techniques were successfully applied and through *association rule* it was possible to extract the temporal and spatial pattern of drought from 1981-2006 by identifying the relationship between climatic and meteorological data. Spatial pattern generated from the above clearly indicates the stations under drought whereas temporal pattern highlights the years experiencing drought.

## 6. References

Dunham Margaret.H. (2004). Basic Data Mining Tasks. Singapore, Pearson Education

Ester, M., A. Frommelt, et al. (1998). Algorithms for Characterization and Trend Detection in Spatial Databases. 4th International Conference on Knowledge Discovery and Data Mining.

Ester, M., S. Gundlach, et al. (1999). Database Primitives for Spatial Data Mining. International Conference on Databases in Office, Engineering and Science, Freiburg.

Harms, S., D. Li, et al. (2002).Efficient rule discovery in a Geo-Spatial Decision Support System

Joshi, K. P. (1997).Analysis of Data Mining Algorithms

Peuquet, D. J. and D. Guo.(2000).Mining Spatial Data Using An Interactive Rule-Based Approach

Tadesse, T., D. A. Wilhite, et al. (2004). "Drought Monitoring Using Data Mining Techniques: A case study for Nebraska, USA." Natural Hazards 33: 137-159.

# Chapter - 9

# Potability Analysis of Ground Water of Thar Desert

**Trapti Sharma & S. Palria**

## Abstract

*Quality of water is a matter of serious concern for arid and semi arid region like Rajasthan. The study area chosen is extreme western arid part of Rajasthan state where water quality and quantity both are very important issues. Due to absence of any perennial river, scanty rainfall and scarcity of surface water resources, ground water is very deep and saline. The groundwater potability assessment was based on 1119 wells water quality data. The purposes of this investigation were (1) to provide an overview of present groundwater quality and (2) to determine spatial distribution of groundwater quality parameters which affect water potability and (3) to map groundwater potability in the study area by using GIS and Geostatistics techniques. Arc GIS 9.3 and Arc GIS Geostatistical Analyst were used for generation of various thematic maps and Arc GIS Spatial Analyst to produce the final groundwater potability map. An interpolation technique, ordinary Kriging, was used to obtain the spatial distribution of groundwater quality parameters. The final water potability map was prepared by spatially integrating the nine thematic grid maps of $NO_3SO_4$, pH, TDS, TH, Cl, Ca, Mg and F.*

***Keywords*** *GIS, Geostatistics, Groundwater quality, Kriging*

## Introduction

Water is vital for the existence of all living organisms, but this valued resource is increasingly being threatened. This is due to exorbitant increase in human populations, economic expansion, decline in groundwater recharge and over-abstraction caused by the rapid increase in the number of wells and tube wells. Thus demand of more water of high quality for domestic and drinking purposes and for other activities. Ground water fulfills the 90% requirement of drinking and 71% for irrigation in Rajasthan. Presently, there is tremendous pressure to exploit groundwater by State and private users, who have access and control over this limited resource. Due to over exploitation of ground water, without adequate recharge, there is unequilibrium in the ground water regime. As a consequence of this, level of ground water is going down and deterioration of water quality is rising. The hydro-geologic environment controls the occurrence, distribution and movement of groundwater. Besides understanding the hydro-geological framework in which groundwater occurs and assessing its utilizable component, an equally important concern is assessing its quality. The quality of ground water is of great importance in determining the suitability of particular ground water for a certain use (public water supply, irrigation, industrial applications, power generation etc.). Today scarcity of water is becoming a major constraint not only for the development of this region but also has emerged as a major problem for the survival of population of this region. The situation is further aggravated in such areas where quality of ground water is poor. Many districts of western Rajasthan like Barmer, Bikaner, Jaisalmer, Churu, Jalore, Pali, western part of Jhunjhunun have saline ground water. As there is no alternative source, people are forced to use saline ground water for drinking purpose. The ground water in these areas has high concentration of various chemical constituents, which are harmful to the health of the people and livestock. The purposes of this study were (1) to provide an overview of present groundwater quality, (2) to determine spatial distribution of groundwater quality parameters which affect water potability and (3) to map groundwater potability in the study area by using GIS and Geostatistics techniques. This study will provide the decision makers with accurate and reliable scientific database regarding the quality of water and type of contamination which is essential for taking policy decisions.

Thus for protecting groundwater quality, data on spatial distribution are very important. Geo-statistical methods are one of the most advanced techniques for interpolation of groundwater quality. The accuracy of interpolation methods for predicting water quality has been proven and analyzed in several studies. Geo-statistical techniques use both, mathematical and statistical methods to predict values at unknown locations and to provide probabilistic estimates of the quality of the interpolation based on the spatial autocorrelation among data points (Burrough and McDonnell 1998, Johnston et al. 2001). These techniques quantify the spatial autocorrelation among measured points and account for the spatial configuration of the sample points around the prediction location. Formation of spatial distribution/ thematic maps using interpolation technique in GIS gives good results and information for the ground water quality mapping (Shankar, 2010).

Kriging is a type of Geostatistical technique which is proven to be the best interpolation technique for the spatial mapping of ground water quality. Ground water quality can also be interpolated by Inverse Distance Weighted or by Co-Kriging methods but they resulted in higher errors as compared to Kriging. Bilgehan Nas Ali Berktay (2010) generate the drinking water quality map using following parameters: chloride, sulphate, hardness, conductivity and nitrate. The construction of the groundwater quality map was carried out using Kriging. Uyan, M. et. al. (2010) applied Geostatistical technique to predict the spatial distribution and groundwater nitrate concentration. They found the spherical model is the best model representing the spatial variability of groundwater nitrate concentrations. M. M. Amin et al. (2010) in their study used Kriging for spatially model the groundwater pollution in Isfahan, Iran. To predict the spatial distribution and uncertainty of some groundwater quality parameters in Iran, Taghizadeh M, et.al (2008) used three interpolation techniques Kriging, cokriging and IDW. In this research Kriging method is used for predicting spatial distribution of Groundwater characteristics such as: TDS, TH, $Cl^-$, Mg, F, $NO_3$, pH, Ca, and $SO_4^{2-}$.

**Study Area:** In India to the west of the Northern Plains lies a large flat sandy area in which very few plants grow. This is the Great Indian Desert or the Thar Desert. It is about 800 Km long and 500 Km wide. It covers the major part of western Rajasthan. The land in this area is partly sandy and partly rocky.

The present study area Western Rajasthan is the region lying on the west of Aravalli hill range (24° 37′ & 30° 11′ north latitude and 69° 29′ & 75° 15′ east longitudes) and covers twelve districts of Rajasthan viz. Barmer, Bikaner, Churu, Ganganagar, Hanumangarh, Jaisalmer, Jhunjhunun, Jodhpur, Jalore, Nagaur, Pali and Sikar. Major part of it comes under the Thar Desert (Fig 1).

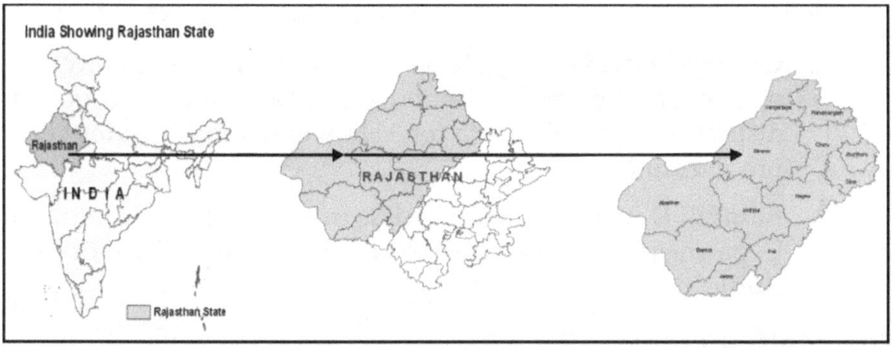


**Figure 1 :** Location map showing India, Rajasthan and study area-Western Rajasthan

Climate of Western Rajasthan is arid and is characterized by low rainfall with erratic distribution, extremes of diurnal and annual temperatures, low humidity and high wind velocity. A major part of the state, comprising the arid zone of the west and the semi-arid zone of the mid-west has an average maximum temperature of $45^0$C. Occasionally, it rises up to $49^0$ C during the months of May and June. The winter season is from December to February with marked decline in minimum temperatures in December and January. The minimum temperatures may fall to below $0^0$ C. Rainfall is scanty in Thar Desert and its surrounding semi arid region. Very little rainfall occurs during the months of July to September. It varies throughout the region. In the extreme west the number of rainy days is even less than ten. Annual potential evapo-transpiration is highest (206.32 cm) at Jaisalmer and the lowest (166.22 cm) at Ganganager in the Western Rajasthan.

Western Rajasthan has basically an agrarian economy. But the sandy soil and lack of water in the desert make growing of crops very difficult. Agriculture in the Western Rajasthan is the primary sector and mainstay of over 80 % of population contributing agriculture activity. Main crops are Maize, Guar, Cotton, Groundnut, Paddy, Kharif Pulses, Wheat, Barley, Gram, Rapeseed and Mustard. Most part of the Western Rajasthan is however covered by arid and semi-arid climatic conditions, which is characterized by low, erratic and uneven distribution of rainfall associated with lack of other water-resources for irrigated farming. Growing of only one rain fed crop in kharif season that too associated with high risk has led to dependence of village community on livestock. Most of the western districts of Rajasthan are irrigated by wells and tube well which further indicate importance of ground water.

## Materials and Methodology Used

## Data Used

Following types of datasets are used for the present study.

1. **Ground Water Quality Data Used :** Ground water quality data was acquired from the Ground Water Department of concern district of year 2007. Data for 1119 wells were considered for present study.
2. **Rainfall Data Used :** Rainfall data from Indian Meteorological Department was taken for the period 2005 to 2007 for present study.

## GIS Methodology

For the analysis of various ground water quality parameters

Geographic Information System (GIS) technique is used. A GIS software package Arc GIS 9.3 and ArcGIS Geostatistical Analyst extension was used for the Ordinary Kriging estimations. Ordinary Kriging was used to obtain the spatial distribution of groundwater quality parameters over the area. Spatial distribution maps or thematic maps for parameters were generated by carrying out point interpolations using Kriging. Thematic maps pertaining to $NO_3$.$SO_4$, pH, TDS, TH, Cl, Ca, Mg and F were interpolated in 1:250,000 scale from the well data. Different classes in thematic maps were categorized according to their concentration range. Area under quality zone was determined from histograms generated by the software for these maps. Flow chart of GIS Methodology used in study is shown in Fig 2.

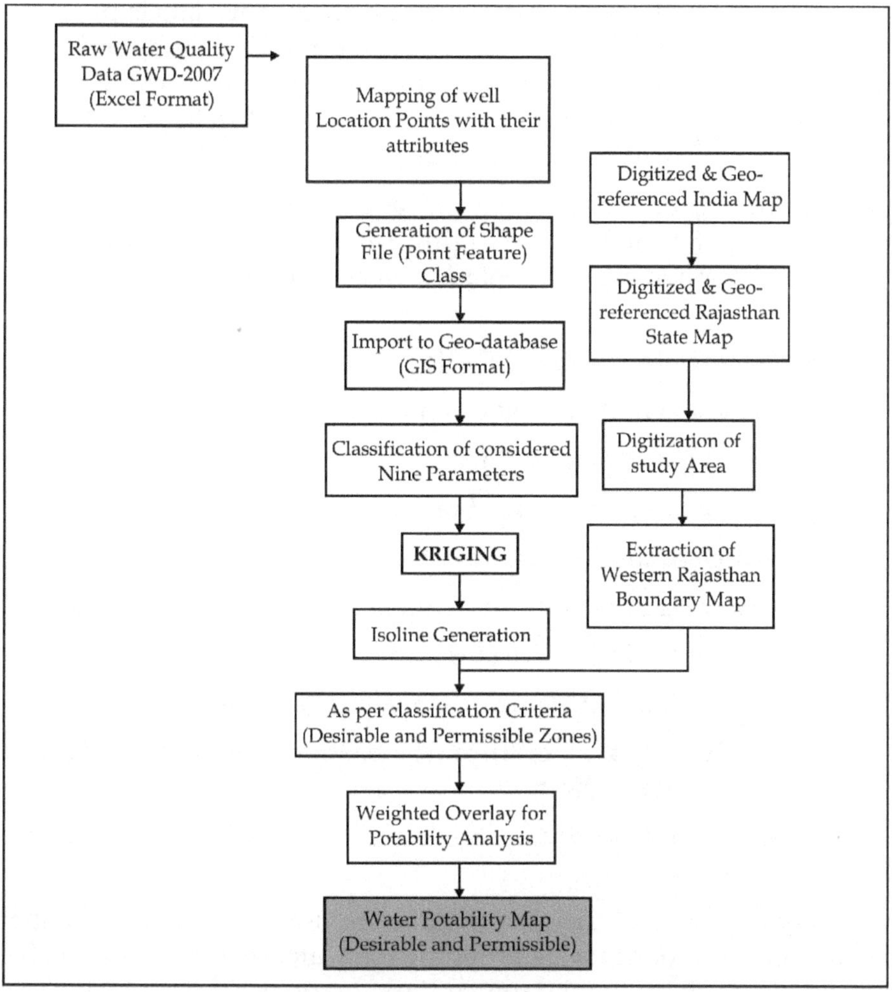


**Figure 2 :** Flow chart showing the methodology adapted for the generation of Maps

The groundwater quality data has been checked by a histogram tool and normal QQ Plots to find out the normality of data. All the parameters show normal distribution pattern, so that a log transformation has not been applied to make the distribution closer to normal. In this situation of water quality data, it is not necessary to remove the trend before modeling the semivariogram/covariance for Kriging. For each water quality parameter, an analysis trend was made, and it was determined that there is no global trend for all parameters. For each water quality parameter, and different semivariogram models were tested. In this work the spherical model is fitted best to the empirical semivariogram. Prediction performances were assessed by cross-validation. Data pertaining to different parameters of water quality was obtained from the Ground Water Department, for the twelve districts. Data obtained in the form of excel spread sheet. This data is provided with Location code (L code) with their latitude and longitude and water quality parameters of all wells. Location code (L code) is an NRIS code for a location (village). After linking, around 1119 wells were located satisfactorily with all parameter values.

Overlaying all the 9 parameters, a raster dataset file was created using the spatial analyst - Kriging Interpolation technique. The conditions applied for interpolation in respect of the Kriging is as follows:

- **Input Data Points** : wells_lc
- **Z-Value Field** : various parameters
- **Kriging method**: Ordinary
- **Semivariogram model**: Spherical
- **Search Radius Type** : Variable
- **Search Radius Settings** : Number of Points - 12, Maximum
- **Output Cell Size** : 0.025691025
- **Output Raster** : Name of analyzing Parameter
- **Analysis Mask of Inputs**: rajasthan_select (boundary of western Rajasthan)
- **Analysis Coordinate System** : Same as Input Coordinate System

**Suitability Analysis:** The parameter-wise analysis alone cannot be taken into account for determining the water quality in its potable sense. For any given sample/source, the value of each parameter is likely to exceed its limits in terms of the permissible range stipulated by BIS. Hence, in a particular sample, even if any one of the parameter violates the permissible limits for water quality, the entire significance of other parameters, regardless of their suitability, becomes void and unsuitable.

The water quality data for the chosen parameters to be overlaid for final analysis with respect to the source, must satisfy the suitability rule in respect of all the parameters. Thus, various zones can be easily created in terms of ground water quality. All the values were ranked with respect to the water quality standards in each of the nine parameters. These ranks which are assigned based on a given parameter value range were subsequently considered for overlay analysis. For each of the parameters, the range of acceptable values was classified separately according to the desirable and permissible limits of that parameter (table 1). The criteria set forth in this stage were based on the following condition:

- Rank 1: Potable
- Rank 2: Non Potable

} Desirable and Permissible Limit

**Table 1** : Percentage Weight of the different parameters for potability

| S.No. | Parameters | Percentage weight | Desirable | Permissible |
|---|---|---|---|---|
| 1. | Magnesium | 7.69 % | 30 | 100 |
| 2. | Fluoride | 7.69 % | 1 | 1.5 |
| 3. | Calcium | 7.69 % | 75 | 200 |
| 4. | Chloride | 15.38 % | 250 | 1000 |
| 5. | Nitrate | 7.69 % | 45 | 100 |
| 6. | pH | 15.38 % | 6.5 | 8.5 |
| 7. | Sulphate | 7.69 % | 200 | 400 |
| 8. | Total Dissolved Solids | 15.38 % | 500 | 2000 |
| 9. | Total Hardness | 15.38 % | 300 | 600 |

## Result and Discussion:

Potability of the ground water depends on the concentration of different chemical constituents. Certain standards for drinking water are prescribed by various agencies like World Health Organization (WHO), BIS, Indian Standards (IS 10500:1991) and Indian Council of Medical Research (ICMR). The ground water is considered potable if the physico-chemical parameters of ground waters lie within the limits prescribed by the above mentioned agencies.

The overall quality of a ground water is always represented by the total dissolved solids as it is a result of all the chemical constituents in the ground water. Therefore, overall ground water potability can also be considered based on the TDS content of ground water. In the present study, the potable and non-potable ground water zones in the Western Rajasthan has been demarcated based on the nine parameters (magnesium, fluoride, calcium, chloride, nitrate, pH, sulphate, total hardness and total dissolved solids). Fig: 3 present the ground water potability maps of Western Rajasthan.

## Ground Water Potability Analysis

Potable water is water that is safe to drink. Potable water should be free from pollution, harmful organisms and impurities. Non-potable water is water which is not fit for drinking, but may still be used for many other purposes, depending on its quality.

**Table 2 :** Area covered by Non-Potable and Potable Ground Water

| Water type | Desirable limits Area (Km$^2$) | Permissible limits Area (Km$^2$) |
|---|---|---|
| Non-potable water | 204571 | 4175 |
| Potable water | 6262 | 202483 |

The analysis of ground water potability of Western Rajasthan is considered with desirable and permissible limits. Map for desirable limits indicates that about 204571 km$^2$ area of Western Rajasthan has non-potable water and 6262 km$^2$ area is under potable water quality as per BIS drinking water standards. The analysis of ground water potability map for permissible limits indicates that about 4175 km$^2$ area of Western Rajasthan has non-potable water and 202483 km$^2$ areas is under potable water quality as per BIS drinking water standards (Table 2 & fig 3)

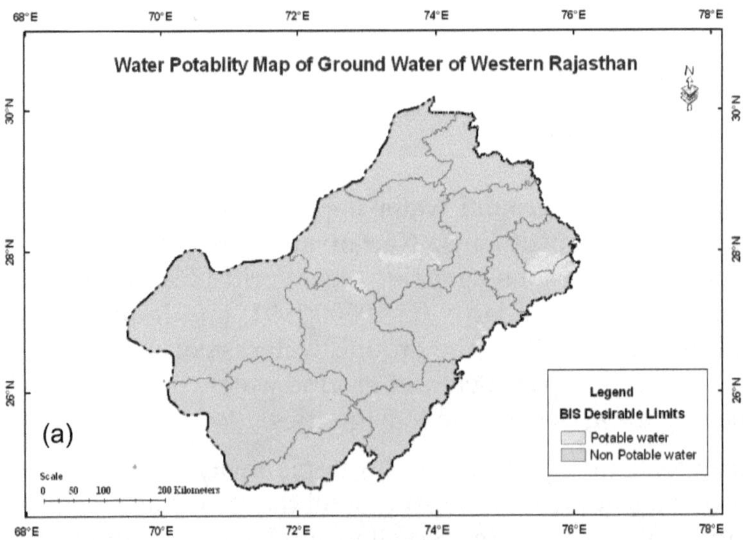


**Figure 3(a) :** Water potability Map (Desirable)

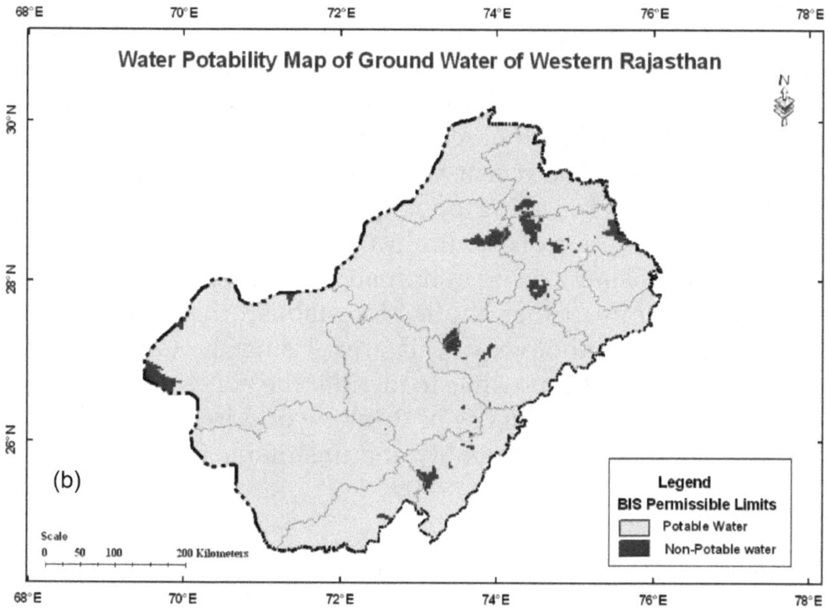


**Figure 3(b) :** Water potability Map : Permissible Limit

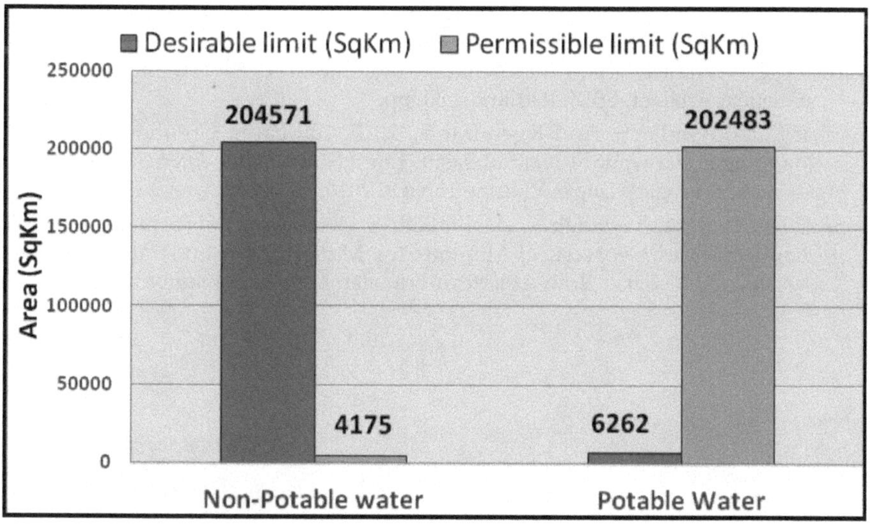


**Figure 4 :** Graphical Representation of Area covered by Non- Potable & Potable Water

## Conclusion

The primary objective of this study was to map and evaluate the groundwater potability in western Rajasthan. Spatial distribution of groundwater quality parameters were carried out through GIS and Geostatistical techniques. These techniques have successfully demonstrated its capability in groundwater potability mapping of western Rajasthan. The final groundwater potability map, shows that approximately 97% of study area is highly saline and not suitable for potability purposes. The recent work clearly revealed that people of Western Rajasthan are using poor quality of water. As a result of continuous consumption of chemically contaminated non-potable water many health problems arising in human and animals. However, there is need to carry out further studies to establish the correlation with hydrogeology and Lithology of this area. Using the GIS and RS and their water database it is possible to identify the zones where quality of water is suitable and can be use of drinking and irrigation and which zones have degraded water quality and unsuitable for various uses.

## References :

Amin M.M., A. Ebrahimi, M. Hajian, N. Iranpanah, and B. Bina, 2010. Spatial Analysis of Three Agrichemicals in Groundwater of Isfahan Using GIS. *Iran. J. Environ. Health.* Sci. Eng., 7(1) : 71-86

Bilgehan Nas Ali Berktay, 2010. Groundwater quality mapping in urban groundwater using GIS. *Environ Monit Assess.* 160 : 215–227.

Burrough, P. A. and McDonnell, R. A., 1998. Principles of geographical information systems. Oxford University Press, Oxford. 333pp.

Johnston, K., Ver-Hoef, J. M., Krivoruchko, K. and Lucas, N., 2001. Using Arc GIS Geo-statistical Analyst. ESRI. Redlands, 300pp.

Shankar,K., Aravindan.S. And Rajendran.S., 2010. GIS based Groundwater Quality Mapping in Paravanar River SubBasin, Tamil Nadu, India. *International Journal of Geomatics and Geosciences.* Volume 1, No 3, 2010.

Taghizadeh, M. R., Akbarzadeh, A., Mahmoodi. S. Heidari, A. and Sarmadian, F., 2008. Application of Geostastical Methods for Mapping Groundwater Quality in Azarbayjan Province, Iran American-Eurasian *J. Agric. & Environ. Sci.,* 3 (5): 726-735.

Chapter - 10

# Agricultural Resources in Arid Western Rajasthan-Issue and Options

**Balak Ram**

## Introduction

Agriculture contributes to 29.93% of India's GDP with a share of 27% of the world's agriculture production. India is ranked 6th in the world. In Rajasthan significant portion of economy of is agrarian. The agriculture sector of the state accounts for 22.5% share of agriculture in GSDP which is estimated to 18.39% in 2009-10. About 80% of the total population resides in rural area and largely dependent on agriculture as the source of their livelihood. The hot arid region of western Rajasthan occupies 61% of the total geographical area and 39.51% population of the state. During the last four decades the burgeoning human and livestock population and increased societal demands for land based products led to significant expansion and intensification of agriculture by putting even marginal lands under plough, excessive exploitation of ground water and importing of surface water for irrigation. The per capita availability of land got reduced from 6.1 ha in 1901 to 1 ha in 2001. Such uneconomic and over utilization of natural resources has caused degradation of land and water resources, emergence of irreversible hazards and damage to micro-environment. As such it is very much essential to have sustained availability of food, fodder and feed; maintain and increase the use capability of land and water resources and make appropriate use of each parcel of land. The present study focus the present status of agriculture in arid western Rajasthan taking an integrated approach of Remote Sensing and GIS technologies, field observations and use of agricultural statistics and researched findings; highlights constraints and potentials are highlighted and options and measures for overall development of agriculture and improvement of ecological environment are suggested.

## Location and Environment

About 61% of the total area of Indian arid zone is concentrated in 12 western districts of Rajasthan with a total geographical area of 208,751 km$^2$. The region is extended from 24° 37′ 00″ to 30° 10′ 48″ north latitudes and between 69° 29′ 00″ and 76° 05′ 33″ east longitudes. This hot arid region is characterized by low and high erratic rainfall, high evaporation loss and extremes of diurnal and seasonal temperatures. The average annual rainfall varies from 456 mm in north east to less than 100 mm in western most part of Jaisalmer district. The coefficient of variability of annual rainfall varies from 40% to 70%. Temperature during long hot summer days goes as high as 50°C while in cold winter it falls below - 6°C. Mean aridity index is 78%. Probability of occurrence of drought varies from 50-60%. Mean moisture index varies from -59.5 in Sikar to -88.9 in Jaisalmer. The length of crop growing period varies from 8-15 weeks (Fig. 1). The mean maximum expected wind velocity is about 30-40 kmph, but can reach as high as 100 kmph during severe dust storms. Mean relative humidity during July and August ranges between 75-80%, but during winter it varies from 46-56%.

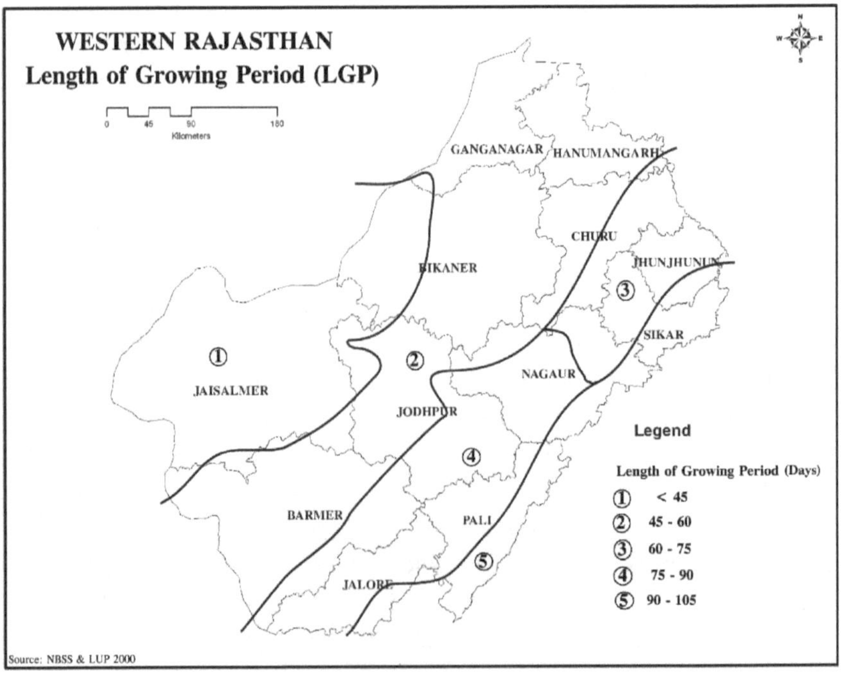


**Figure 1 :** Length of Growing Period (LGP)

Dominant landforms are the sandy plains with varying degree of hummocks and sand dunes. These are encountered with hills and outcrops (of mainly granite, rhyolite, sandstone and limestone), saline depressions and buried channels. Light textured sandy soils cover major part of this region. Desert soils are mostly sterile with a poor fertility status, poor water holding capacity, very low in organic matter, weaker in exchangeable bases, high in pH values and low in nitrogen and phosphate. The soils have high infiltration rate with unsaturated hydraulic conductivity (Dhir, 1977). The vegetation of this region is quite sparse with limited number of xerophytic plants and thorny bushes. The drainage, except the ephemeral Luni river system, is mostly internal. Ground water is deep, scares and over exploited. About 45% area has saline to very saline water and 40% moderately brackish ground water. The region has been divided into 7 agro-ecological sub-regions and 21 agro-ecological zones (Faroda et. al. 1999). As per 2011 census, the region has 27.12 million human populations and 29.11 million livestock population as per 2007 livestock census. The density of human population varies from 17 in Jaisalmer to 361 in Jhunjhunun district and that of livestock from 74 to 293 in Sikar district. The region has 14,374 villages and 86 towns. There are 62 Krishi Upaj Mandis, 17 Quality Control Labs. and 6 Government. Seed Farms in the region.

## Method and Materials

Recently prepared remote sensing and GIS based district wise land use maps at CAZRI have been used as base material to have complete scenario of the spatial distribution of irrigated and rainfed croplands. District wise reports of natural resource appraisal are consulted to assess the agricultural situation in different parts of the region and the factors responsible for development or deterioration of agriculture. District wise statistics on land use, cropping pattern, crop production and irrigation for the year 2007- 08 are analyzed and correlated with maps statistics to have better understanding of present agricultural situation. Data on land use, cropping pattern, crop production and irrigation as well as human and livestock populations from 1960-61 are analyzed to work out changes/trend over the last 48 years. Based on the Food Security Atlas of Rural Rajasthan the Food Security Index for different districts as well as related parameters Rajasthan are mapped for Arid Rajasthan. Through integration of present findings and available researched data/reports strategies are suggested for sustainable development of agriculture.

## Results and Discussion

### Agricultural land use

Out of the 208751 km$^2$ total geographical area of western Rajasthan the total reporting area is 208228 km$^2$. Out of this the total cultivated area comes to 66.56% or 13.86 mha. This includes 52.67% net sown area, 6.29 current fallow, 7.6% other fallow lands. The total cropped area comes to 64.15% and double cropped area 11.48%. Area under other land use categories includes 17.69% culturable waste, 0.02% miscellaneous. tree crops & groves, 4.9% permanent pasture and other grazing lands, 4.58% barren and uncultivable land, 4.9% land put to non-agricultural uses and 2.95% forest as per land use statistics 2007-08 (Anon. 2010). District wise details are given in table 1. Through digital interpretation of IRS P6 LISS-III data from 2004 to 2009, CAZRI has prepared land use maps of the 12 arid districts on 1:250, 000 scale using Arc GIS 9.2 version software. These maps show the spatial distribution of agricultural lands based on the intensity of cultivation. The irrigated and rainfed croplands in western Rajasthan are shown in table 2 and fig. 2. The statistics of irrigated and rainfed areas are shown in table 4. Churu district occupies highest (88.4%) rainfed agricultural area while Jaisalmer has least one (21.1%). Of the gross irrigated area through canals 45.7% is shared by Ganganagar, 34.7% by Hanumangarh, 12.4% by Bikaner and 7.2% by Jaisalmer districts respectively. Through ground water irrigation Jhunjhunun district has 42.36% and Sikar 32.15%. Churu, Jaisalmer and Barmer have lowest concentration of 3.75%, 5.01% and 6.1% respectively (Balak Ram and Chauhan, 2009). As per land use statistics 2007-08, the region has 13.69% (2.85 mha) net irrigated area and 20.1% (4.18 mha) gross irrigated area. There are 237490 irrigation wells and 108342 tube wells. Hanumangarh and Nagaur district have highest concentration of tube wells. Due to ground water degradation about 91438 wells have gone dry. The district wise details are given in table 3. Out of the total gross irrigated area 45.56% (1903548 ha) is contributed by canals. The intensity of cultivation varies from 67% in Jodhpur to 92% in Churu district (Fig. 3). Cropping intensity varies from 106 in Barmer to 153% in Jhunjhunun district (Fig. 4). The irrigation intensity varies from 106% in Pali to 182% in Hanumangarh district (Fig. 5). The percentage of net irrigated area to net sown area varies from 4.02% in Churu to 70.61% in Ganganagar district. The level of agricultural development (Varghese and Mordia, 2002) varies from 24.65% in Barmer district to 46.88% in Ganganagar district. District wise details of these parameters are given in table 4.

**Table 1 :** Agricultural Land Use (%) in Western Rajasthan 2007-08

| District | Total reporting area (ha) | Culturable waste | Other fallow land | Current fallow | Net sown area | Total cropped area | Area sown more than once |
|---|---|---|---|---|---|---|---|
| Barmer | 2817332 | 7.08 | 10.89 | 7.97 | 58.70 | 62.34 | 3.64 |
| Bikaner | 3040476 | 27.16 | 10.60 | 5.81 | 41.88 | 47.17 | 5.30 |
| Churu | 1385898 | 0.72 | 2.96 | 4.25 | 84.13 | 104.53 | 20.40 |
| Ganganagar | 1093352 | 4.21 | 9.46 | 7.63 | 88.28 | 96.4 | 30.12 |
| Hanumangarh | 970340 | 0.49 | 2.68 | 4.28 | 84.43 | 127.17 | 42.73 |
| Jaisalmer | 3839154 | 63.92 | 2.95 | 1.54 | 14.65 | 16.61 | 1.96 |
| Jalor | 1056602 | 2.97 | 7.95 | 6.80 | 64.04 | 80.91 | 16.87 |
| Jhunjhunun | 591536 | | 3.84 | 3.72 | 71.49 | 109.54 | 38.04 |
| Jodhpur | 2256405 | 1.80 | 14.31 | 12.57 | 55.60 | 61.10 | 5.50 |
| Nagaur | 1764380 | 0.81 | 4.64 | 9.16 | 71.96 | 82.97 | 11.01 |
| Pali | 1233079 | 3.22 | 9.45 | 5.67 | 50.20 | 57.27 | 7.07 |
| Sikar | 774244 | 1.27 | 5.41 | 5.62 | 67.73 | 95.54 | 27.81 |
| Total | 20822798 | 3682847 (17.69) | 1582684 (7.60) | 1309057 (6.29) | 10966682 (52.67) | 13356253 (64.14) | 2389511 (11.47) |

**Table 2.** Distribution of irrigated and rainfed area in western Rajasthan (Based on remote sensing data and field check)

| District | Irrigated area (ha) | % | Rainfed Area (ha) | % | Total | % |
|---|---|---|---|---|---|---|
| Barmer | 170727 | 6.10 | 1687697 | 59.45 | 1858424 | 65.55 |
| Bikaner | 306594 | 11.30 | 1782965 | 65.40 | 2089559 | 76.70 |
| Churu | 51912 | 3.75 | 1225290 | 88.4 | 1277202 | 92.15 |
| Ganganagar | 362316 | 33.14 | 606788 | 55.50 | 969104 | 88.64 |
| Hanumangarh | 683478 | 33.10 | 581528 | 40.09 | 126506 | 73.19 |
| Jaisalmer | 192439 | 5.01 | 813472 | 21.10 | 935911 | 26.11 |
| Jalor | 277030 | 26.22 | 565164 | 53.49 | 842194 | 79.71 |
| Jhunjhunun | 25425 | 42.36 | 247223 | 41.70 | 272648 | 84.06 |
| Jodhpur | 190521 | 8.34 | 1403415 | 61.42 | 1593936 | 69.76 |
| Nagaur | 367170 | 20.72 | 1174815 | 66.31 | 1541985 | 87.03 |
| Pali | 130559 | 10.54 | 715158 | 57.73 | 845717 | 68.27 |
| Sikar | 249150 | 32.15 | 362815 | 49.40 | 611965 | 81.55 |

*Source: CAZRI, Jodhpur*

**Table 3 :** Irrigated area and status of irrigation wells and tube-wells (2007-08)

| District | Net irrigated area (%) | Gross Irrigated area (%) | Wells in use | Tube wells | Wells and tube-wells | Wells out of use |
|---|---|---|---|---|---|---|
| Barmer | 4.53 | 6.98 | 22266 | 2811 | 25077 | 2466 |
| Bikaner | 7.79 | 12.35 | 5014 | 7351 | 12365 | 189 |
| Churu | 4.96 | 6.42 | 6131 | 2023 | 8154 | 00 |
| Ganganagar | 51.00 | 79.93 | 00 | 13903 | 13903 | 00 |
| Hanumangarh | 37.85 | 68.96 | 2564 | 20625 | 23189 | 41 |
| Jaisalmer | 3.22 | 5.00 | 920 | 1400 | 2320 | 33 |
| Jalor | 24.53 | 29.31 | 46027 | 6743 | 52770 | 19983 |
| Jhunjhunun | 39.01 | 49.34 | 42367 | 1802 | 44169 | 3987 |
| Jodhpur | 9.15 | 13.51 | 10246 | 25579 | 35825 | 9202 |
| Nagaur | 15.11 | 21.94 | 25740 | 14393 | 40133 | 22470 |
| Pali | 13.16 | 13.94 | 30085 | 2448 | 32533 | 20542 |
| Sikar | 31.43 | 41.79 | 46130 | 7264 | 53394 | 12525 |
| Total | 2850149<br>13.65 | 4184615<br>20.05 | 237490 | 106342 | 343832 | 91438 |

*Source: Agricultural Statistics, Rajasthan, 2009-10*

**Table 4 :** Intensity of cultivation, cropping intensity and irrigation intensity (%) in western Rajasthan (2007-08)

| District | Intensity of cultivation | Cropping intensity | Irrigation intensity | % net irrigated area to net sown area | Index level of development |
|---|---|---|---|---|---|
| Barmer | 76 | 106 | 154 | 5.84 | 24.65 |
| Bikaner | 72 | 113 | 159 | 9.89 | 35.08 |
| Churu | 92 | 124 | 129 | 4.02 | 33.62 |
| Ganganagar | 79 | 145 | 157 | 70.61 | 46.88 |
| Hanumangarh | 92 | 151 | 182 | 38.19 | 33.61 |
| Jaisalmer | 77 | 113 | 155 | 7.96 | 28.60 |
| Jalor | 81 | 126 | 119 | 30.96 | 33.78 |
| Jhunjhunun | 90 | 153 | 126 | 50.52 | 35.67 |
| Jodhpur | 67 | 110 | 148 | 10.10 | 39.71 |
| Nagaur | 84 | 115 | 134 | 19.18 | 31.54 |
| Pali | 76 | 114 | 106 | 15.83 | 40.61 |
| Sikar | 86 | 141 | 133 | 41.52 | 32.58 |
| Total | 79 | 122 | | | |

*Source: Agricultural Statistics, Rajasthan, 2009-10,*

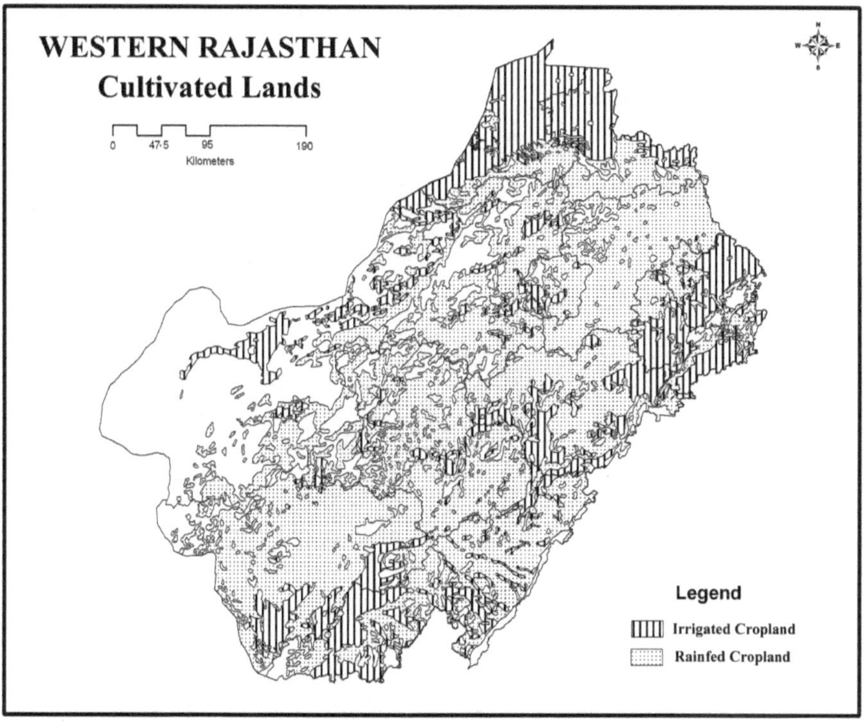


**Figure 2 :** Cultivated Lands

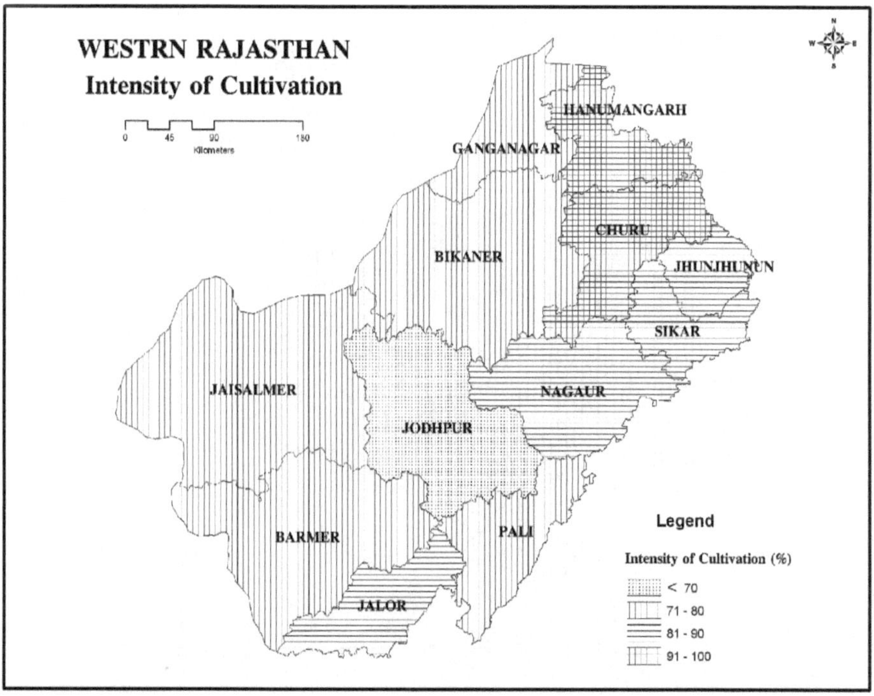


**Figure 3 :** Intensity of Cultivation

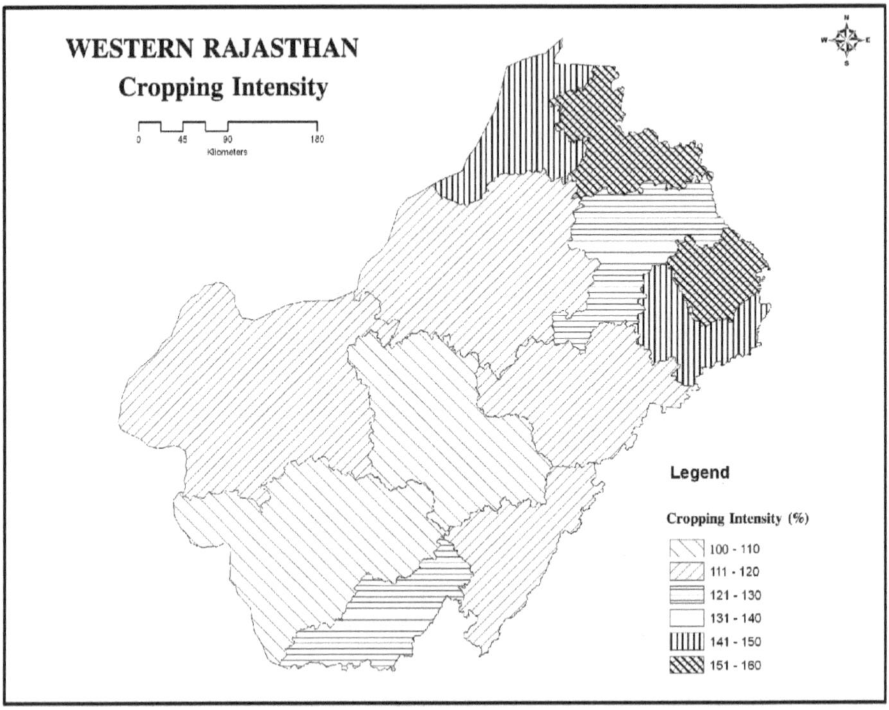


**Figure 4 :** Cropping Intensity

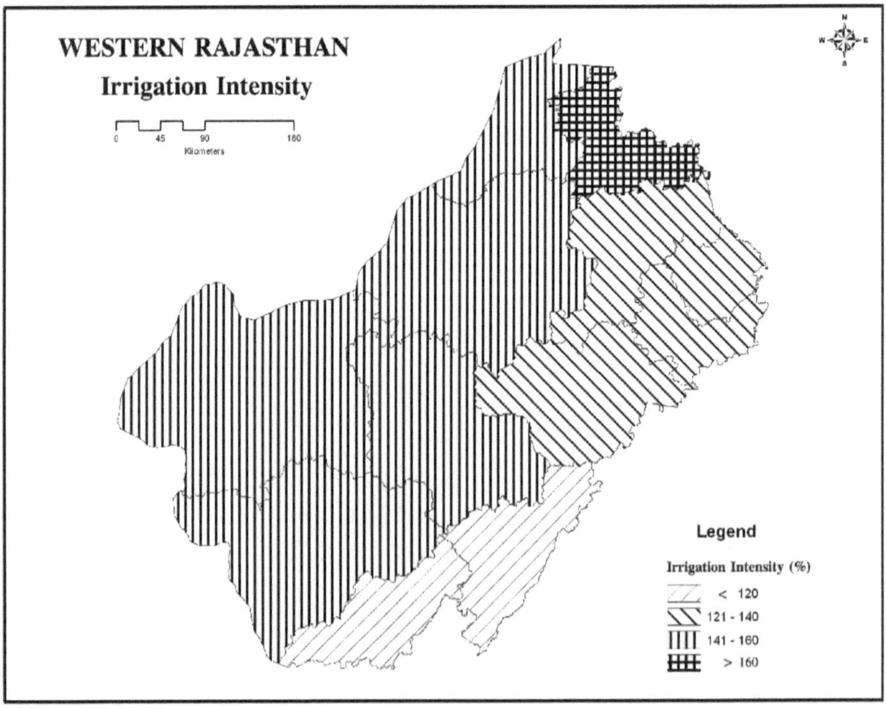


**Figure 5 :** Irrigation Intensity

## Cropping pattern

In spite of the hot desertic conditions, western Rajasthan occupies 74.5% of the total cropped area under bajra crop in Rajasthan state, guar 45.8%, moth 99.75%, moong 77.19%, sesame 44.41%, groundnut 75.53%, mustard 33.04%, cumin 97.61%, isabgol 98.52% and methi 61.98% as per agricultural statistics 2009-10 (Table 5). Bajra, guar, moth, moong and sesame are most important rainfed kharif season crops produced throughout the region covering 28.31%, 17.74%, 9.31%, 5.22% and 1.62% of the total cropped area in western Rajasthan. Jowar and cowpea are lesser significant crops. Jowar (1.87%) is mainly produced in Pali, Nagaur and Jodpur districts and cowpea (0.80%) in Sikar, Jhunjhunun and Nagaur districts. Wheat, mustard, gram, groundnut and cotton are important irrigated crops constituting 7.19%, 5.36%, 4.89%, 1.51%, 2.78% respectively. Others are barley (0.83%), cumin (1.45%) isabgol (1.65%) and paddy (0.25%). Cumin and isabgol are mainly taken in Barmer, Jalor, Nagaur and Jasisalmer region; barley in Sikar, Ganganagar and Hanumangarh districts and paddy absolutely in Ganganagar and Hanumangarh districts. Castor (0.64%) and Taramira (0.13%) are mainly taken under conserved moisture. Castor is mainly produced in Jalor, Jodhpur and Barmer districts. Similarly gram is taken both under irrigation and conserved moisture. Producing areas of other crops are methi (Sikar, Bikaner, Nagaur, Jhunjhunun), onion (Jodhpur, Pali, Sikar, Jhunjhunun), mehandi (Pali) and red chillies (Jodhpur). Besides, coriander, garlic, fennel, linseed, sugarcane, potato and sweet potato are also taken in small and fragmented pockets. Area under fodder and vegetable & fruits crops (0.8%) is extremely less looking into the demand in the region. The total fertilizer consumption during 2009-10 was 343134 tons which includes 67.8% nitrogen, 28.08% potassium and 4.02% phosphate. Ganganagar and Hanumangarh districts alone consume 47.4% of the total fertilizer.

## Crop production

Out of the total crop production in the region i.e. 11.27 million tonnes, 50.88% is contributed by cereals, 12.57% by pulses, 14.75% by oilseed and 21.8% by other commercial and cash crops as per agricultural statistics 2008-09. District wise production of above group of crops is given in Table 6. Bikaner, Churu and Nagaur have highest production of pulse crops while Ganaganagar, Jodhpur, Hanumangarh, Bikaner and Nagaur have dominancy over oilseed production. Hyper arid Jaisalmer district contribute only 1.1% of the total crop production in

**Table 5 :** Area (%) of important crops in western Rajasthan (2009-10)

| District | Bajra | Moong | Moth | Sesame | Groundnut | Cotton | Guar | Wheat | Gram | Mustard | Cumin | Isabgol |
|---|---|---|---|---|---|---|---|---|---|---|---|---|
| Barmer | 24.84 | 6.70 | 18.24 | 1.74 | 0.32 | N | 15.19 | 1.73 | N | 2.31 | 34.57 | 32.0 |
| Bikaner | 6.42 | 0.23 | 22.47 | 1.60 | 38.02 | 1.36 | 26.58 | 6.36 | 17.82 | 4.14 | 0.97 | 1.75 |
| Churu | 10.59 | 4.44 | 22.39 | 0.51 | 10.97 | N | 12.18 | 2.58 | 24.37 | 1.99 | 0.43 | 0.92 |
| Ganganagar | 0.11 | 3.16 | 0.14 | 1.30 | 2.42 | 43.31 | 6.29 | 22.98 | 12.20 | 29.10 | N | N |
| Hanumangarh | 1.39 | 1.47 | 3.20 | 2.63 | 2.16 | 46.25 | 8.10 | 24.69 | 22.66 | 12.30 | N | N |
| Jaisalmer | 4.20 | 0.35 | 0.03 | 0.41 | 4.47 | 0.03 | 13.13 | 1.22 | 7.90 | 4.74 | 5.22 | 9.69 |
| Jalor | 8.52 | 12.96 | 2.56 | 10.21 | 1.26 | 0.18 | 2.27 | 3.52 | N | 8.57 | 30.32 | 18.29 |
| Jhunjhunun | 6.85 | 2.12 | 0.17 | 0.05 | 0.75 | N | 2.49 | 9.04 | 8.33 | 10.73 | N | N |
| Jodhpur | 14.93 | 13.82 | 13.45 | 14.56 | 21.25 | 2.92 | 5.49 | 6.04 | 0.02 | 10.15 | 18.56 | 14.38 |
| Nagaur | 12.20 | 38.19 | 16.33 | 7.51 | 6.90 | 4.79 | 3.53 | 8.77 | 2.27 | 6.37 | 8.12 | 22.53 |
| Pali | 2.06 | 14.62 | 0.20 | 54.20 | 0.77 | 1.12 | 1.97 | 2.77 | 0.02 | 1.44 | 1.81 | 0.37 |
| Sikar | 7.89 | 1.94 | 0.80 | 0.22 | 10.68 | 0.04 | 2.78 | 10.30 | 4.71 | 8.16 | N | 0.07 |
| Total (ha) | 3860720 | 712353 | 1269638 | 247731 | 206566 | 379022 | 2418775 | 979966 | 666354 | 730878 | 197766 | 224339 |
| % to TCA | 28.31 | 5.22 | 9.31 | 1.82 | 1.51 | 2.78 | 17.74 | 7.19 | 4.89 | 5.36 | 1.45 | 1.65 |
| State % | 74.5 | 77.19 | 99.75 | 44.41 | 63.36 | 85.28 | 45.80 | 36.75 | 75.35 | 33.04 | 97.01 | 98.52 |

N = not available

**Table 6 :** Production (%) of cereals, pulses, oilseed and other crops in western Rajasthan (2008-09)

| District | Cereals | Pulses | Total Food grains | Oilseeds | Others | Total |
|---|---|---|---|---|---|---|
| Barmer | 58.74 (2.5) | 11.50 (1.9) | 70.24 (2.3) | 12.29 (1.8) | 17.47 (1.7) | 239122<br>100.0 (2.1) |
| Bikaner | 30.31 (5.3) | 20.29 (14.4) | 50.59 (7.1) | 14.56 (8.8) | 34.84 (14.2) | 1003643<br>100.0 (8.9) |
| Churu | 47.57 (6.5) | 34.57 (19.1) | 82.14 (9.0) | 08.60 (4.1) | 09.27 (3.0) | 783590<br>100.0 (7.0) |
| Ganganagar | 48.71(18.6) | 05.93 (9.2) | 54.64 (16.7) | 17.63 (23.2) | 27.73 (24.7) | 2189343<br>100.0 (19.3) |
| Hanumangarh | 52.23 (18.3) | 07.46 (10.6) | 59.69 (16.8) | 07.79 (9.4) | 35.52 (26.6) | 2009280<br>100.0 (17.8) |
| Jaisalmer | 09.67 (0.2) | 28.46 (2.5) | 38.14 (0.7) | 47.68 (3.6) | 14.18 (0.7) | 125623<br>100.0 (1.2) |
| Jalor | 40.78 (3.0) | 08.75 (2.6) | 49.52 (2.9) | 39.95 (10.1) | 10.53 (1.8) | 420981<br>100.0 (3.7) |
| Jhunjhunun | 68.25 (10.8) | 12.07 (7.7) | 80.32 (10.2) | 13.48 (7.3) | 06.20 (2.3) | 904457<br>100.0 (8.1) |
| Jodhpur | 42.78 (6.0) | 09.31 (5.3) | 52.09 (5.9) | 23.20 (11.3) | 24.70 (8.1) | 810018<br>100.0 (7.2) |
| Nagaur | 53.68 (12.7) | 19.17 (18.4) | 72.85 (13.8) | 10.93 (8.9) | 16.22 (9.0) | 1356753<br>100.0 (12.0) |
| Pali | 51.03 (2.6) | 10.24 (2.2) | 61.27 (2.6) | 20.54 (3.7) | 18.19 (2.2) | 297473<br>100.0 (2.7) |
| Sikar | 68.49 (13.5) | 07.68 (6.1) | 76.17 (12.0) | 11.44 (7.8) | 12.39 (5.7) | 1127152<br>100.0 (10.0) |
| Total | 5732983<br>(50.88) | 1415989<br>(12.57) | 7148972<br>(63.45) | 1661675<br>(14.75) | 2456788<br>(21.80) | 11267435<br>(100.00) |

*Figures in bracket show percentage to total area under a crop* arid Rajasthan.The production of rice during the year was 11640 tons, bajra 2485617 tons, jowar 89185 tons, moong 290644, moth 324287, cowpea 52094, sesame 62604, groundnut 329213, castor 140508, cotton 636542, guar 1113500, red chillies 6150, wheat 2522734, barley 527668, gram 744485, mustard 1116592, taramira 16886, cumin 40418, methi 33595 and isabgol 67699 tons respectively. Nagaur district tops in bajra, moong and isabgol production; Pali in jowar, sesame and mehandi; Churu in moth; Sikar in cowpea and methi; Bikaner in groundnut, guar and gram; Ganganagar and Hanumangarh in wheat, mustard, barley and cotton production; Jalor in castor and cumin; and Jodhpur in chillies production. Barmer district though have significant area under bajra, cumin, isabgol and castor but production remains low due to extremely arid conditions. The production of rainfed crops is entirely dependent on the availability of moisture and other climatic conditions and as such the production use to fluctuate at distinct level each year. Therefore, during moderate and severe drought years the production goes down at very low level in spite the cropped area is more.

## Salient features of arid rajasthan:

1. Restricted crop growing period and choice of crops.
2. The extent of rainfed cropped area varies from year to year depending on amount, distribution and intensity of rainfall.
3. The intensity of cultivation decreases from east to west with the declining rainfall pattern.
4. Ground water is scarce, deep and brackish to saline. Availability of surface water depends on rainfall since there are no perennial streams and lakes.
5. In better rainfall area of Sikar and Jhunjhjunun districts towards eastern margin, the entire sand dunes are cultivated whereas in lesser rainfall areas in extreme west only lower part of the sand dunes is cultivated.
6. Few rabi crops like mustard, taramira and gram are taken on the conserved moisture. 'Khadin' cultivation is an important crop production system in water scarce Jaisalmer region.
7. Production of rainfed crops is badly affected during drought years. Even the acreage of ground water irrigated crops gets affected.
8. Production of crops taken over the conserved moisture is comparatively low.

## Changes in Agricultural Land Use, Cropping Pattern and Crop Production

The human population in the region has increased from 7.64 million in 1961 to 27.12 million in 2011 and that of livestock population from 13.72 million in 1961 to 29.11 million in 2007. This burgeoning pressure and societal demands on one hand and division of holdings on the other led to expansion and intensification of agriculture through reduction in fallowing practices, putting more land from culturable waste category to cultivation and development of irrigation. Expansion of cultivated area is mainly taken place in Bikaner, Jaisalmer, Churu and Barmer districts where as intensification of agriculture has taken place in Sikar, Jhunjhunun, Ganganagar and Hanumangarh districts. During the period from 1960-61 to 2007-08 (Malhotra et. al., 1983 and Anon. 2010) the net sown area has increased by 33.64%, total cropped area by 60.44% and double cropped area by 20 times. On the other hand culturable waste lands are declined by 21.81%, other fallow lands by 36.13% and current fallow by 17.15% respectively. The net irrigated area has increased by 366.92% (from 0.61 to 2.85 mha).

The development of irrigation and introduction of sprinkle system brought significant change in cropping pattern particularly in canal irrigated area. Cumin, isabgol, methi, fennel and castor were introduced as new crops after eighties. Similarly large scale cultivation of gram and taramira on the conserved moisture was taken up. The mixed cropping practice like bajra+moth+guar or bajra+moong+guar is almost withdrawn. The changes in area and production of important crops from 1970-71 to 2008-09 are given in table 7. The acreage of rice, wheat, groundnut, mustard and cotton has increased by 153.5, 117.3, 935.0, 1988.6 and 289.7 per cent respectively. Jowar, barley and gram have registered marginal increase. The acreage of sesame has declined by 5.3%. Similarly the production of wheat, barley, groundnut, mustard and cotton has increased manifolds while the production of sesame, jowar and rice has declined. The acreage of bajra though slightly declined over the years to give way for other crops but its production has increased by 26.4%. As such the productivity of almost all the crops has increased by adoption of new farming technologies and high yielding varieties although the acreage of few rainfed crops has declined.

**Table 7 :** Trend in area and production of principal crops in western Rajasthan

| Crop | Area (000 ha) | | | Production (000 tons) | | |
|---|---|---|---|---|---|---|
| | 1970-71 | 2009-10 | Change (%) | 1970-71 | 2008-09 | Change (%) |
| Rice | 13 | 33 | 153.8 | 23 | 12 | -47.8 |
| Jowar | 229 | 255 | 11.4 | 93 | 89 | -4.3 |
| Bajra | 4179 | 3851 | -7.8 | 1966 | 2486 | 26.4 |
| Wheat | 405 | 880 | 117.3 | 525 | 2523 | 380.6 |
| Barley | 112 | 113 | 0.9 | 127 | 528 | 315.7 |
| Kharif pulses | 1614 | 2093 | 29.7 | 432 | 667 | 54.4 |
| Gram | 679 | 666 | -1.9 | 438 | 744 | 69.9 |
| Groundnut | 20 | 207 | 935.0 | 4 | 329 | 8125.0 |
| Sesame | 262 | 248 | -5.3 | 69 | 63 | -8.7 |
| Mustard | 35 | 731 | 1988.6 | 42 | 1117 | 2559.5 |
| Cotton | 97 | 378 | 289.7 | 133 | 637 | 378.9 |

*Source: Agricultural Statistics, Rajasthan 2009-10*

*CAZRI, 1982*

## Food Security Analysis

The ultimate objective of the agriculture development is to get self sufficiency in food, feed, fodder and milk and meat production and insure the availability of food to the poorest of the poor citizen. Based on the Food Security Atlas of Rural Rajasthan (IHD, 2009), various indicators are analyzed (Table 8) to find out the food security in arid Rajasthan. In per capita value of agricultural output only Ganganagar and Hanumangarh districts have moderate value while Churu, Jodhpur and Pali have extremely low value. In share of agricultural labour force only Ganganagar gets high and Sikar moderate rank. In food availability index too Ganaganagar district is secure while Hanumangarh and Jhunjhunun are moderately secure. In food security outcome index, Ganganagar, Hanumangarh and Sikar are secure. Barmer district is extremely insecure in most of the parameters. In food access index only Jhunjhunun district is secure while Pali and Jodhpur are severely insecure. The status in term of Food Secuity Index (FSI) shows Pali and Barmer as extremely insecure districts, Jhunjhunun as secure and Ganganagar, Hanumangarh and Sikar as moderately secure districts. The left over Bikaner, Jaisalmer, Jalor and Nagaur districts show moderate status in food security.

**Table 8 :** Food Security Parameters in Western Rajasthan

| District | Per capita value of agrl. output | Food availability index | Food access index | Status in term of FSI | Food security outcome index |
|---|---|---|---|---|---|
| Barmer | Extremely low | Extremely insecure | Moderately secure | Extremely insecure | Extremely insecure |
| Bikaner | Very low | Severely insecure | Moderately insecure | Severely insecure | Moderately secure |
| Churu | Extremely low | Severely insecure | Moderately secure | Moderately secure | Moderately insecure |
| Severely insecure | Severely insecure | Severely insecure | Severely insecure | Severely insecure | Severely insecure |
| Hanumangarh | Moderate | Moderately secure | Moderately secure | Moderately secure | Secure |
| Jaisalmer | Extremely low | Extremely insecure | Moderately insecure | Severely insecure | Moderately insecure |
| Jalor | Very low | Moderately insecure | Moderately insecure | Severely insecure | Moderately insecure |
| Jhunjhunu | Very low | Moderately secure | Secure | Secure | Moderately secure |
| Jodhpur | Extremely low | Severely insecure | Severely insecure | Severely insecure | Moderately secure |
| Nagaur | Very low | Severely insecure | Moderately insecure | Severely insecure | Moderately insecure |
| Pali | Extremely low | Severely insecure | Severely insecure | Extremely insecure | Severely insecure |
| Sikar | Very low | Moderately insecure | Moderately secure | Moderately secure | Secure |

## Development Measures

The major challenges for the sustainable development of agriculture in arid Rajasthan are increase the productivity and production stability of rainfed crops, achieve food and fodder security, increase water use efficiency, reversal of land degradation, minimization of risk and intensive labour absorption. The important measures in this direction are to boost agricultural productivity in dry lands, promotion of low/zero tillage, in-situ moisture conservation through raised bed, ridge, furrow, mulching etc.; develop climate hardy crops; improve water use efficiency; increase and maintain green cover by reducing losses to agricultural lands due to grazing; eco-restoration of grazing land; use of sustainable and low-emission technologies; develop organic farming module to be disseminate; improve land condition for agriculture; green manuring and organic farming; use of solar pump for drip irrigation; harnessing potentials of crop productivity and allowing advance decision making at the farm level; improve animal feeding technology modules; assessment of epidemiology of climate sensitive disease of animals; make availability of certified quality seeds; rodent control and reduction in fluoride level in drinking water.

Other measures are direct income support for farmers, protected markets, recognize small and marginal farmers a special group, enhancing value of their produce and production, integrated pest management, farm tools and equipment popularization, improved germplasm of sheep, cattle , buffaloes and camel for enhancing milk, wool and meat production, area specific mineral supplement compound feed block and bypass of protein technology, field based diagnostic kit and value addition of milk and other farm produce.

## Judicious Use and Management of Ground Water

For judicious use and efficient management of ground water and increase of water use efficiency, the important measures are to minimize loss in conveyance, minimize loss from the point of delivery, minimize loss in water application, minimize operational loss, growing of less water requiring crops, appropriate scheduling of water application, use of drip and sprinkler system of irrigation, artificial recharge, provide incentives for water conservation and artificial recharge, ground water modelling and suitable legislation should be promulgated for the controlled use of water.

## Development of Sustainable Rainfed Agriculture System

The important strategies for developing such lands are wind erosion control, development of agro-forestry models, field boundary plantation, in situ water harvesting and soil moisture conservation through contour furrows, contour bunding and vegetative barriers; adoption of suitable tillage and sowing practices, soil fertility management, taking suitable crop varieties matching with the rainfall pattern (Bhati,1997), crop-fallow rotation, mulching, developing short duration and drought tolerant varieties and strengthening local capacity to manage agriculture and natural resources. Improve cultural practices, controlling wind erosion through '*kanabandi*' , strip cropping for soil and water conservation, weed management, integrated nutrient management, integrated pest management, wider spacing for rainfed bajra, development of disease and pest resistant genotype of oilseeds and developing eco-friendly plant protection schedules for various crops are also needed (RAU, 2006)

Other measures are to cultivate and promote indigenous technical knowledge (ITK) of desert people, develop rainfed farming packages of cropping system under different situation of onset of monsoon, develop reliable weather forecast system, creation of alternate/ supplement employment, availability of quality seeds, effective penetration of agriculture extension programme, poverty alleviation and women empowerment, management of crop residue, developing integrated pest management, developing post harvest technologies and value addition of traditional crops and by-products.

## Improvement and Value Addition of Biomass Productivity

The strategies include maintenance of tree on non-arable land, around house, along boundaries and in other community sites; intercropping on arable land monocropping on arable farm woodlots and development of silvi-pasture with suitable blend of trees, shrubs and grasses.

## Modernization of Animal Husbandry, Dairying and Livestock Production

Enhancement of fodder production, creation of adequate fodder bank, availability of quality fodder and feed, development of livestock based infrastructure facilities, dairy development, meat production and make this profession more attractive with social status are important sectors for development. The important strategies could be the improvement of carrying capacity of pastures through regeneration, development and management; development of silvi-pasture with suitable blend of tree, shrub and grass species, adequate production of fodder during lean period, production of feed blocks and improvement of nutritious value of fodder, establishment of breeding farm for cattle, sheep, goat and camel; genotype improvement of local breeds, providing adequate veterinary facilities (including mobile dispensary) as per livestock population over the region; effective mechanism for collection chilling, processing and marketing of milk and its by-products; mobilization and need based training to pastoralist and credit facility. Social stigma should be removed with people's involvement for developing meat industry.

## Development of Alternate Land Use Systems

The important alternate land use system for arid region are agro-forestry, agri-horticulture, agri-horti-pasture, agri-silvi-pasture, silvi-pasture, horti-pasture, horti-silvi-pasture and cultivation of medicinal and aromatic plants.

## Drought Proofing and Management

The strategies includes choice of crops, plant population

management, preparation of rainfall distribution model for different situations, weed management, life saving irrigation, establishment of fodder and seed bank at tehsil level, adequate vaccination programme, inclusion of agro-industrial bye-products and non-conventional feed, putting more area for cultivation of fodder crops and optimize crop production.

## Conclusion

Nearly two third area of western Rajasthan is under agriculture which includes 13.6% irrigated and 52.0% dry lands. The intensity of cultivation decreases from east to west with the declining rainfall. Irrigated area through ground water is mainly concentrated in alluvial plains, interdunal plains and valley fill regions. Canal irrigated area in mainly concentrated in Ganganagar, Hanumangarh and Bikaner districts. Bajra, moth, guar, moong and sesamum are important rainfed crops. Wheat, mustard, gram, groundnut, cumin and isabgol are important irrigated crops. The total crop production comes to 11.3 million tonnes. During the last 48 years the irrigated and double cropped area is significantly increased. On the other hand long fallowing practices has reduced due to division of holdings and increase of population. Introduction of sprinkler system not only increased the area under irrigation but also helped to grow crops on the undulating sandy terrain. Production has registered increasing trend due to irrigation, fertilization, plant protection and adoption of modern technologies. Mixed cropping system has also declined. More emphasis is now being given to cash crops. Agricultural development taken place in areas where irrigation came up. Due to uncertainty and instability of crop production, the rainfed areas are highly insecure in food security. Excessive irrigation, over exploitation of ground water and constant irrigation with saline/ sodic water caused sharp decline in water level and development of salinity/sodicity/waterlogging. Degradation of land and creation of new wastelands has also taken place. The region still has scope for enhancing the productivity of dry lands; livestock based integrated farming system; modernization of animal husbandry, dairying and livestock production; development of silvi-pasture and economic utilization of native flora.

## References

Anon. Rajasthan Agricultural Statistics at a glance for the year 2009-10. Commissionerate of Agriculture, Rajasthan, Jaipur, p. 147, 2010.

Balak Ram and Chauhan, J, S. , 2009. Cartographic Assessment of Land Use Changes in arid Rajasthan using Remote Sensing and GIS. *Indian Cartographer*, 29 : 123-135.

Bhati, T.K. 1997. Management of dry land crops in arid ecosystem. In: Desertification control in the arid ecosystem of India for sustainable development (Eds. Surendra Singh and Amal Kar), pp.298-307, Agro-botanical Pub. (India), Bikaner.

Dhir, R.P. 1977. Western Rajasthan Soils. In: Desertification and its Control. ICAR, New Delhi, pp.102-115.

Faroda, A.S. Joshi, D.C. and Balak Ram, 1999. Agro-ecological zones of north-western hot arid region of India. CAZRI, Jodhpur, p.24.

Gupta, J.P. and Ojaswi, P.R. 1996. Agricultural resources of western Rajasthan and their management for sustainable development, CAZRI, Jodhpur, p. 70.

IHD. 2009. Food Security Atlas of Rural Rajasthan. Institute of Human Development, New Delhi, p.124.

Malhotra, S.P., Gupta, B.S., Goyal, D. and Taimani, V. Population, Land, Crops and Livestock Statistics of Arid Zone of Rajasthan. CAZRI, Jodhpur, p.148, 1983.

RAU, 2007. Research Highlights-2006, RAU, Bikaner, p.28.

Varghese, K.A. and Mordia, A. Classification of districts of Rajasthan according to level of development using multivariate indicators. http./www. nabard,org/databank/ rajasthan

# Chapter - 11

# Demography and Occupational Structure of Thar Desert

**Mahesh Kumar Gaur and Mrs. Narvada Sharma**

## Abstract

*Population of Rajasthan as per Census 2011 is 68,621,012 persons which is much higher than census 2001 that records 56,507,188. It includes 12 districts of Rajasthan. Western Rajasthan covers about 61% area of the total geographical area and 39.51% population of the total population of the Rajasthan. The growth rate of population is much higher than that of the Rajasthan State. In the census 1981 Western Rajasthan had registered a phenomenal increase of 35.48 per cent over 1971 population. Regional pattern of growth rate in western Rajasthan shows that extreme hot arid region district (Barmer) is on top having 32.55 per cent growth rate whereas Indira Gandhi Canal Region (Ganganagar) district has least population growth of 10.06 per cent only. In 1981, the average urban growth rate had been 54.52 per cent. In 2001, the rural density of population was 83 persons per sq. km. The highest rural density of population in the region was 322 persons per sq. km. in Nawalgarh tehsil and the lowest was 6 persons per sq. km. in Jaisalmer tehsil. The highest increase in rural density of population was 129 persons per sq. km. in Shri Madhopur tehsil and the lowest increase was 3 persons per sq. km. in Jaisalmer tehsil. Urban density of population shows varying flexibility and variability. The change in literacy of western Rajasthan between 1981 and 2011 is 40.94 per cent increase in literacy. The highest increase is 50.23 per cent in Jhunjhunun district and the lowest increase of 31.09 per cent is observed in Jalore district during 2011 over 1981. Therefore, to follow the theoretical approach in demographic study and sectoral shifts would require comprehensive knowledge of quantitative techniques. Hence quantification should be used to bring precision in description and also to reach at a concrete conclusion. This study is a special reference to demographic variations in Western Rajasthan. Distribution, growth rate, literacy, sex ratio, and occupational structure in population have been made to delineate and demarcate the spatial patterns and variations in demographic development. Present work is based on census data from 1901 to 2011. While presenting the general account of demographic elements in general and sectoral shifts in particular. The study also elaborates the inter-regional variations.*

## Introduction

India's population is now 1.21 billion, which is bigger than the combined populations of the US, Indonesia, Brazil, Pakistan and Bangladesh. India launched the latest census exercise in 2010. Some 2.5 million officials visited households in about 7,000 towns and 600,000 villages. Population of Rajasthan as per Census 2011 is 68,621,012 persons which is much higher than census 2001 which recorded 56,507,188. Child sex ratio at 883 in the state which is below the national average of 914 and is among those states with lowest child sex ratio. Even the sex ratio in the Rajasthan at 926 is below the national average of 940, a rise by seven points since 2001 to record the highest figures in the country since 1971. But the current sex ratio is a jump by four points since the last census. The rising population in the state is also reflected in the population density which now stands at 201 persons per sq. km. against figures of 165 and 129 in 2001 and 1991 respectively. Significantly, Rajasthan is the largest state in the country. On the significant side, Rajasthan has also shown an improvement in the female literacy rate that now stands at 52.66 against that of 43.9 in 2001. But even this figure stands way below the national average of 65.46. In fact, this is the lowest female literacy rate in the country and is closely followed by Bihar at 53.33.

## Study Area

The study area (western Rajasthan) is a dry and sandy plain. This part of India lies along western side of Arawali hills in Rajasthan. According to geologists this part of Rajasthan was an extended part of sea. This situation of the study area prevailed during the Eocene and Pleistocene period. The geographical area of western Rajasthan is about 208,228 sq. km. Extension of the study area from north to south and east to west are 640 km. and 300 km. respectively. It includes 12 districts of Rajasthan (Sriganganagar, Hanumangarh, Jhunjhunu, Sikar, Bikaner, Churu, Barmer, Jaisalmer, Jodhpur, Nagaur, Pali and Jalore.) Western Rajasthan covers about 61% area of the total geographical area and 39.51% population of the total population of the Rajasthan.

The topography of western Rajasthan is not result of superficial or biogenic agencies, but is attributed to geological processes largely in the nature of the sheet movements leading to pen plantation, rapid changes in the drainage system, enormous accumulation of loose rocky materials, deepening of water table and consequent depletion of vegetation, and thereby accentuation of desert conditions. Most spectacular amongst the land forms are the dunes. These are present in 58 percent area of the region. The area of the occurrence of dunes lies in the northern and western half of the region but even here it is not a contiguous feature and does have intlerludes of dune free corridors such as the areas from Pokhran through Bap to Bikaner. A variety of dune formations such as coalesced parabolic, longitudinal, transverse, barkhan and obstacle have been recognized.

Climate of this region is characterized by extremes of temperature, severe drought accompanied by high wind velocity, low relative humidity, and evaporation far exceeding precipitation and too scanty of rainfall. The variations in the annual rainfall are large. The winter is quite cold, temperature sometimes falling blow the freezing point. During winter, clean skies, fine weather and low humidity, large diurnal variations of temperature and light north-easterly winds are the features of weather. These features are changed only when western disturbances pass over the region and cloudy conditions and sometimes even rain may occur. Both the normal values of maximum and minimum temperatures during this season follow roughly the latitudinal pattern. The soils of the area are broadly classified into three groups viz. desert soils, grey-brown soils and alluvial soils.

## Data Base

1. In case of western Rajasthan demographic information, the database is primarily the Census of India. The main sources for the year 1991 and 2011 needed for the present study are 12 district handbooks of the regions.
2. Data pertaining to land use, crops and animal husbandry are taken from District Statistical Abstracts (2011) of 12 districts, published by Directorate of Economic and Statistics, Rajasthan, Jaipur.
3. Other variables are taken from primary census abstracts and 'village directory' of Sriganganagar, Hanumangarh, Jhunjhunu, Sikar, Bikaner, Churu, Barmer, Jaisalmer, Jodhpur, Nagaur, Pali and Jalore.
4. Base map is generated from Census of India, 2001 and 2011.

## Demographic Indicators of Development

The society in arid zone is patriarchial, patrilineal and patrilocal and the agnatic ties have greater influence on the proximity of relationships. The human settlement in the region started several centuries ago. Most of the villages are 500 to 800 years old. Despite this population growth was slow. For example, some data for a part of arid zone shows that over a period of 250 years, the population increased only by two and half times. The situation persisted right upto the beginning of the century. However, with the year 1921 - a period of rapid growth - population increased by more than two fold in four decades from 1921 to 1961. The growth rate of population is much higher than that of the Rajasthan State. For the first time, the growth rate has shown a decline during the decade 1981 to 1991 and this is ascribed to awareness created by family planning programme.

Population, in its various aspects, has two-fold relevance to development: it is integral part of the resource base of the region and in itself, at least partly, an outcome of development. Development can never be thought independent of man, his capabilities and his aspirations.

The inter-relation between man and development is not a static one. It is a dynamic relationship both in temporal and spatial perspectives. The changing numbers and characteristics of population through time, and the spatial variations therein, are vital to the whole process of socio-economic advancement. (Table 1)

**Table 1** : District-wise Number of Literates and Literacy Rate in Western Rajasthan, 2011

| State/ District | Number of Literates | | | Literacy rate | | |
|---|---|---|---|---|---|---|
| | P | M | F | P | M | F |
| Rajasthan | 38970500 | 24184782 | 14785718 | 67.06 | 80.51 | 52.66 |
| Ganganagar | 1206320 | 719999 | 486321 | 70.25 | 79.33 | 60.07 |
| Hanumangarh | 1057534 | 637735 | 419799 | 68.37 | 78.82 | 56.91 |
| Bikaner | 1300806 | 797119 | 503687 | 65.92 | 76.90 | 53.77 |
| Churu | 1165255 | 709852 | 455403 | 67.46 | 79.95 | 54.25 |
| Jhunjhunun | 1385546 | 827466 | 558080 | 74.72 | 87.88 | 61.15 |
| Sikar | 1679913 | 1016609 | 663304 | 72.98 | 86.66 | 58.76 |
| Nagaur | 1801102 | 1131927 | 669175 | 64.08 | 78.90 | 48.63 |
| Jodhpur | 2075029 | 1295900 | 779129 | 67.09 | 80.46 | 52.57 |
| Jaisalmer | 314345 | 214540 | 99805 | 58.04 | 73.09 | 40.23 |
| Barmer | 1210278 | 800983 | 409295 | 57.49 | 72.32 | 41.03 |
| Jalor | 842793 | 554463 | 288330 | 55.58 | 71.83 | 38.73 |
| Pali | 1103704 | 680962 | 422742 | 63.23 | 78.16 | 48.35 |
| Western Rajasthan | **15142625** | **9387555** | **5755070** | **65.43** | **78.69** | **51.2** |
| % age of Rajasthan | 38.86 | 38.82 | 38.92 | | | |

*Source: Census of India 2011, Government of India.*

Population characteristics of an area are undisputedly a manifest expression of the level of development. Usually fertility and mortality rates of population are used as indicators of demographic development.

Census data 2001 defines five major indicators which are essential for levels of demographic development.

1. Population growth rate
2. Density of population
3. Literacy
4. Urbanization and sex ratio
5. Occupational Structure

## 1.0 Population Growth Rate

In the census 1981 Western Rajasthan had registered a phenomenal increase of 35.48 per cent over 1971 population. Despite this steep rise in population, the area still retains its sparsely populated character. In the census 2001, the desert region experienced an increase of 29.68 per cent over 1991 population whereas in 2011, there is a decline in the percentage decadal growth rate of population. It is just 20.92 per cent census. Barmer and Jalore are the two districts which have witnessed insignificant decline whereas Jhunjhunun and Pali have recorded almost fifty per cent of population growth rate during 2001-2011. Regional pattern of growth rate in western Rajasthan shows that extreme hot arid region district (Barmer) is on top having 32.55 per cent growth rate whereas Indira Gandhi Canal Region (Ganganagar) district has least population growth of 10.06 per cent only. High population growth rate of Barmer may be attributed to the increased economic and employment opportunity in the district due to oil and gas exploration. This increase is a result of developmental activities including the construction of border roads and trans-border railway line, laying of petroleum pipelines, allotment of land, adaption of improved agro-horticultural practices, large scale sale and purchase of land, mining activities, setting of ancillary industrial units and hotels for Cairns Energy officials and staff, etc. The development of pasture lands, and vocational training activities for sustenance and drinking water facilities through government and corporate efforts have resulted into inward migration of population to the Barmer district.

Whereas, Ganaganagar experienced decreased and erratic flow of water supply in the Indira Gandhi Canal during the same period. As a result, population either moved to other economically and agriculturally potential areas or started searching employment opportunities in sectors other than agriculture.

A comparative study between the growth rate in 1971-81 and 2001-11 shows that there is -17.77 per cent less growth rate as compared to 1971. In 1981, the average urban growth rate had been 54.52 per cent. The noticeable reason leading the highest rate in Anupgarh (493.73 per cent) and the lowest in Churu tehsil (19.01 per cent) was taking up of agro-horticultural activities. In 2001, the average urban growth rate is 23.74 per cent and highest has been observed in Gharsana (100 per cent) and lowest in Siwana and Raipur (-100 per cent for each). Both the highest and lowest occurred due to changes incorporated in the census definition. The highest positive change of 39.37 per cent has been recorded in Degana tehsil of Nagaur district and the highest negative of -410 per cent in Vijaynagar tehsil of Ganaganagar district. (Table 2; Fig. 1 & 2)

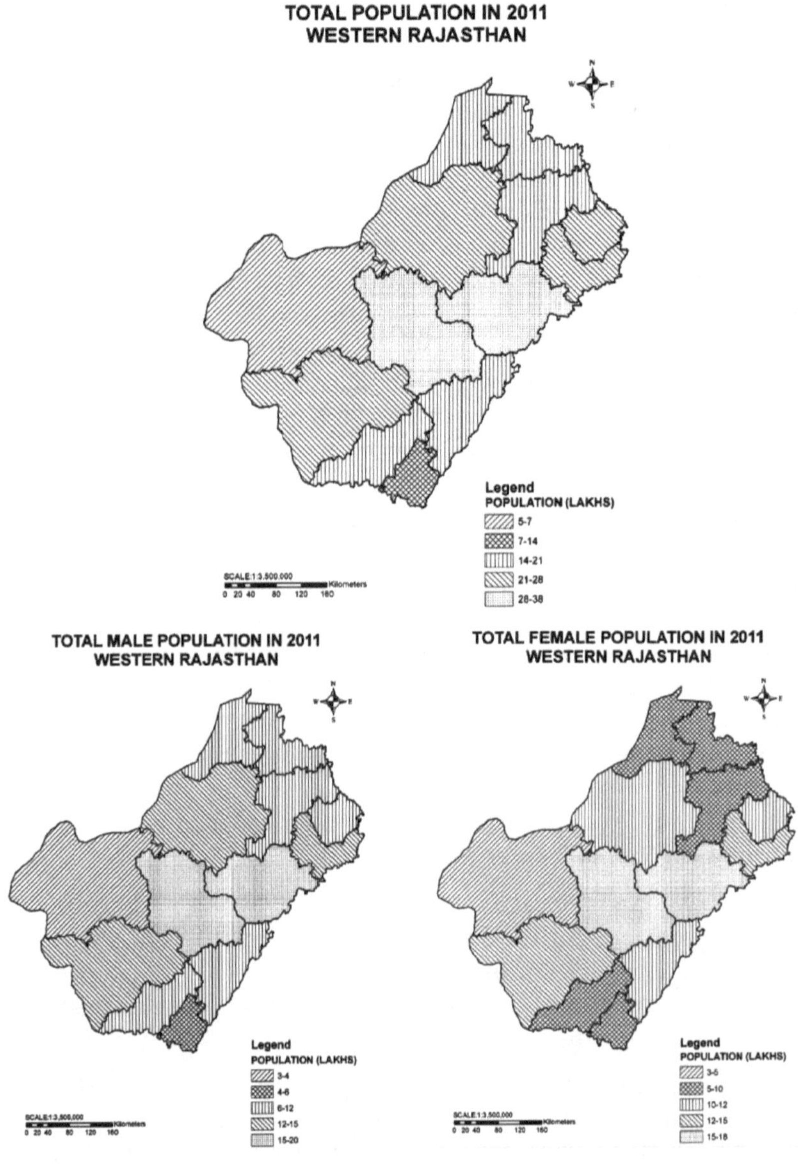


**Figure. 1 :** Population distribution in Western Rajasthan

**Table 2 :** Population Structure of Western Rajasthan, 1981 - 2011

| State/ District | Population, 2001 | | | Population in 2011 | | | Percentage Decadal Growth Rate of Population | | | Sex Ratio (No. of Females per 1000 Males) | | | Population density per Sq. Km | | |
|---|---|---|---|---|---|---|---|---|---|---|---|---|---|---|---|
| | **Person** | **Male** | **Female** | **Person** | **Male** | **Female** | **1981-1991** | **1991-2001** | **2001-11** | **1991** | **2001** | **2011** | **1991** | **2001** | **2011** |
| **Rajasthan** | **56,473,122** | **29,381,657** | **27,091,465** | **68621012** | **35620086** | **33000926** | **28.44** | **28.33** | **21.44** | **910** | **922** | **926** | **129** | **165** | **201** |
| Ganganagar | 1,788,487 | 955,027 | 833,460 | 1969520 | 1043730 | 925790 | 18.25 | 27.53 | 10.06 | 865 | 873 | 887 | 176 | 224 | 179 |
| Hanumangarh | 1,517,390 | 800,796 | 716,594 | 1779650 | 933660 | 845990 | 44.6 | 24.34 | 17.24 | 891 | 895 | 906 | 96 | 120 | 184 |
| Bikaner | 1,673,562 | 885,722 | 787,840 | 2367745 | 1243916 | 1123829 | 42.7 | 38.18 | 24.48 | 885 | 889 | 903 | 44 | 61 | 78 |
| Churu | 1,922,908 | 986,867 | 936,041 | 2041172 | 1053375 | 987797 | 30.84 | 24.6 | 20.35 | 937 | 948 | 938 | 92 | 114 | 148 |
| Jhunjhunun | 1,913,099 | 983,158 | 929,941 | 2139658 | 1097390 | 1042268 | 30.61 | 20.9 | 11.81 | 931 | 946 | 950 | 267 | 323 | 361 |
| Sikar | 2,287,229 | 1,172,129 | 1,115,100 | 2677737 | 1377120 | 1300617 | 33.81 | 24.11 | 17.04 | 946 | 951 | 944 | 238 | 296 | 346 |
| Nagaur | 2,773,894 | 1,421,455 | 1,352,439 | 3309234 | 1698760 | 1610474 | 31.69 | 29.33 | 19.25 | 942 | 951 | 948 | 121 | 157 | 187 |
| Jodhpur | 2,880,777 | 1,509,563 | 1,371,214 | 3685681 | 1924326 | 1761355 | 29.12 | 33.77 | 27.69 | 891 | 908 | 915 | 95 | 126 | 161 |
| Jaisalmer | 507,999 | 278,973 | 229,026 | 672008 | 363346 | 308662 | 41.73 | 47.45 | 32.22 | 807 | 821 | 849 | 9 | 13 | 17 |
| Barmer | 1,963,758 | 1,035,813 | 927,945 | 2604453 | 1370494 | 1233959 | 28.27 | 36.83 | 32.55 | 891 | 896 | 900 | 51 | 69 | 92 |
| Jalor | 1,448,486 | 736,029 | 712,457 | 1830151 | 937918 | 892233 | 26.52 | 26.78 | 26.31 | 942 | 968 | 951 | 107 | 136 | 172 |
| Pali | 1,819,201 | 917,320 | 901,881 | 2038533 | 1025895 | 1012638 | 16.63 | 22.39 | 11.99 | 956 | 983 | 987 | 120 | 147 | 165 |
| **Western Rajasthan** | **22,496,790** | **11,682,852** | **10,813,938** | **27115542** | **14069930** | **13045612** | **31.23** | **29.68** | **20.92** | **907** | **919** | **923** | **118** | **149** | **174** |

*Source: Census of India 2011, Government of India.*

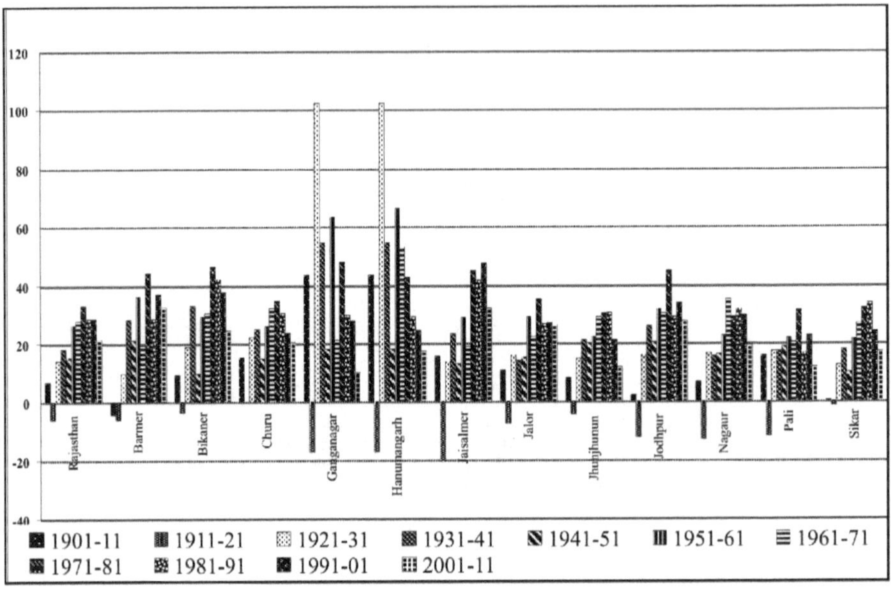


**Figure. 2 :** Percentage Decadal Variation in Population for State and Deistricts :

In 2001, there were in all 99 urban centers as against 91 in 1981. So, there was an increase of nine urban centers. Out of 86 tehsils in the Rajasthan Desert, 26 did not have any urban center including the tehsil headquarters. These were Tibi, Poongal, Lunkaransar, Kolayat, Khajuwala, Chatatrgarh, Buhana, Jayal, Kheensar, Osian, Bhopalgarh, Shergarh, Luni, Fatehgarh, Sheo, Baytoo, Siwana, Gudha Malani, Ramsar, Chouhtan, Sayla, Ahore, Bhagora, Raniwara, Raipur and Rohat.

## 2.0 Density of population

Despite high growth rate, the arid region by and large, still remains a sparsely populated region. Density of Western part of Rajasthan is very thin. Even though, it occupies about 61 per cent area of the state but average density is only 174 which is much less than the state average of 201 in 2011 census. Population density of Jaisalmer in 2011 is 17 persons per sq.km whereas it was just 9 in 1991. Sikar district maintains highest density of 361 persons per sq. km., followed by Jhunjhunun (346 persons per sq. km.). In spite of scarcity of irrigation facilities and sandy tract, reasons ascribed to comparatively dense population are availability of better quality of medical and educational facilities in these districts; mining of various minerals and adoption of improved agro-horticulture practices. Density of Jaisalmer, Bikaner, and Barmer is less than 100. Looking at their size, density of these districts appears to be very thin and sparse. Beyond these town, density of these districts is almost devoid of any human population. (Fig. 3)

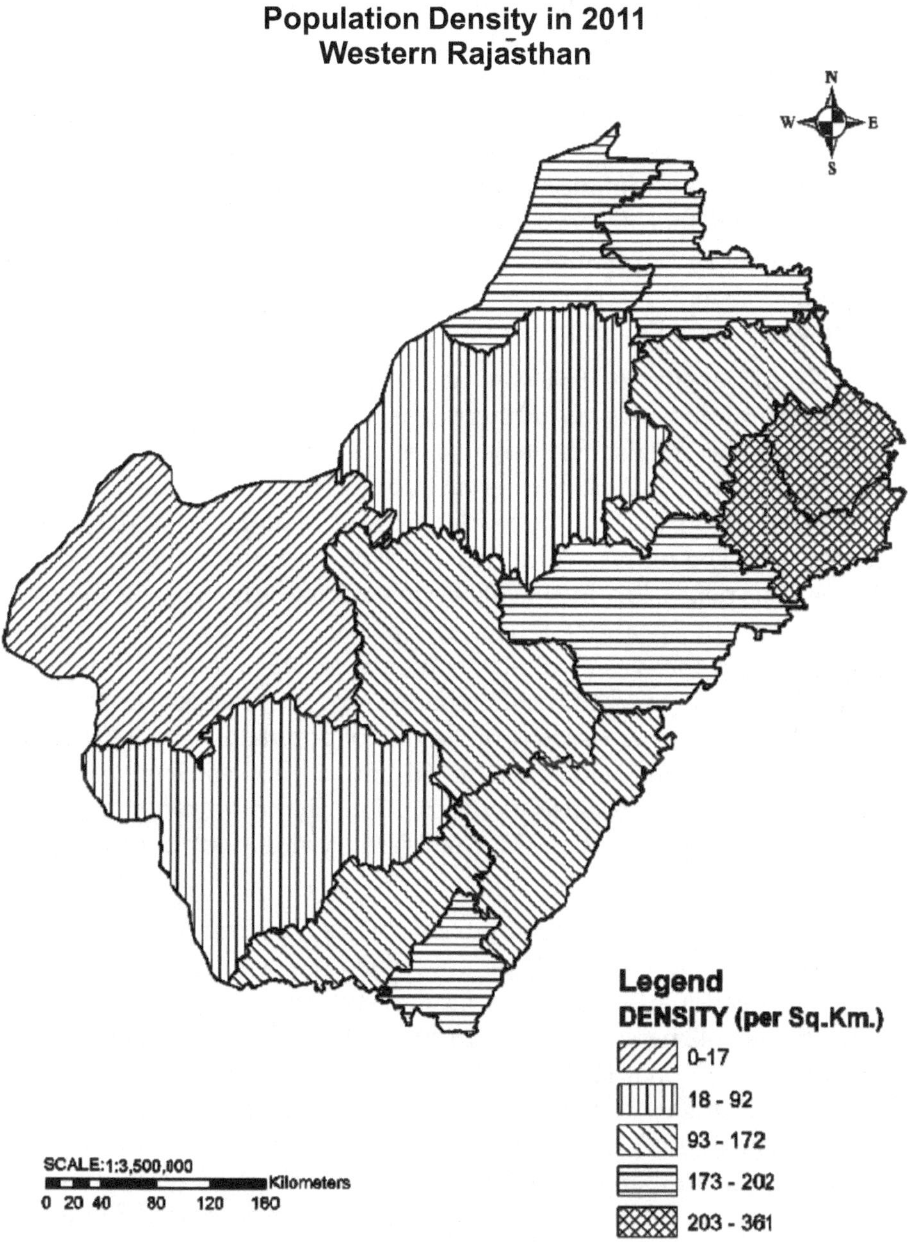


**Figure. 3 :** Population density in Western Rajasthan

Situation is better in comparison to 1981 and 1991. In 1981, the density of population was 56 persons per sq. km. which was quite low as compared to the State average of 100 and national average of 220 persons per sq. km. The highest density of population of 282 persons per sq. km. was reported for Sriganganagar tehsil and the lowest in Jaisalmer tehsil (4 persons per sq. km.).

Census 2001 presented a better picture. The density of population in the western arid region was 108 persons per sq. km. which was quite below to the state average of 165 persons per sq. km. The highest density of population in the region was 619 persons per sq. km. in Jodhpur tehsil and the lowest 9 persons per sq. km. in Jaisalmer tehsil. The change in the density of population in the western Rajasthan between 1991 and 2001 depicted that 62 persons per sq. km. increased as compared to 1981 density of population, while the highest increase in density of population was 257 persons per sq. km.in Jodhpur tehsil and the lowest 5 persons per sq. km. in Jaisalmer tehsil.

Rural density in 1981 was only 39 persons per sq. km. and the highest rural density was in Khetri tehsil (190 persons per sq. km.) and the lowest was in Jaisalmer tehsil (4 persons per sq. km.). Despite high rural growth rate in 2001, the region by and large still remains a sparsely inhabited region. In 2001, the rural density of population was 83 persons per sq. km. The highest rural density of population in the region was 322 persons per sq. km. in Nawalgarh tehsil and the lowest was 6 persons per sq. km. in Jaisalmer tehsil. The change in the rural density of population in the western Rajasthan between 1981 and 2001 depicted an average change of 48 persons per sq. km. increase as compared to 1981. The highest increase in rural density of population was 129 persons per sq. km. in Shri Madhopur tehsil and the lowest increase was 3 persons per sq. km. in Jaisalmer tehsil. The Shekhawati region has the highest rural density due too high soil fertility and moderate availability of water.

Urban density of population shows varying flexibility and variability. In fact, 26 tehsils do not have any urban centers including the tehsil headquarters. The average density of urban population in 2001 census was 4142 persons per sq. km. while the highest was 14103 persons per sq. km. in Nohar tehsil and the lowest was 356 persons per sq. km. in Danta Ramgarh tehsil. The average change in urban density of population in 1981 and 2001 shows that 1765 persons per sq. km. increased as compared to 1981, while the highest increase was 5166 persons per sq. km. in Hanumangarh tehsil and decreased 89 persons per sq. km. in Jaitaran tehsil.

## 3.0 Literacy

Literacy depends upon socio-economic structure of the society. Therefore, it is a manifest of indicator of other independent variables. At the same time literacy helps in creating greater awareness, increases wages and earnings and social development. Therefore, it can also be classified as a process variable responsible for social and economic transformation. The Rajasthan literacy scene, as compared with other states is not satisfactory. During the 1981-1991 decade, 53.33 lakh persons were added to the ranks of the literates. The figure in the previous decade, 1971-81, was only 32.71 lakhs. And the number of illiterates increased for the first time - from 259.01 lakhs in 1981 to 212.03 lakhs in 1991. In the censuses of 1961, 1971 and 1981, although the percentage of literacy increased every time, the number of illiterates increased simultaneously. Number of illiterates decreased for the first time in 1991.

Within the state, desert districts during 1981-91 also recorded improvement in the ratio of literates to total population - in the case of both males and females. The average literacy of western Rajasthan was 24.49 per cent. The highest literacy was observed in Jodhpur tehsil (41.24 per cent) and the lowest in Chouhtan tehsil of Barmer district (6.67 per cent). The gap in the literacy ratio for males narrowed down between districts while this increased in the case of females. The increase in percentage points for both males and females were higher in Jhunjhunun, Pali and Sikar districts than the Rajasthan average. The best performance in terms of improvement in the ratio of literates to population was in Jhunjhunun district. The highest ratio was in Jaisalmer (14.87 percentage points) district. The highest increase in the literacy ratio for females was in Ganganagar (11.92 percentage points) and Jhunjhunun (11.85 per centage points) districts. The most dismal picture regarding the increase in the female literacy ratio came from Jalore district (2.50 percentage points) district. Although Jalore had risen above the Rajasthan average with regard to male literacy ratio, its position pertaining to the females was the lowest. (Table 3)

Thus, among the various indicators of population quality, achievements in the sphere of literacy are the most important in the context of developing areas (Gosal and Gopal Krishan, 1981). In 2001, average literacy of western Rajasthan further improved and was 47.21 per cent. The highest literacy percentage was observed in Buhana tehsil (64.03 per cent) and the lowest in Poogal tehsil of Bikaner district (28.38 per cent). The change in literacy of western Rajasthan between 1981 and 2011 is 40.94 per cent increase in literacy. The highest increase is 50.23 per cent in Jhunjhunun district and the lowest increase of 31.09 per cent is observed in Jalore district during 2011 over 1981. (Fig. 4)

The average male literacy of western Rajasthan during 2011 census is 78.69 per cent while female literacy is just 51.20 per cent. The highest male and female literacy are 87.88 and 61.15 per cent respectively and both are in Jhunjhunun district. The lowest male and female literacy 71.83 and 38.73 per cent respectively and both are in Jalore district. Female literacy is function of general literacy. The other contributing factors are growth of industries. Presence of marginal workforce is

**Table 3 :** Districts arranged in descending order of growth rate of population 19061-1971 to 2001-2011

| Rank | State/ District | 1961-71 | State/ District | 1971-81 | State/ District | 1981-91 | Rank | State/ District | 1991-01 | State/ District | 2001-11 |
|---|---|---|---|---|---|---|---|---|---|---|---|
| 1 | Hanumangarh | 53.02 | Ganganagar | 48.08 | Bikaner | 41.84 | 1 | Jaisalmer | 47.52 | Barmer | 32.55 |
| 2 | Kota | 42.98 | Bikaner | 46.34 | Jaisalmer | 41.73 | 2 | Bikaner | 37.71 | Jaisalmer | 32.22 |
| 3 | Banswara | 37.71 | Jaisalmer | 44.84 | Jaipur | 38.73 | 3 | Barmer | 36.90 | Jodhpur | 27.69 |
| 4 | Nagaur | 35.00 | Jodhpur | 44.82 | Kota | 35.89 | 4 | Jaipur | 35.06 | Jaipur | 26.91 |
| 5 | Pratapgarh* | 32.91 | Kota | 44.58 | Sikar | 33.81 | 5 | Jodhpur | 34.04 | Banswara | 26.58 |
| 6 | Bundi | 32.42 | Barmer | 44.41 | Nagaur | 31.69 | 6 | Dausa | 32.40 | Jalor | 26.31 |
| 7 | Jaipur | 32.19 | Hanumangarh | 42.90 | Dausa | 30.84 | 7 | Dhaulpur | 31.19 | Dungarpur | 25.39 |
| 8 | Churu | 32.07 | Jaipur | 40.58 | Alwar | 30.79 | 8 | Karauli | 30.41 | Bikaner | 24.48 |
| 9 | Bharatpur | 31.07 | Pratapgarh* | 36.27 | Jhunjhunun | 30.61 | 9 | Alwar | 30.31 | Kota | 24.34 |
| 10 | Udaipur | 30.37 | Jalor | 35.20 | Churu | 30.23 | 10 | Sirohi | 30.13 | Dausa | 23.75 |
| 11 | Dungarpur | 30.30 | Banswara | 35.03 | Banswara | 30.19 | 11 | Banswara | 29.62 | Udaipur | 23.63 |
| 12 | Jodhpur | 30.17 | Churu | 34.81 | Ganganagar | 29.36 | 12 | Nagaur | 29.38 | Pratapgarh* | 22.84 |
| | | | Rajasthan | 32.97 | | | 13 | Kota | 28.52 | Dhaulpur | 22.78 |
| 13 | Bikaner | 30.14 | Udaipur | 32.81 | Jodhpur | 29.12 | | Rajasthan | 28.41 | | |
| 14 | Jhunjhunun | 29.12 | Sikar | 32.09 | Hanumangarh | 29.03 | 14 | Pratapgarh* | 27.61 | Alwar | 22.75 |
| | Rajasthan | 27.83 | | | | | 15 | Ganganagar | 27.59 | Sirohi | 21.86 |
| 15 | Sawai Madhopur | 27.69 | Pali | 31.39 | Karauli | 28.66 | | | | Rajasthan | 21.44 |
| | | | | | Rajasthan | 28.44 | 16 | Sawai Madhopur | 27.55 | Bharatpur | 21.39 |

| | | | | | | | | | | | |
|---|---|---|---|---|---|---|---|---|---|---|---|
| 16 | Alwar | 27.64 | Bundi | 30.83 | Barmer | 28.27 | 17 | Bharatpur | 27.22 | Karauli | 20.94 |
| 17 | Sikar | 27.11 | Jhunjhunun | 30.39 | Dhaulpur | 28.10 | 18 | Udaipur | 27.09 | Churu | 20.35 |
| 18 | Jhalawar | 26.81 | Dausa | 29.73 | Dungarpur | 28.07 | 19 | Jalor | 26.81 | Baran | 19.82 |
| 19 | Dhaulpur | 26.37 | Nagaur | 29.04 | Baran | 27.30 | 20 | Dungarpur | 26.65 | Sawai Madhopur | 19.79 |
| 20 | Baran | 26.35 | Karauli | 29.00 | Sawai Madhopur | 27.22 | 21 | Bhilwara | 26.39 | Jhalawar | 19.57 |
| 21 | Tonk | 25.74 | Chittaurgarh | 29.00 | Bharatpur | 27.15 | 22 | Ajmer | 26.17 | Bhilwara | 19.27 |
| 22 | Karauli | 25.72 | Dungarpur | 28.78 | Jalor | 26.52 | 23 | Baran | 26.08 | Nagaur | 19.25 |
| 23 | Chittaurgarh | 25.05 | Sawai Madhopur | 28.03 | Bundi | 25.85 | 24 | Bundi | 24.98 | Ajmer | 18.66 |
| 24 | Dausa | 24.22 | Sirohi | 27.90 | Pratapgarh* | 25.45 | 25 | Hanumangarh | 24.39 | Rajsamand | 17.89 |
| 25 | Jalor | 22.10 | Dhaulpur | 27.28 | Tonk | 24.42 | 26 | Tonk | 24.27 | Tonk | 17.33 |
| 26 | Bhilwara | 21.81 | Baran | 26.78 | Udaipur | 23.97 | 27 | Sikar | 24.14 | Hanumangarh | 17.24 |
| 27 | Ganganagar | 21.04 | Alwar | 26.21 | Jhalawar | 21.91 | 28 | Churu | 23.51 | Sikar | 17.04 |
| 28 | Pali | 20.40 | Bharatpur | 26.05 | Bhilwara | 21.55 | 29 | Jhalawar | 23.34 | Chittaurgarh | 16.09 |
| 29 | Sirohi | 20.30 | Jhalawar | 25.85 | Sirohi | 20.66 | 30 | Pali | 22.46 | Bundi | 15.70 |
| 30 | Barmer | 19.24 | Ajmer | 25.49 | Ajmer | 20.66 | 31 | Jhunjhunun | 20.93 | Pali | 11.99 |
| 31 | Jaisalmer | 18.82 | Rajsamand | 25.34 | Chittaurgarh | 19.85 | 32 | Chittaurgarh | 20.44 | Jhunjhunun | 11.81 |
| 32 | Rajsamand | 17.67 | Tonk | 25.22 | Rajsamand | 18.12 | 33 | Rajsamand | 19.96 | Ganganagar | 10.06 |
| 33 | Ajmer | 17.53 | Bhilwara | 24.22 | Pali | 16.63 | | | | | |

*Source: Census of India 2011, Government of India.*

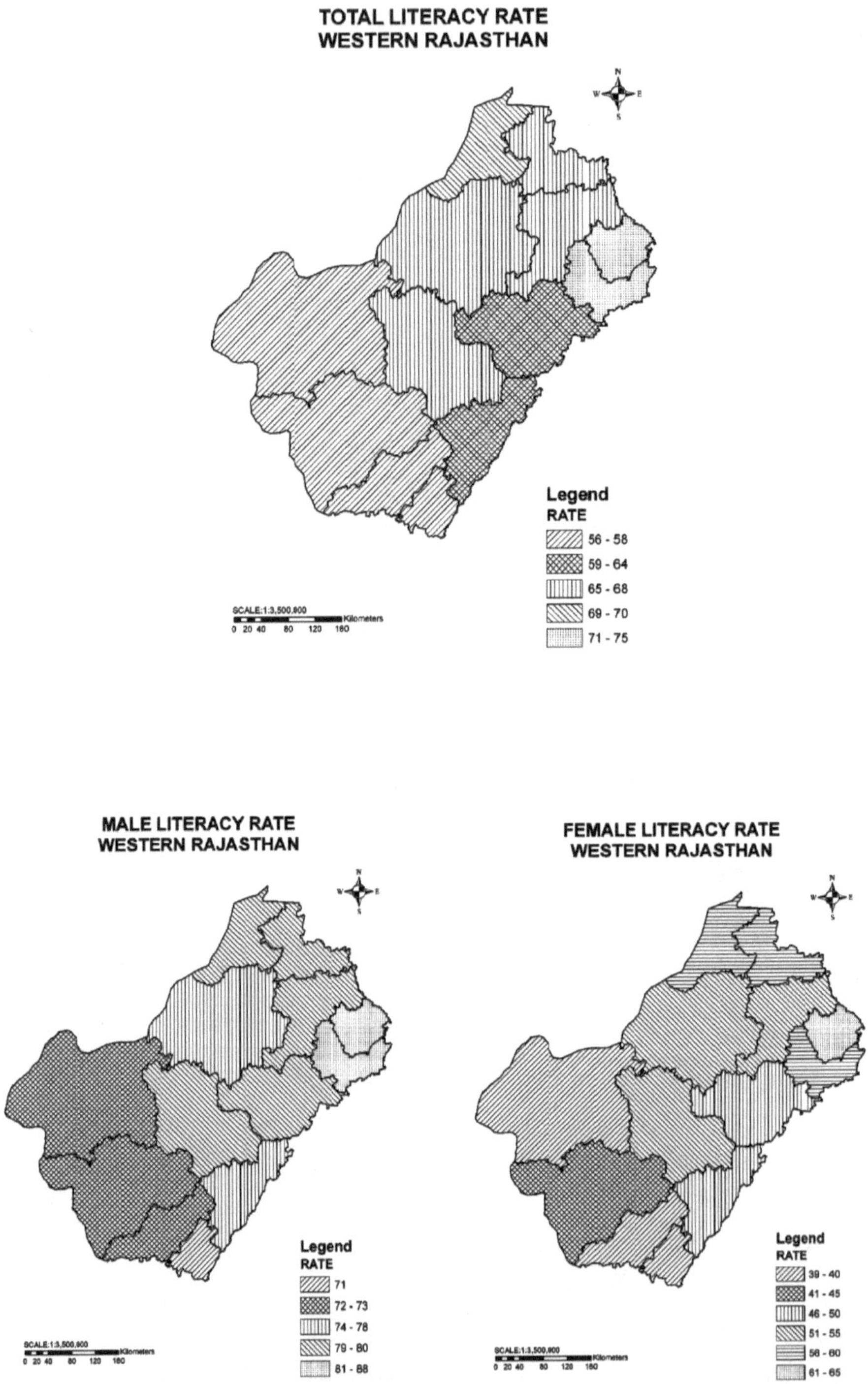


**Figure. 4 :** Literacy rate in Western Rajasthan

deterrent to female literacy. The availability of the labour force reduces the work burden on female children in those families that can afford wage labour and facilitate female children to attend school. The sex ratio contributes in female literacy.

The average rural literacy of western Rajasthan was 45.09 per cent in 2001 census. The highest rural literacy 64.03 per cent was recorded in Buhana tehsil and the lowest 28.38 in Poogal tehsil. The average change in rural literacy showed 28.52 per cent increase in 2001 as compared to 1981 rural literacy. The highest increase was 49.93 per cent in Taranagar tehsil and the lowest increase of 12.73 per cent in Poogal tehsil. Similarly, the average rural male literacy was 57.19 per cent, while female literacy was 32.10 per cent only in 2001 census. The highest rural male and female literacy were 73.55 per cent in Buhana tehsil and 53.96 per cent in Udaipurwati tehsil respectively. The lowest rural male literacy of 39.50 per cent was observed in Poogal tehsil and the lowest female literacy of 14.99 per cent in Shergarh tehsil.

The average urban literacy of the arid zone was 59.69 per cent. The highest urban literacy increase was 72.26 per cent in Khetri tehsil and the lowest increase of 47.09 per cent in Sanchor tehsil. Similarly, the average urban literacy showed 19 percent increase. The highest urban literacy increase was 153 percent in Nawalgarh tehsil and the lowest increase 30.35 percent in Parbatsar tehsil. The average urban male and female literacy were 81.43 per cent in Khetri tehsil and 62.17 per cent in Ganganagar tehsil respectively. The lowest urban male literacy of 61.24 per cent and the lowest urban female literacy of 31.33 per cent and both in Sanchor tehsil.

## 4.0 Urbanisation and Sex Ratio

**Urban Population :** The degree of urbanization in a region is important measure of socio-economic progress. Towns are the centres for new innovations, technology, modern traits of life and growth of new ideas and from where they spread to the countryside. The degree and rate of diffusion of these traits determine the speed of transformation of the countryside in social as well as economic terms. In return the demand which the villages make from the towns for a variety of services and the contribution which villages make in supplying goods and materials to the towns play their own role in accelerating the process of urbanisation.

The average urban population is 23.01 percent, while the highest is 80.48 percent in Jodhpur tehsil and the lowest is 3.69 percent in Degana tehsil.

The western Rajasthan has in all 99 urban centres according to 2001 census as given below in table.

| S.No. | Name of district | No. of towns |
|---|---|---|
| 1 | Ganganagar | 13 |
| 2 | Hanumagarh | 6 |
| 3 | Bikaner | 3 |
| 4 | Churu | 15 |
| 5 | Jhunjhunu | 13 |
| 6 | Sikar | 11 |
| 7 | Nagaur | 14 |
| 8 | Jodhpur | 6 |
| 9 | Jaisalmer | 2 |
| 10 | Barmer | 2 |
| 11 | Jalore | 3 |
| 12 | Pali | 11 |

**Sex-Ratio:** Sex ratio is very sensitive to socio-economic development. The sex bias discrimination in lower age groups and sex differential mortality in reproductive ages are two important factors responsible for lower sex ratio in the arid zone of Rajasthan. Western Rajasthan shares a general trend of sex ratio being adverse to women prevailing in other parts of the country. In 1981 Jaisalmer tehsil - the least populated tehsil - had recorded the lowest sex ratio of 791 and the highest was 1023 in Fatehpur tehsil of Sikar district. The sex-ratio over the entire region was 925. In 2001, again the lowest sex ratio 775 was observed in Jaisalmer tehsil, while Desuri tehsil had the highest 1036 sex ratio. The sex-ratio over the entire region was 925. The change in sex ratio of Western Rajasthan between 1981 and 2001 was 4 females per 1000 males increased as compared to 1981. The highest increase was 65 females per 1000 males in Desuri tehsil and the highest decrease was -51 in Barmer tehsil. Table 4 shows status of sex ratio in western arid region between 1901 to 2011. (Fig. 5)

Sex ratio of 987, in the Pali district, is highest in 2011 census whereas it is lowest in lowest in Jaisalmer district (849 females per 1000 males). In fact, Jaisalmer district has consistently shown lower sex ratio in comparison to other districts of the arid zone.

Furthermore, it is evident from the census data of 1991, 2001, and 2011 that sex ratio in the arid zone is positively influenced by sex ratio among children, large households, joint family system and female work participation. It is negatively influenced by easily available manpower that generally comes from scheduled caste population and weaker

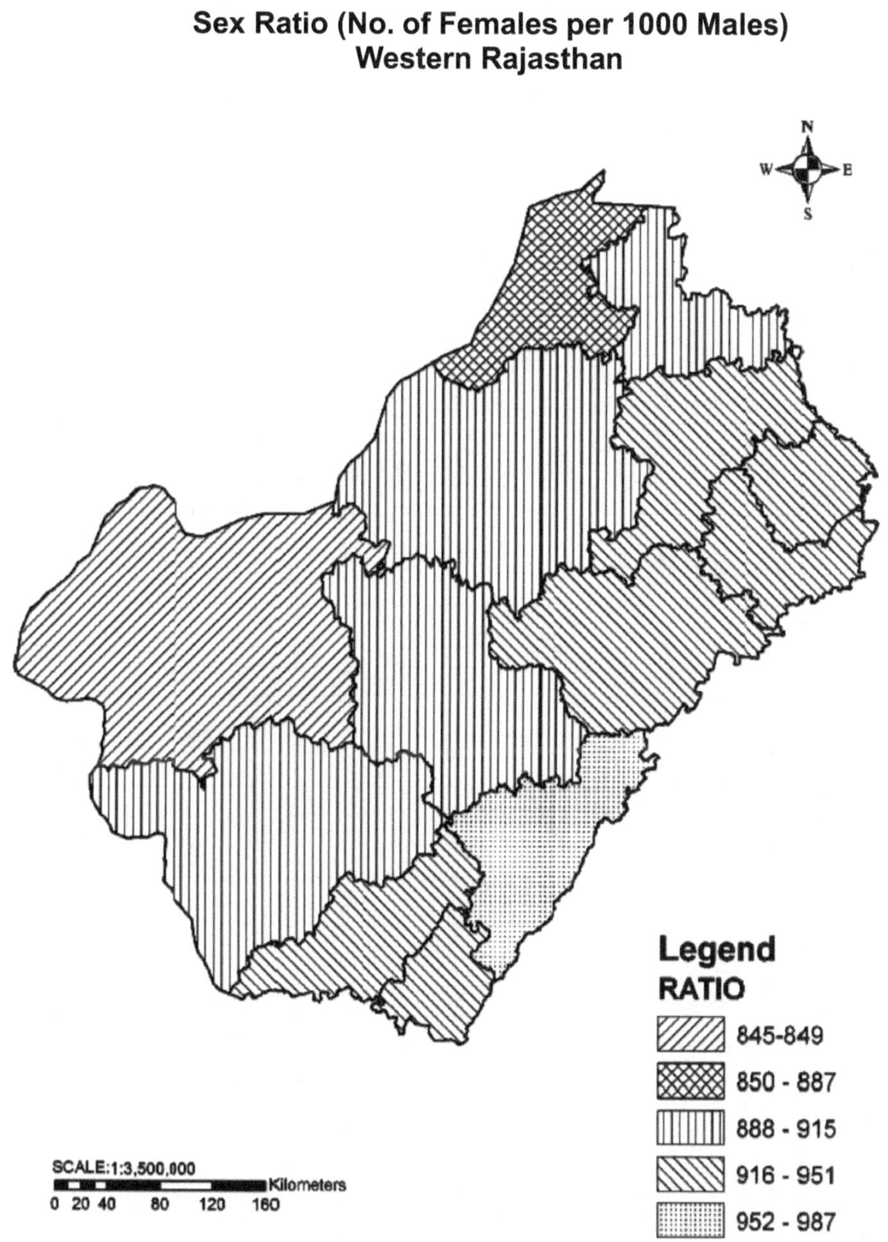


**Figure. 5 :** Sex ratio in western Rajasthan

sections of the society. The literacy level is also negatively associated with lower sex ratio. The medical termination of additional births by literates restricts additional possibility of female births and therefore, is responsible for low sex ratio. Industrialisation attracts male migration and further reduces the sex ratio.

**Table 4 :** District-wise Sex Ratio (females per 1,000 males) : 1901-2011

| | Sex-Ratio Since 1901 (Number of females per 1000 males) | | | | | | | | | | | |
|---|---|---|---|---|---|---|---|---|---|---|---|---|
| **State/ District** | **1901** | **1911** | **1921** | **1931** | **1941** | **1951** | **1961** | **1971** | **1981** | **1991** | **2001** | **2011** |
| **Rajasthan** | 905 | 908 | 896 | 907 | 906 | 921 | 908 | 911 | 919 | 910 | 921 | 926 |
| Ganganagar | 853 | 818 | 857 | 797 | 814 | 828 | 843 | 863 | 859 | 865 | 873 | 887 |
| Hanumangarh | 853 | 818 | 857 | 797 | 814 | 848 | 840 | 887 | 892 | 891 | 894 | 906 |
| Bikaner | 918 | 907 | 891 | 895 | 869 | 920 | 916 | 908 | 899 | 891 | 896 | 903 |
| Churu | 932 | 922 | 926 | 924 | 908 | 945 | 936 | 946 | 953 | 937 | 948 | 938 |
| Jhunjhunun | 884 | 906 | 878 | 878 | 881 | 956 | 943 | 928 | 956 | 931 | 946 | 950 |
| Sikar | 877 | 899 | 886 | 913 | 920 | 972 | 964 | 961 | 963 | 946 | 951 | 944 |
| Nagaur | 914 | 928 | 900 | 920 | 912 | 936 | 945 | 942 | 958 | 942 | 947 | 948 |
| Jodhpur | 888 | 891 | 870 | 885 | 885 | 900 | 888 | 900 | 909 | 891 | 907 | 915 |
| Jaisalmer | 870 | 837 | 808 | 851 | 829 | 817 | 802 | 810 | 811 | 807 | 821 | 849 |
| Barmer | 874 | 880 | 860 | 891 | 873 | 868 | 868 | 887 | 904 | 891 | 892 | 900 |
| Jalor | 898 | 916 | 911 | 910 | 921 | 918 | 919 | 932 | 942 | 942 | 964 | 951 |
| Pali | 946 | 938 | 934 | 954 | 943 | 946 | 943 | 950 | 946 | 956 | 981 | 987 |
| **Western Rajasthan** | 892 | 888 | 882 | 885 | 881 | 905 | 901 | 910 | 916 | 908 | 918 | 923 |

## 5. Occupational Structure

Economic structure shapes socio-economic, cultural and political landscape of a region. It is the strongest factor and decides various aspects of social life. Most of the planning is directed to intervene in economic process to change the socio-economic structures and to achieve desired objectives. Employment sector, ownership of resources, income generation and its redistribution are some important dimensions to be analysed.

Aridland economies are generally dominated by sedentary occupations such as agriculture and animal husbandry. Employment opportunities are comparatively less. Sectoral shift in employment is very slow. Work participation is generally high. Joint family system prevails to absorb the disguised unemployment.

Main and Marginal Workers: In the working population male workers continue to predominate. In arid western Rajasthan, the proportion of male main workers is 43.77 percent which is higher than the National and State average. Obviously, the female worker's proportion is much lower, being only a little less than 16.81 percent as per 2001 census. It is not surprising that among the total male population of the region one third female population is a part of mainstream of the working population. The distribution of main workers shows that Sadulshahar ranks first in male and Sanchor in female main workers with percentages of 52.79 percent and 40.45 percent respectively. The lowest male main workers are 34.02 percent in Shergarh and female are 3.68 percent in Churu.

Regional pattern of male main workers in Western Rajasthan shows that Ganganagar and Hanumangarh region is on top having an average of 48.31 percent followed by 43.16 percent in dry desert region of Bikaner, Churu, Sikar and Jhunjhunun.

The distribution of marginal male workers shows that the highest percentage is in Sanchor (18.02 percent) and the lowest in Raniwara (2.48 percent). The female marginal workers distribution is the highest in Lachhmangarh (33.45 percent) and the lowest is in Jodhpur (4.22 percent). The average ratio of male and female marginal worker in Western Rajasthan is 6.35 percent and 17.63 percent respectively. In the marginal working population female workers continue to predominate.

Therefore, marginality is exceptionally higher in literate population. The low female work participation increases marginality. Similarly dryland farming is associated with high marginality. Therefore, the villages with good irrigation, having power connections and with higher female work participation have low level of marginality among workers.

Female work participation is a function of work participation, sex ratio, industrialisation and sex ratio among children. Generally, the female work participation is higher among lower sections of the society that is characterised with higher sex ratio. Majority of the female work force finds job in agriculture. Therefore, female work participation rate is found higher in agriculturally developed areas.

## Conclusion

The arid zone of Rajasthan presently has 27.11 million people and this number is 20 million more than it had in year 1961. However, agricultural production has not kept pace with it in terms of production and even less in terms of productivity. It is seen that though the area under the crop has increased somewhat during the past few years, the production and productivity have not increased appreciably. Since, the irrigated area constitutes only 13.65 per cent of the total cropped areas, and population has increased a lot more, food production has not increased with the demand. This is reflected in per capita availability of grains for some of the arid districts. All the districts except Ganganagar are showing a decline in per capita availability, the trend being more serious in Barmer and Jodhpur districts. Situation is further compounded by the declining groundwater table, frequent droughts and increase in the cattle population. Biotic resources of the desert are under severe strain. Thus, the arid zone is facing a serious crisis.

## Acknowledgement

Authors acknowledge assistance rendered by Mr. Bahul Vyas is duly acknowledged.

## References

Gaur, Mahesh. 1992. Environmental Crisis: Population, Pollution and Resources, in Third World, New Delhi. vol. (8). 1992

Gaur, Mahesh, Hemlata Gaur & B.B.L. Sharma. 1999. Man & Natural Resource Bases Relationship in an Arid Environment (A study through Remote Sensing Technique). UGC National Seminar at Churu, Dec. 1999.

Gaur, Mahesh, Hemlata Gaur & B.B.L. Sharma. 2000. Urbanization & the Environment (A case Study of Rajasthan State). In: Internat. Conf. on Asian Urbanization on 21st Century Challenges of Urbanization, Chennai, Jan 5-9, 2000.

Gaur, Mahesh. 2001. Impact of Human Activities on Land degradation & Desertification in the Thar desert-Using Remote Sensing Technique. In: Land Degradation and Desertification. V.C. Jha (Ed.), Rawat Publications, Jaipur. 2003:386-402.

Gosal, G.S. & Gopal Krishan. 1984. Regional Disparities in the level of Socio-Economic Development in Punjab. Vishal Publications, Kurukshetra.

Sharma, J.R. & R.B. Singh.1988. Indian Desert:Population, Rsources and Development, Anupama Publications, Delhi,

Census of India, 2001. www.censusindia.net

# Chapter - 12

# Water Management in Dry lands (In-Field Water Harvesting for Sustainability)

**Divya Mishra, Jyotsana Chuchra and JR Sharma**

## 1.0 Introduction

Population in dry lands of the world is increasing; more fresh water is required in order to meet the growing demands. To meet this challenging situation developing new technologies for identification of potential sites for rainwater harvesting (RWH) is an important step towards maximizing water availability in semi-arid areas.

Dry lands are typically defined as those areas of the world where the potential average yearly loss in the form of evapotranspiration (1500-2000 mm/y) exceeds average yearly moisture gain in the form of precipitation (100-420 mm/y). Rajasthan, which is located in the arid zone of India, has been the worst affected of drought. Around 26 million people of 23,406 villages in 26 districts (of the total 32 districts) of the state are affected. Livelihood security in western arid Rajasthan is strongly dependent on rainfall distribution and land management practices among small land holding farmers. Depending upon the local environment conditions water harvesting provides supplementary supply, an alternative supply or the only feasible improved supply, especially in the rural areas of arid and Semi-Arid countries. The potential very much depends upon climate, hydrology, landscape and vegetation and many other factors.

Remote sensing derived parameters facilitate spatial modeling. High temporal resolution remote sensing data has been found useful in deriving the land use/cover required for such study as it captures the variation both in spatial and temporal domains.

Water Harvesting is the process of intercepting storm water runoff from a surface (rooftop, land surface) and putting it to beneficial use. Rainwater harvesting captures precipitation and uses it as close as possible to where it falls. It is the deliberate collection and concentration of rainwater from a surface known as catchment and its storage in physical structures or within the soil profile. Rainwater can also provide affordable water for household use, agriculture, environmental flows and prevention of flood damage.

The process mimics intact and healthy ecosystems, which naturally infiltrate rainwater into the soil and cycle it through myriad life forms and thus follows its natural path to productivity. It is the feasible option for improving the living conditions of many millions of people currently facing serious water scare problems.

The Principles of Rainwater Harvesting:

1. Begin with long and thoughtful observation.
2. Start at the top or high point of watershed and mark the way down.
3. Start with small and simple.
4. Spread and infiltrate the flow of Water.
5. Always plan for an overflow route, and manage that overflow water as resource.
6. Maximize beneficial relationships and efficiency by "stacking functions".
7. Continually reassess the system : The "Feedback Loop"

## 1.1 Need to focus on rain water harvesting

Rainfall in semi-arid areas with fine-textured soils is mainly lost through evaporation from the soil surface (Es) and runoff (R). Under these conditions Es can be 60% to 70% of the annual rainfall (Bennie and Hensley, 2001), and R can vary between 8% and 49% of the annual rainfall depending on the prevailing conditions (Haylett, 1960; Du Plessis and Mostert, 1965; Bennie et al., 1994; Hensley et al., (2000) and Botha et al., 2003).

One of the problems has been lack of tangible scientifically verified information, which can be used to demonstrate to policy makers, areas where rainwater harvesting can be applied. The focus is management of the rural water using low-tech practices, both for domestic purposes and for crop production. Small-scale strategies are the most effective and the least expensive which help the farmers in utilizing their resources, for rainwater harvesting in a sustainable manner, in their fields for different purposes.

Therefore, there is urgent need to focus on rainwater harvesting as it is much cheaper to harvest rainwater then to exploit groundwater. Because collecting rainwater is the only way of recharging water sources. Millions of liters of water can percolate into the ground, rather than evaporating and going waste.

## 1.2 Advantages of rain water harvesting

1. Rain Harvesting is independent of any centralized system, so it not only leads to conservation of water but also energy as the input required to operate the centralized is bypassed.
2. Reduce flooding and lessen soil erosion and thereby preventing contamination of groundwater.
3. Generates an impressive array of resources: provide drinking water, generate high quality irrigation water, enhance soil fertility, increases local water resources, prevents contamination of ground water and reduces demand for ground water.
4. Restoring the productivity of land which suffers from inadequate rainfall.
5. Increasing yield of rainfed farming.

Basically Rain water is harvested for two main purposes:

Firstly, rainwater is harvested and stored in some artificial structures made in the ground or for ready use in containers above or below ground.

Secondly, it is stored in the soil to recharge groundwater and for withdrawal later with the construction of structures like pits, trenches, dug wells, hand pumps and recharge wells.

Rainwater harvesting is defined as a method for inducing, collecting, storing and conserving local surface runoff for agriculture in arid and semi-arid regions (Boers & Ben-Asher 1982).

## 1.3 In-Field water harvesting

In-field approach to RWH has a high potential for improving land conservation. In-field rainwater harvesting is a method of collecting surface runoff from a small catchment area.

The advantages of In-field rainwater harvesting systems are:

- Simple design and cheap to install, therefore easily replicable and adaptable.
- Higher runoff efficiency than medium or large-scale water harvesting systems; no conveyance losses.
- Erosion control function
- Can be constructed on almost any slope, including almost level plains.

The need for information in user-friendly formats and which is easy to update, query, manage and utilize, has popularized the use of GIS in this field. So keeping in mind the requirement of the farmers of our country various efficient strategies have been proposed for locals who wants to harvest rainwater in a safe, productive, and sustainable way. It allows for the decentralized control and community management of water. The aim is to realize the maximum effectiveness for the least effort and cost. The study provides in spatial domains the expansive opportunities for rain water harvesting.

## 2.0 Study Area

Jodhpur and Barmer are among the most water insecure districts in the state. The contributing factors for this dismal scenario include non-existence of perennial rivers, erratic rainfall pattern and excessive salinity. The unpredictability of rainfall has badly affected agriculture production in most areas of the both districts and has resulted in a threat to the survival and livelihood of the rural population due to almost complete crop failure. Therefore, rainwater harvesting will meet twin objectives of long-term water conservation and generation of extra income for farmer.

Fifty villages of Jodhpur and Barmer districts of Rajasthan state (Latitude: 27°00' N, Longitude: 74°00' E) of India was used to provide data for development, testing and validation. Study area is covering about 18375.26 km2 area. Latitudes and longitudinal extend of Jodhpur are 26°18' N, 73°04' E and Barmer are 25°45' N, 71°25' E. Thirty four villages of tehsil Luni in district Jodhpur and sixteen villages of tehsil Pachpadra in district Barmer were selected based on potential of rainwater harvesting.

There are scattered open Hills and sand dunes in the north, west and south. The average height of the district above the mean sea level is between 250-300 meters. The general slope is towards the West. Soil type is mainly sandy and the weather is dry. The groundwater table levels runs very deep and with scanty rainfall.

## 3.0 Role of Earth Observation and GIS

Remote Sensing information is used to derive latest land cover, slope and terrain conditions. IRS- P6 (LISS IV) satellite imagery of path 202 and row 44, 65 with date of pass December 25, 2006 has been used to prepare the land use/ land cover and geomorphology of study area which was needed for runoff estimation using GIS. Soil maps are required for identification of soil type. Amount of runoff generated depend upon size of the soil particle. Infiltration rate is high in coarser soil particles like sand and loam as compared to fine soil particles like clay. Therefore, the volume of runoff generated varies according to soil type of an area.

Daily rainfall data from year 1998 to 2011 is used to study rainfall pattern in study area. Rainfall is required to compute runoff generated from individual fields. The knowledge of rainfall characteristics (intensity and distribution) for a given area is one of the pre-requisites for designing a water harvesting system. Runoff volume estimation was done using the complete database from the Access sheets and GIS derived output. The GIS derived output was obtained by after merging the curve number map and village map with field boundaries. These combinations were used as input with rainfall data for runoff estimation.

SRTM/Aster DEM is used for delineation of the drainage for the whole area, slope and the determination of lowest point in agricultural fields. Lowest points are required to identify site for construction of water harvesting structure. Slope identifies the rate of maximum change in Z value from each cell. For each cell, slope calculates the maximum rate of change in value from selected cell to its neighbours. Slope classes were generated. The slope map was merged with the village shapefile. It identifies the steepest downhill slope for a location on a surface. The slope gradient of the drainage line is a key factor for the selection of water harvesting locations in order to get the maximum storage capacity in the channel. It also affects the amount of runoff. The maximum change in elevation over the distance between the cell and its eight neighbours identifies the steepest, downhill descent from the cell.

GIS was useful in preparing the primary data, which involved the village boundaries for each village. The cadastral maps for some villages of Luni and Pachbadhra district were obtained from settlement office, Jodhpur. Cadastral Maps were digitized at a scale of 1:4000 (Map 1).

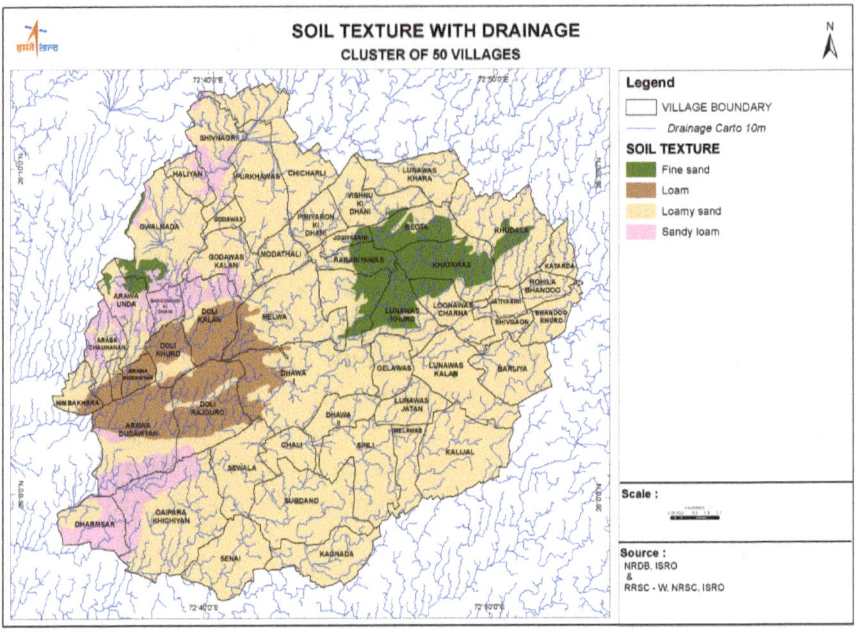


**Map 1:** Soil Texture with Drainage Map of Cluster of 50 villages

## 4.0 Methodology

This study aims to encourage the farmers to conserve and utilize available water resources. The methodology used to access the surface runoff contributing to rainwater harvest is as follow:

## 4.1 Calculation of runoff coefficient

The runoff coefficient, C, is a dimensionless ratio intended to indicate the amount of runoff generated by a watershed given an average intensity of precipitation for a storm. The intensity of runoff is proportional to intensity of rainfall; calibration of the runoff coefficient has almost always depended on comparing the total depth of runoff with the total depth of precipitation,

$C = R/P$

Where:

R = Total depth of runoff (L), and

P = Total depth of precipitation (L).

The runoff coefficient represents the fraction of rainfall converted to runoff.

Amount of rainwater harvested can be calculated based on:

### 4.2.1 Based on Annual Rainfall

The formula for the calculation of total amount of rainwater available that can be harvested is:

Rainwater Harvested (L) = Annual rainfall (mm) x 0.8 x Area (sq. m.) x Runoff coefficient

Annual rainfall is the total amount of rainfall that an area receives in a year. Here we calculate the volume of rainwater that can be harvested in a year. During the rainfall in the initial phase some of the water goes as soil moisture & seepage and after that runoff begin depending upon intensity of rain. This runoff factor depends on various factors like the type & size of catchment, the slope, land use, the soil characteristics, the texture of soil, its depth etc. Based on these factors run off coefficient is decided as per field observations in an area. This runoff coefficient is a dimensionless figure and can be estimated from the individual field data.

On the basis of Table 1, the runoff coefficient factor is decided (Murthy, 2003; CGWB, 2007). These coefficients have been adjusted for the agricultural land (no crop, early crop and late crop) and considering the soil factors:

- Soil Compaction
- Crop Canopy
- Stored Soil Moisture and Soil Infiltration

This coefficient is multiplied with the annual rainfall (mm) and the drainage area (sq. m), which gives the estimate of the amount of water (L) that can be harvested in that particular area. We assume that around 20 per cent of water is loss due to various factors like evaporation, seepage etc.

The runoff coefficient depends on different factors as:

a) Type of vegetation of the area

b) Slope Range (%) and

c) Soil texture

The coefficient is based on two years' experience at the pilot site, the intensity of rain observed and last ten years daily rainfall data of the area (Table 1).

**Table-1 :** Values of Runoff Coefficient Factor (C) for different Soil Conditions

| Type of Vegetation | Slope Range (%) | Runoff Coefficient (C) | | |
|---|---|---|---|---|
| | | **Coarse** (Sand, Loamy Sand, Sandy Loam) | **Medium** (Silty Loam, Loam, Silt, Sandy Clay Loam, Clay Loam, Silty Clay Loam) | **Fine** (Sandy Clay, Silty Clay, Clay) |
| Dense scrub | 0-2 | 0.1 | 0.3 | 0.4 |
| | 2-5 | 0.1 | 0.3 | 0.4 |
| | 5-10 | 0.16 | 0.35 | 0.50 |
| Open scrub/Open land | 0-2 | 0.10 | 0.30 | 0.40 |
| | 2-5 | 0.10 | 0.30 | 0.40 |
| | 5-10 | 0.25 | 0.36 | 0.55 |
| Agricultural Land (No crop) | 0-2 | 0.28 | 0.47 | 0.58 |
| | 2-5 | 0.30 | 0.50 | 0.60 |
| | 5-10 | 0.40 | 0.60 | 0.70 |
| Agricultural Land (Early crop stage) | 0-2 | 0.17 | 0.31 | 0.33 |
| | 2-5 | 0.20 | 0.33 | 0.37 |
| | 5-10 | 0.30 | 0.39 | 0.40 |
| Agricultural land (Late crop stage) | 0-2 | 0.24 | 0.38 | 0.40 |
| | 2-5 | 0.25 | 0.42 | 0.45 |
| | 5-10 | 0.35 | 0.48 | 0.50 |

## 4.2.2 Based on Event Rainfall

The amount of water that can be harvested in a particular event / day is:

Rainwater Harvested (L) = Rainfall (mm) x R factor from Table 2 x Area (sq.m.) x Runoff coefficient

This concept takes into account the last 48 hrs rainfall along with its intensity. The following Table 2 describes R factor of amount of rainfall occurrence.

**Table 2 :** Rainfall factor values for different intensity rainfall

| Intensity of rainfall (mm) | No wetness (Last 48 hrs. < 20mm rainfall ) | Wetness (Last 48 hrs. >20mm rainfall) |
|---|---|---|
| <20 | 0.10 | 0.20 |
| 20-30 | 0.40 | 0.60 |
| >30 | 0.50 | 0.80 |

## 5.0 Identification of potential RWH sites:

Aster data have sinks and peaks which are often errors due to the resolution of the data or rounding of elevations to the nearest integer value. Using fill tool all sinks and peaks in Aster data are filled.

Fill = <input raster> <output raster>

The function iterates until all sinks within the specified Z limit are filled. As sinks are filled, others can be created at the boundaries of the filled areas, which are removed in the next iteration. Sinks is filled to ensure proper delineation of flow accumulation. If the sinks are not filled, a derived drainage network may be discontinuous.

Then correct Aster is used as input to flow direction. The direction of flow is determined by the direction of steepest descent from each cell. This is calculated as change in z-value / distance * 100

Flow direction = <input_fill> <output_name>

The distance is calculated between cell centers. Therefore, if the cell size is 1, the distance between two orthogonal cells is 1, and the distance between two diagonal cells is 1.414. If the descent to all adjacent cells is the same, then the neighborhood is enlarged until a steepest descent is found. If all neighbors are higher than the processing cell, it will be considered noise, filled to the lowest value of its neighbors, and have a flow direction toward this cell. The output of flow direction is used in flow accumulation.

Flow Accumulation tool determine how much rain has fallen within a given watershed. The output of Flow Accumulation would then represent the amount of rain that would flow through each cell, assuming that all rain became runoff and there was no interception, evapotranspiration, or loss to groundwater. Finally a Con operation is used to extract drainage.

Drainage = Con (Flow accumulation > 75, 1, 0)

Final drainage extracted from Aster data is used to identify the lowest point for construction of rainwater harvest structure. Site shape file was digitized as point feature using derived drainage for all the villages in study area.

## 6.0 Development of IT based tools for facilitating farmers

The use of IT provides a powerful, efficient and effective mechanism to offer rainwater harvesting planning in a way that is simple to use and has low investment requirements. A desktop based application is made for the user with basic hardware and software requirements.

This software demonstrates the sustainable strategies for rainwater harvesting. It gives the complete idea about farm level rainwater harvesting. It is semi-automated software program for quantitative analysis of sustainable rainwater harvesting techniques. This software is farmer centric hence it has extremely intuitive GUI.

The software module displays the village wise cadastral fields. The drainage map is overlaid on the fields for the user to identify the slope and lowest point in the field. The farmer has the option to see the drainage map on the field by turning off / on check box at any point of time. This will typically be the site where most of the rainwater falling on his field will converge and may serve as the site for construction of rainwater storage tank.

The software has been designed to calculate amount of In-field Rainwater Harvesting Potential based on site/field conditions like the rainfall intensity, current land use, slope etc. The farmer provides input regarding his village name, type of vegetation and slope by making selection in dropdown box. As the user farmer selects any village from the combo box, the map is zoomed to that village and the name of the village as label appears on the screen. Various tools are provided to make map interactive. Software assists in identification of corresponding field by highlighting field on satellite imagery and cadastral shape file.

## 7.0 Results

The volume of rainwater harvested can be calculated annually or event based. This numerical value is the excess of rainwater that can be harvested. The value is obtained by considering all the factors that affects the runoff. There are many factors like soil seepage, slope range, vegetation, amount of rainfall and soil infiltration rate. The combined effect of all these factors is taken into account and the runoff is calculated. Soil texture map can also be overlaid on the fields for runoff calculation. The purpose of the application is to get an estimate of the amount of excess of rainwater that can be stored into the desired structures.

### Annual based

Based on the input provided by the user with respect to slope, vegetation and soil characteristics the runoff coefficient is calculated and multiplied with the annual rainfall (mm) and the drainage area (sq-m), which gives the estimate of the amount of water (L) that can be harvested in that particular area. We assume that around 20 per cent of water is lost due to various factors like evaporation, seepage etc.

## Event based

In addition to the input parameters like slope, vegetation and soil characteristics, last two days rainfall (48 hours) and particular day intensity of rainfall is also provided to calculate the runoff coefficient and estimate of the amount of water (L) that can be harvested in that particular area.

Total Volume is calculated independent of the event selected. Total volume is the total rainfall an area receives without any prior losses. It is basically calculated by just multiplying the rainfall value to the area, will give the total volume of rainfall an area receives. The percentage of rainwater harvested of total volume of Rainfall is displayed as result.

## 7.1 Size and type of cistern

Water consumption in a house is throughout the year and water availability from rainfall is for a limited period restricted to number of rainy days in a year. To make rainwater available in non-rainy days, storage device need to be designed with an optimum capacity to suit the need / requirement. Optimum size of the storage device and cost effective methods to store water are the key issues for viable rainwater harvesting system. The storage structures may vary from permanent masonry tanks, Ferro-cement tanks to plastic or metal tanks. The capacity of storage structure can be decided by considering parameters such as catchment area like surface runoff etc. water usage and space availability.

## 7.2 Tank size and type calculation

There are three tank types calculators viz., circular, rectangular and Trapezoidal depending upon the amount of water and utilization, site suitability and choice of the farmer.

1) Circular Tank: It is basically suitable for collection of small amount of water and sandy areas.

Circular Tank Volume:

Volume = $\pi * r * r * h$

where,

$\pi = 3.142$, r = Radius, h = Height of the cylinder

2) Rectangular Tank: It is basically suitable in areas with stones.

Rectangular Tank Volume

Volume = $l * w * h$

where,

l = Length, w = Width, h = Height

3) Trapezoidal Tank: These structures are used to impound larger runoff volumes from a larger area. It is simple to design and construct and requires minimum maintenance. They are basically used to grow annual crops in dry areas. All those crops which are drought tolerant like sorghum are preferred as it can withstand temporary water logging.

Trapezoidal Tank Volume

Volume = ½ (l*b1+l*b2)* h

where

b1 = Base 1, b2 = Base 2, l = Length, h = Height

Side Slopes are based on the angle of repose of the soil. Side slope considered for different soil types are: Loam: 1.5:1(horizontal to vertical), Sand: 2:1(horizontal to vertical)

## 8.0 Utilization

Irrigation: Since the largest need for irrigation water in our area occurs during the time of lowest rainfall and highest temperature, a rainwater system designed to meet this need will have to capture water prior to the summer.

The size of the storage system may be prohibitive for using rainfall for the sole source of irrigation water in large or water-intensive landscapes. A low water demanding landscape is required. Depending on the water requirement of the crops and the water harvested in the area the type of crop is decided.

Domestic: Harvesting rainwater can reduce the use of drinking water for landscape irrigation. Water harvested from rain can be used for washing, drinking. Although at home water is needed daily but rains only for a certain period of time. So this harvested water can be used.

## 8.1 Yield of crops achieved at pilot site during 2010 and 2011 and comparison

Pilot studies were conducted to study the difference in yield during normal rain fed conditions and with harvested rainwater as supplementary source of irrigation. The difference achieved in the yield is illustrated in Table 3.

**Table 3:** Yield During Normal Conditions in and with Supplementary Irrigation

| Sl. No. | Name of the Crop | Yield Achieved In 2010-11 at Pilot site (in kg/ha/year) | Normal Yield (in kg/ha) | Yield With SI as claimed by research organisations by experiments & published in research papers (in kg/ha) |
|---|---|---|---|---|
| 1 | Pearl millet (Variety No. 666) | 1798 -2010<br>1330-2011 | 600 - 700 | 2000 - 2500 |
| 2 | Pearl millet (Variety from ICRISAT, Hyderabad) | 1273-2010<br>---- | New variety of seed tried | ---- |
| 3 | Moog (Green gram) | 634-2010<br>558-2011 | 200 - 250 | 600 - 800 |
| 4 | Moth bean (Kidney bean) | 441-2010<br>360-2011 | 200 - 250 | 600 - 800 |
| Second season crop based on soil moisture and irrigation through harvested rain water 2010-11 | | | | |
| 5 | Mustard | 490-2010 | First time in the area | ----- |
| 6 | Taramera | 520-2010 | --Do-- | |

Rainwater harvesting is clearly of enormous benefit to agricultural use. It results in increase in productivity by increasing yields and reducing cropping risk.

Hence, it is clear from the fore-going discussion that earth observation based study provides expansive opportunities for RWH. Also it has an in-built capacity to fulfill requirement of the farmers to devise various efficient strategies to harvest rainwater in a safe, productive, and sustainable way. Further, it offers decentralization of control and an opportunity for the community management of water to realize the maximum efficiency with minimum endeavor and investment.

## Conclusion

For sustainable agricultural development farmers of dry areas needs to utilize water harvesting strategies. This study indicates that Rainwater Harvesting is a promising source of supplementary irrigation in areas where rainfed agriculture is a major challenge and the constraints in further expansion of irrigated area in the country. Rainwater harvesting and efficient water use are inevitable options to sustain rainfed agriculture in future. The water harvesting strategies adopted should suit its purpose, accepted by local population and should be sustainable in local environment.

## References


Adato, M. 1987, Runoff Agriculture for Arid and Semi-Arid Lands. Prospects based on Research at the Desert Runoff Farms in the Negev Desert of Israel. Desert Runoff Farms Unit. Bersheba, Israel.

Bancy M. Mati, Maimbo Malesu and Alex Oduor, Promoting rainwater harvesting eastern and southern Africa

Barnes, D., Bliss, P.J., Gould, B.W. and Valentin, H.R. 1983. Water and wastewater engineering systems. Pitman, London.

Boers, Th.M. 1994. Rainwater Harvesting in Arid and Semi-arid Zones. Doctoral Thesis, Wageningen Agricultural University, Wageningen, the Netherlands.

Boers, Th. M., and J. Ben-Asher. 1982. A review of rainwater harvesting. *Agricultural water management* 5:145-158.

Bunch, R. 1989. Encouraging farmer's experiments. In: Chambers, R., Pacey, A. and Thrupp, L. (eds) Farmer First: Farmer Innovation and Agricultural Research. Intermediate Technology Publications, London, p. 55-59.

Cosgrove, W.J and F.R. Rijsberman. 1998. Creating a vision for water, life and the environment. *Water Policy* 1(1) 115-22.

Critchley, W. and Siegert, C. 1991. Water Harvesting Manual. FAO Paper AGL/MISC/17/91, FAO, Rome

Dieter Prinz1, Water Harvesting past and future, Universität Karlsruhe (TH)

Gould J, Nissen-Petersen E. 1999. Rainwater catchment systems for domestic rain: design construction and implementation. London: Intermediate Technology Publications. p. 335

Gupta, K; Deelstra, J; and Sharma K D. 1997, Estimation of water harvesting potential for a semiarid area using GIS and Remote Sensing. IAHS Publ. no. 242

IAHS. 1997. Remote Sensing and Geographic Information Systems for Design and Operation of Water Resources Systems (Proceedings of Rabat Symposium S3, April 1997). IAHS Publ. no. 242, 1997

Lancaster, Brad, 2008 Rainwater Harvesting for Drylands and Beyond.Vol 1 and Vol 2

Pacey, A. and Cullis, A. 1986, 1991, Rainwater Harvesting. The collection of rainfall and runoff in rural areas. Intermediate Technology Publ., London

Prinz, D., Water Harvesting - Past and Future, Universität Karlsruhe (TH) Institute für Wasserbau und Kulturtechnik 76128 Karlsruhe

Prinz, D., Oweis Theib and Oberle Annette. Rainwater Harvesting for Dry Land Agriculture - Developing a Methodology Based on Remote Sensing and GIS.

Prinz, D., W. Tauer, and Th. Vögtle, 1994. Application of remote sensing and geographic information systems for determining potential sites for Water Harvesting. Proceedings, FAO Expert Consultation on Water Harvesting for Improved Agricultural Production. Cairo, 21 - 25 Nov. 1993. FAO, Rome, 135-144

Shatta, A. and F. Attia 1994, Environmental Aspects of Water Harvesting. In: FAO, Water Harvesting For Improved Agricultural Production. Expert Consultation, Cairo, Egypt 21-25 Nov. 1993, p. 257-270, FAO, Rome.

Suhas P Wani, Johan Rockström, Theib Oweis, Rainfed Agriculture: Unlocking the Potential, ICRISAT, India SEI, Sweden and ICARDA, Syria.

Tauer, W. and Humborg, G. 1992, Runoff Irrigation in the Sahel Zone: Remote Sensing and Geographic Information Systems for Determining Potential Sites. Markgraf, Weikersheim

Texas Water Development Board. 2002. Water for Texas - 2002. Austin (TX): Texas Water Development Board. p. 155

The Texas Manual on Rain Water Harvesting, 2005. Austin, Texas

Theib Oweis, Ahmed Hachum, and Jacob Kijne. Water Harvesting and Supplemental Irrigation for Improved Water Use Efficiency in Dry Areas

Zohar, Y.; Aronson, J.A. and Lövenstein, H. 1987. Fuelwood and Fodder Crops in Rainfall Harvesting Systems Under Arid Conditions. A Research Progress Report, ARO VolcaniCenter.Bet Dagan, Israel.

Websites

American Rainwater Catchment Systems Association,

www.arcsa-usa.org [Accessed 10 Feb 2010]

Organization of American States, Rainwater Harvesting from Rooftop Catchments,

www.oas.org/usde/publications/Unit/oea59e/ch10.htm [Accessed 15 Feb 2010]

Food and agricultural organization,

www.fao.com [Accessed 10 March 2010]

www.rainwaterharvesting.org/whatiswh.htm [Accessed 18 March 2010]

## Chapter - 13

# Role of Space Technology in Studying Land Degradation and Poverty Nexus in Dry lands: A Case Study of Pali Development Block

**Himani Singh**


## Abstract

*This study evaluates the potential of remote sensing techniques for assessing the relationship between land degradation and poverty in Pali development block of Rajasthan. Due to climatic conditions that prevail in this area of the country, less facilities and management practices to conserve land and its fertility becomes an important factor for intensifying poverty. The other countenance of this is that it is poverty which leads to land degradation. This paper assesses the impact of poverty and land degradation on each other. The desertification intensity and BPL density is calculated and relation is seen using Chi square test. The goodness of fit is seen up to 5 per cent level of significance for land degradation and demographic data. The desertification in agricultural area is also considered. The total degradation in the block calculated is 8.39 per cent. It is found that village with highest BPL density has more degradation intensity particularly in agricultural area, proving the agricultural based economy of the dry lands in Rajasthan. An example study is also done at village level. Using cadastral map and satellite image classification, fields that require immediate attention have been identified. Demographic and degradation maps are also prepared.*


*Key words: Remote sensing, Poverty, Land degradation, Desertification, Dry lands, chi square*

## Introduction

The source of basic needs on earth is environment and natural resources. With increasing population over time, the resource consumption and its pattern have changed. The fact is that the dependency of poor people on natural resources is more in comparison to people having fair economical. In Asia, poverty has mainly been a rural phenomenon and nearly three-fourth of the poor live in rural areas, with majority of them dependent on natural resources for employment and income, and their this dependency has become the reason for them degrading the environment. The land is very important natural resource to life on earth which has seen the most degradation due to human practices. In arid and semi-arid parts of the country, where climate also become harsh, the land degradation becomes more challenging. People in this part of country don't have much choice beyond farming. According to Reardon and Vosti (1995), poverty contributes to land degradation as result of poor household's presumed short-term perspective and inability to invest in natural resource conservation and improvement. Poverty is also hypothesized to contribute to population growth.

But apart from this one sided impact of one factor on another, the other theory about this relation between land degradation and poverty is that land degradation brings scarcity of basic needs for people especially poor that are more dependent on natural resources. Land degradation can contribute directly to poverty, separately from its impact on agricultural productivity, by reducing the availability of other valuable goods and services important to poor households (for example, fuel wood, construction materials, wild foods, and medicinal plants) and by increasing the demands on labor needed to forage for such goods.

The intersection of poverty, low agricultural productivity, land degradation (or more generally, natural resource or environmental degradation), and rapid population growth in some developing regions has contributed to a commonly held hypothesis of a downward spiral of mutually reinforcing linkages among these factors. (John Pender, 2004). This hypothesis says that it is land degradation resulting in poverty by affecting agriculture productivity adversely. This downward spiral is not inevitable, as it is influenced by many other factors, particularly policies, institutions, and technologies.

Earlier research in India by Jodha (1986), Reddy and Chakravarty (1999); Narain, Gupta, and van't Veld (2005), and in Nepal by Adhikari (2003) reflects that people under poverty line in these countries profoundly depends on common pool resources for consumption and income, and are in most cases more dependent on this income than wealthier households. Though it does not mean that poorer are larger users of common pool resources in total. Consistent with the hypothesis that poverty causes degradation, there is evidence from several studies in developing countries that poor rural households discount the future heavily, at higher rates than wealthier ones. These studies include those of Pender (1996) in rural India. Contrary to this evidence, some have argued that poor people have a short-term perspective is belied by evidence from many case studies that the poor will often act to reduce consumption and preserve their assets in the face of drought and famine (Moseley 2001; Gray and Moseley 2005)

To link the availability and consumption of natural resources with poverty and other social factors, knowledge of land use, land cover and changes occurring on it and population related factors over a period of time is required. Remote sensing technique has been widely used for mapping and assessing natural resources availability and their status in a geographical area. This technique has also been used for poverty mapping of the country. The same technique is being used in present study for finding linkages between poverty and land degradation.

## Objective

- To generate Population and Poverty map on Scale 1:35,000 for Pali Development Block.
- To create digital database for Poverty and Integration with Land Degradation status Map.
- To generate digital database at the block level.
- To study effect of poverty on land degradation at village level using cadastral map of Sowaniya and LISS III data for three seasons. (Kharif, Rabi and Zaid)

## Study Area

The study area is a development block in Pali district of Rajasthan. It has total of eighty-two villages and one town under it. Area under study lies in a transition zone i.e., semi-arid and sub-humid climate occurs here. It represents agro climatic characteristics. The block is affected with degradation mainly in agriculture areas. Due to decreasing rate of production many people from villages have moved toward town for working in industries. This has not only increased health risk to them but also has created food insecurity. Poverty factor is taken as a study parameter to see effect of land degradation on it. (Fig. 1)

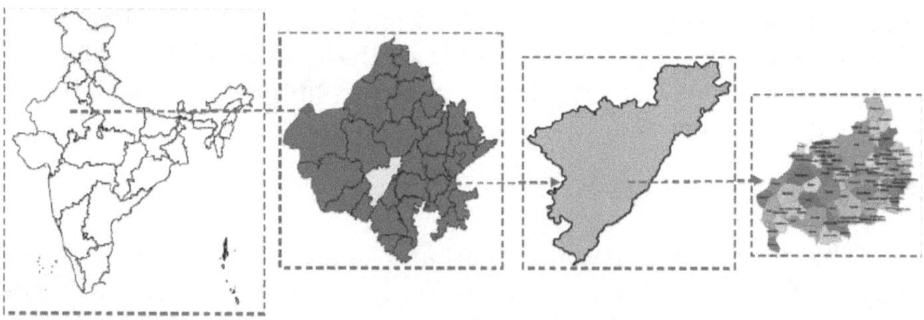

**Figure 1 :** Location map of study area

## Materials and Methods

### Data and base map preparation

IRS LISS III Multi season data of three seasons (Kharif, Rabi, Zaid for year 2005-2006) for image interpretation was used. Features for Base map preparation from topomaps were digitized and further updated using satellite image. (Fig. 2)

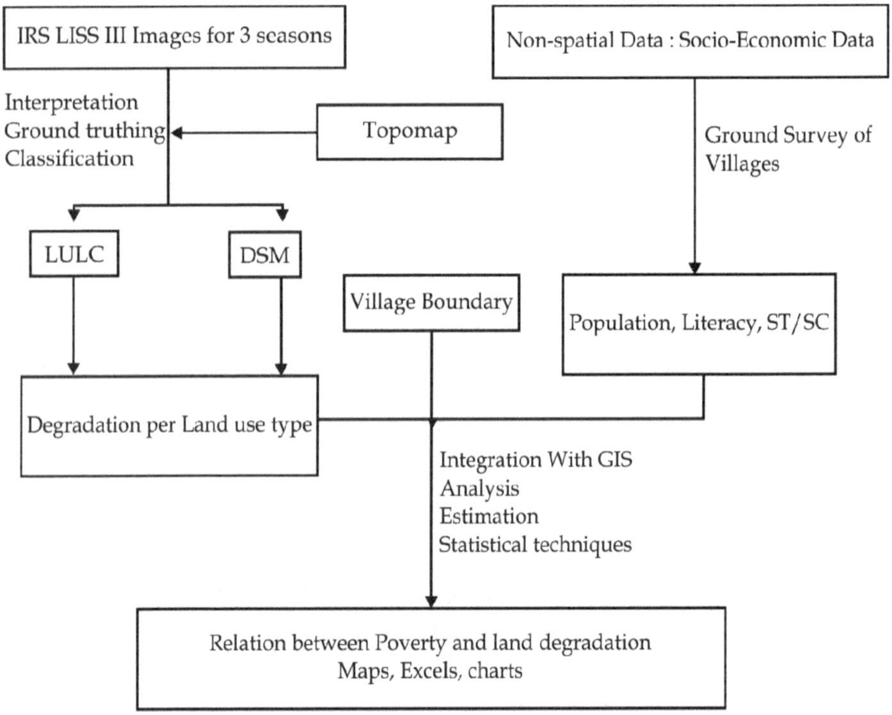


**Figure 2 :** Methodology

### Ground data collection and verification

Town and villages both were visited and supportive details were collected. Land use, agricultural pattern, education system and irrigation facilities were discussed with people of the field area. This collected information later then archived in softcopy information which also included all census data.

### Analysis

The desertification status map and landuse classified map was obtained to find the degradation per landuse in area of interest. Village wise area statistics was calculated. This was also carried out for degraded land.

Statistical Techniques: Chi square was carried out for finding goodness of fit between land degradation with other variables like poverty, literacy rates etc.

## Results

Degraded land area under different types and Poverty Density/ Rates with other few variables was calculated using coverage and database generated. Population Below Poverty Line in Pali Development Block has been 22,938 persons whereas total degraded land was 110.90 sq. km. Degradation percent of Pali development Block has been 8.39 percent. Gura Girdhari village has maximum number of BPL population whereas Khatikhera village was with maximum of land degradation per square kilometer. Maximum agricultural land degradation 34.30 per cent was found in Bhaton ki dhani.

**Table 1** : Goodness of Fit between land degradation and other socio-economic parameters

| Variables with Land Degradation | Calculated value | Result |
|---|---|---|
| Degradation & Illiterate population | 34.42 | Significant at 5% level of significance |
| Degradation & literate population | 28.89 | Significant at 5% level of significance |
| Degradation & % SC population | 20.36 | Non-Significant |
| Degradation & SC Female population | 32.04 | Significant at .02% level of significance |
| Degradation & SC Male population | 35.96 | Significant at .02% level of significance |
| Degradation & ST population | 21.66 | Non-Significant |
| Degradation & Main working population | 62.45 | Non-Significant |
| Degradation & Marginal working population | 16.33 | Significant at 5% level of significance |
| Crop area degradation & BPL | 34.89 | Significant at 5% level of /significance |
| Degradation & BPL | 8.47 | Non-significant |
| Degradation & % Working Population | 3.38 | Non-Significant |
| Degradation & % Working population male | 9.78 | Significant at 0.10% level of significance |
| Degradation & % Working population female | 7.5 | Non- significant |

The maps at 1:35,000 scale has also been generated to judge the scenario and relation between poverty and other factors. (Fig 3, 4, 5, 6 & 7)

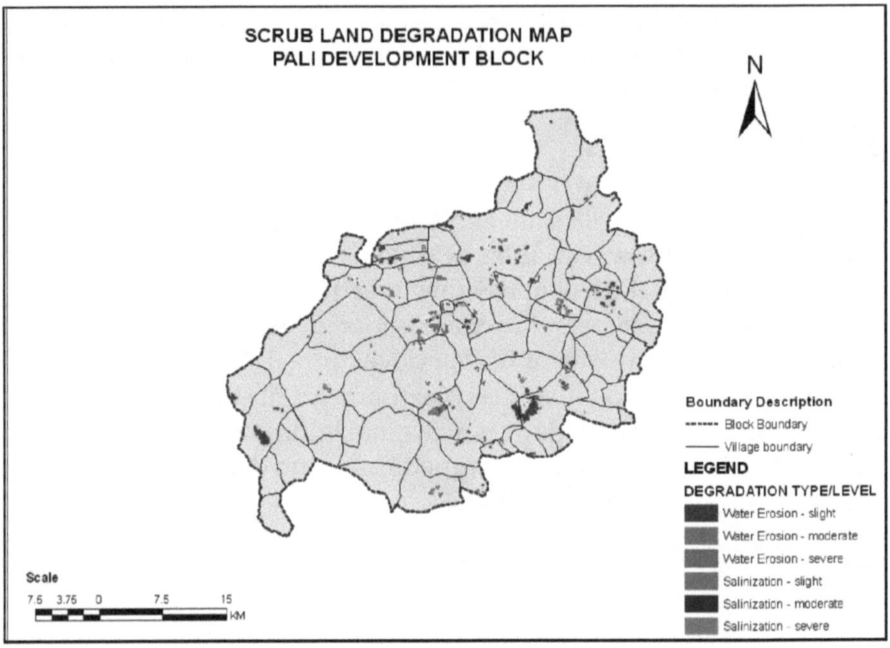


**Figure 3 :** Scrub Land Degradation Map Pali Development Block

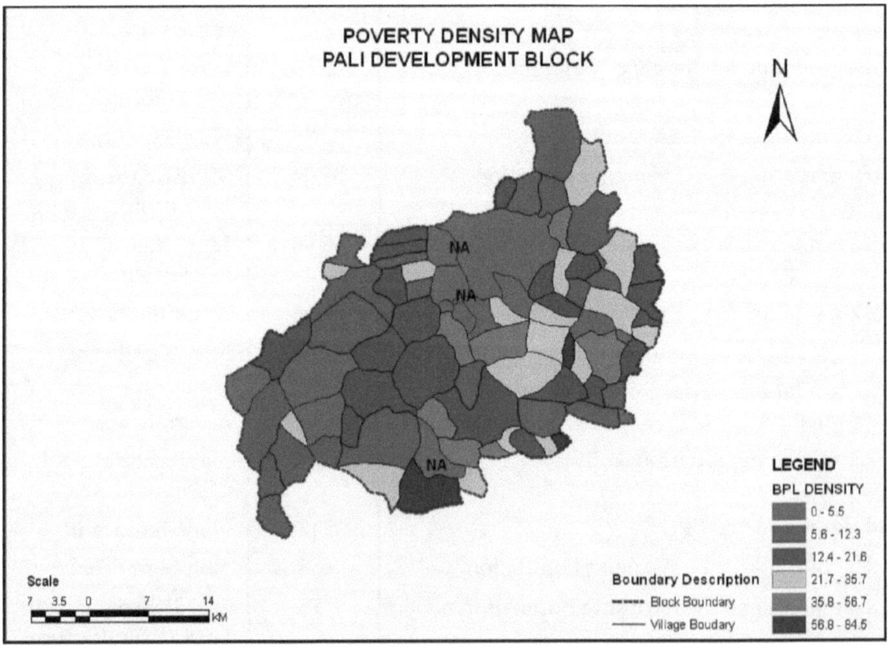


**Figure 4 :** Poverty Density Map Pali Development Block

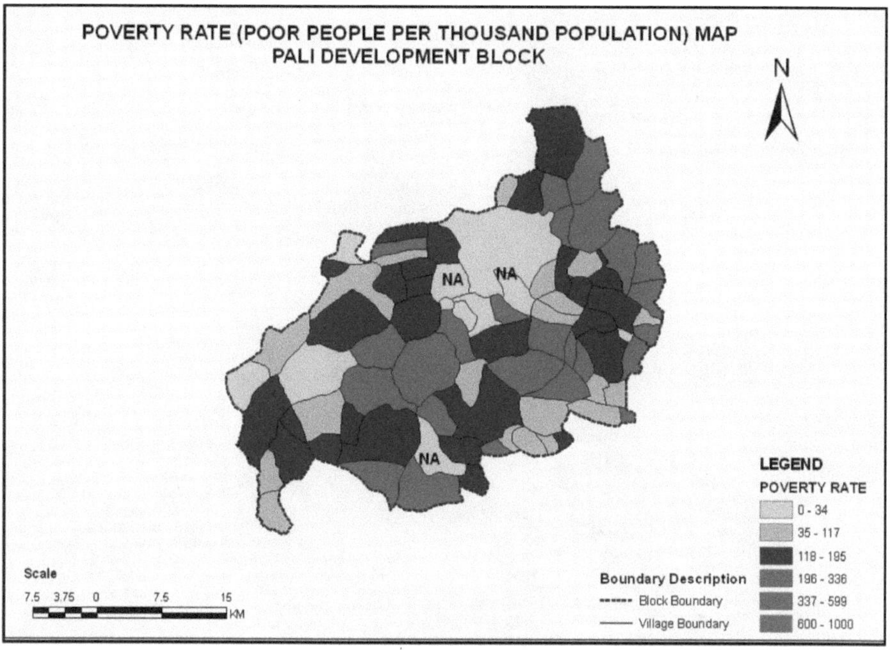


**Figure 5 :** Poverty Rate (Poor People Per Thousand Population) Map Pali Development Block

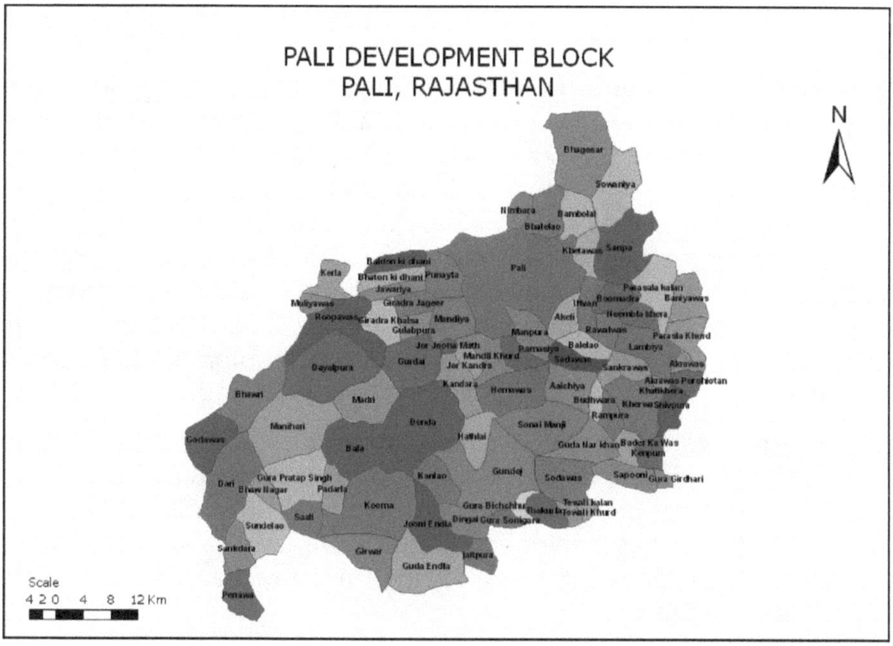


**Figure 6 :** Pali Development Block Pali, Rajasthan

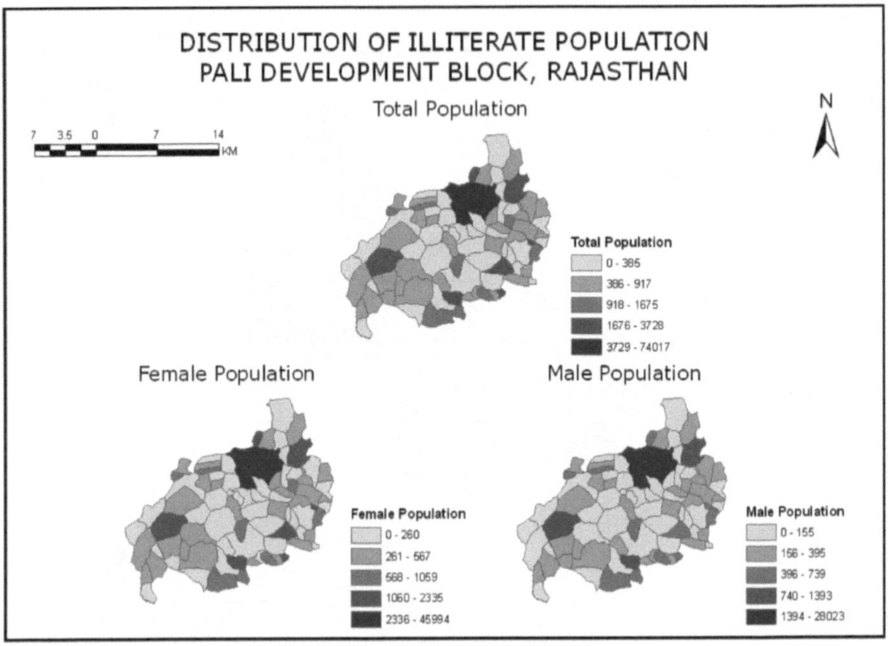


**Figure 7 :** Distribution of Illiterate Population Pali Development Block, Rajasthan

## Discussion

The casual relationship between poverty and land degradation is not simple. Natural resources are important for livelihood of the poor. The result presented thus far shows that degradation especially in crop lands has some linkages with BPL population is not just by chance. Illiteracy has also some impact on degradation as this factor determines the agriculture practices used by the people. The other important finding is that the schedule class has also significant impact on land degradation. The ignorance towards this class by wealthier class prevents them to use better agriculture practices and to access common pool of natural resources.

More vegetation cover is used for fuel wood where women population is high. Poor have less access to fertilizers and other chemicals. For not being able to earn livelihood through other ways than agriculture they generally do not leave land fallow. Non-poor population also does intensive agricultural practices, which cause degradation of land.

## References

Adhikari, B. 2005. Poverty, Property Rights, and Collective Action: Understanding the Distributive Aspects of Common Property Resource Management. *Environment and Development Economics* 10(1): 7-31

Cavendish, W. 2000. Empirical Regularities in the Poverty-Environment Relationship of Rural Households: Evidence from Zimbabwe. *World Development* 28 (11): 1979-2003

Ellis, F. 2000. Rural Livelihoods and Diversity in Developing Countries. New York: Oxford University Press

Jodha, N. (1986). Poverty Debate in India: A Minority View. Economic and Political Weekly, November 1988, 2421-28.

Jodha, N. S. 1986. Common Property Resources and Rural Poor in Dry Regions of India. *Economic and Political Weekly* 21:1169-81

Lu, D. Moran.E.2007. Mapping and monitoring land degradation risks in the western Brazilian amazon using multitemporal Landsat TM/ETM images. Land degradation and development.Wiley InterScience.

Moseley, W. G. 2001. African evidence on the relation of poverty, time preference and the environment. *Ecological Economics* 38: 317-326.

Nachtergaele, Freddy O.1994. Land Degradation Assessment Indicators and the LADA project. Land and Water development Division, FAO. Scape.

Nkonya, E. et al. 2000. "Linkage between Land management, land Degradation and poverty in sub Saharan Africa: A case study of Uganda", IFPRI report 159

Pender, J. et al. 2004. Linkages between poverty and land management in rural Uganda: Evidence from the Uganda National Household Survey, 1999/2000.Environment and Production Technology Division Discussion Paper 122. Washington, D.C.:International Food Policy Research Institute.

Reardon, T. 1995. Links between rural poverty and the environment in developing countries: Asset categories and investment poverty. *World Development* 23 (9): 1495-1506.

Reddy, S. R. C. et al.1999. Forest dependence and income distribution in a subsistence economy: Evidence from India. *World Development* 27 (7): 1141-1149.

# Chapter - 14

# Study of Spatial Distribution of Groundwater Quality with Geostatistics in Western Rajasthan

**Mrs. Trapti Sharma & S. Palria**

## Abstract

*Groundwater is one of the major sources of exploitation in arid and semi -arid regions of Rajasthan. To protect groundwater, its data on spatial distribution are important along with quantity. Analysis of quality of groundwater is also very important, because the physical and chemical characteristics of groundwater determine its suitability for agricultural, industrial and domestic usages. The importance ground water increases many fold for arid tracts like Western Rajasthan. Here quality of water is also matter of serious concern. Geostatistics methods are most advanced techniques for interpolation of Groundwater quality. The present study appraises the groundwater quality of western Rajasthan for agricultural purpose with GIS technology. Kriging Raster Interpolation technique of spatial analyst module in Arc GIS software has been used to generate the spatial distribution map of ground water for agricultural purpose. In western Rajasthan only 28% (58109 km$^2$) of ground water falls under the category of good and is suitable for agricultural use.*

***Keywords:*** *Spatial distribution, groundwater quality, Kriging, raster, interpolation, Geostatistics, Geographical Information System (GIS).*

## Introduction

Water is prime resource for fulfilling diverse needs of man and it is most important ecological component, without which no ecological system can survive. The diverse uses of water include irrigation, domestic and human consumption, agriculture and industrialization. With the rapid increasing population and simultaneously decrease of surface water resources, the importance of ground water increases. In

Rajasthan, owing to meager surface water resources, ground water is only source of dependence and survival. Over 71% irrigation and 90% drinking water supply schemes are based on ground water resources. Varied hydrogeology and vast areas with typical groundwater quality problems characterize Rajasthan state. If we compare quality of water of Rajasthan with India, Rajasthan state accounts for 51% of the fluoride and 42% of the saline affected areas in the entire country. Due to arid climate and complex geology, the ground water is characterized by multiple quality problems (Ozha and Golani, 2005).

The importance ground water increases many fold for arid tracts like Western Rajasthan. Here quality of water is also matter of serious concern. In general, groundwater quality in western Rajasthan showed a considerable deterioration possibly due to over-exploitation besides natural factors. Geological regime is the main natural factor contributing in chemical contamination of water. In this region ground water is very deep and saline. The major contaminants that govern the water quality in western Rajasthan are salinity, sodicity and alkalinity for waters to be used for agriculture (Ozha and Sharma, 2011).

The analysis based on Geographical Information System has indicated that the primary sources for the various chemical constituents in the ground water are the rock forming minerals in the different geological formations of Western Rajasthan. Many studies have proved the capabilities of GIS in generating the hydro-chemical pattern maps and their analysis. To study groundwater quality, data on spatial distribution are very important. Geo-statistical methods are one of the most advanced techniques for the interpolation of Groundwater quality. The accuracy of interpolation methods for predicting water quality has been proven and analyzed in several studies. Geo statistical techniques use both, mathematical and statistical methods to predict values at unknown locations and to provide probabilistic estimates of the quality of the interpolation based on the spatial autocorrelation among data points (Burrough and McDonnell 1998, Johnston et al. 2001). These techniques quantify the spatial autocorrelation among measured points and account for the spatial configuration of the sample points around the prediction location. Formation of spatial distribution/ thematic maps using interpolation technique in GIS gives good results and information for the ground water quality mapping (Shankar., 2010).

Kriging is a type of Geostatistical technique which is proven to be the best interpolation technique for the spatial mapping of Ground water quality. Ground water quality can also be interpolated by Inverse Distance Weighted or by Co-Kriging methods but they resulted in higher errors as compared to Kriging. In this study the construction of the groundwater quality map was carried out using Kriging. Uyan, M. et. al., 2010 applied Geostatistical technique to predict the spatial distribution and groundwater nitrate concentration. They found the spherical model is the best model representing the spatial variability of groundwater nitrate concentrations. In this paper Kriging method is used for predicting spatial distribution of Groundwater characteristics for agriculture purpose. Goyal and Chaudhary, 2010 has analyzed the GIS based Spatio-temporal changes in groundwater depth and irrigation water quality on the basis of EC in Kaithal district of Haryana state.

## Study Area

The present study area Western Rajasthan is the region lying on the west of Aravalli hill range ($24^{\circ} 37'$ & $30^{\circ} 11'$ north latitude and $69^{\circ} 29'$ & $75^{\circ} 15'$ east longitudes) and covers twelve districts of Rajasthan viz. Barmer, Bikaner, Churu, Ganganagar, Hanumangarh, Jaisalmer, Jhunjhunun, Jodhpur, Jalore, Nagaur, Pali and Sikar. Major part of it comes under the Thar Desert (Fig 1).

**Fig 1:** Location map showing study area - Western Rajasthan

Climate of Western Rajasthan is arid and is characterized by sporadic rainfall with erratic distribution patterns, extremes of temperatures, formidable water quality problems, low humidity and high wind velocity. Summer begins in the month of March and the temperature starts rising progressively through April, May and June. In the Western Rajasthan, mainly at Bikaner, Phalodi, Jaisalmer and Barmer, the maximum daily temperature varies between 40°C to 45°C. The winter season is from December to February with marked decline in minimum temperatures in December and January. The minimum temperatures may fall to below 0° C.

## Rainfall

There is a very rapid and marked decrease in rainfall west of the Aravalli range, making Western Rajasthan the most arid part. The average annual rainfall in this part ranges from less than 10 cm in northwestern part of Jaisalmer (lowest in the state), to 20 to 30 cm in Ganganagar, Bikaner and Barmer districts, 30 to 40 cm in Nagaur, Jodhpur, Churu and Jalore districts and more than 40 cm in Sikar, Jhunjhunun and Pali districts and along the western fringes of the Aravalli range. The number of rainy days during monsoon period varies widely in different places, with a lowest of 10 in Jaisalmer district.

## Agriculture

Western Rajasthan has basically an agrarian economy. Most of its population lives in small villages and Dhanies. Major part of the Western Rajasthan is however covered by arid and semi-arid climatic conditions, which is characterized by low, erratic and uneven distribution of rainfall associated with lack of other water-resources for irrigated farming. Growing of only one rain fed crop in kharif season that too associated with high risk has led to dependence of village community on livestock.

Agriculture in the Western Rajasthan is the primary sector and mainstay of over 80 % of population contributing agriculture activity. Main crops are Maize, Guar, Cotton, Groundnut, Paddy, Kharif Pulses, Wheat, Barley, Gram, Rapeseed and Mustard. The gross cultivated area varies from one region to another region due to variation in land capability and irrigation facilities.

## Irrigation

The irrigated area is unevenly spread all over the region. While some district are more fortunate to have a proportionately large area under assured irrigation, other have very little of it and some have almost negligible. Ganganagar district has maximum area under irrigated farming (22.74%). The maximum canal irrigation is in Ganganagar district followed by Bikaner. Churu is the district having the least canal-irrigated area. Most parts of Pali & Jalore are more dependent on Ground Water irrigation. Western Rajasthan also suffers with problems like water-logging, soil salinity etc due to canal irrigation (Table 1).

**Table 1:** Different Sources of Irrigation (1995-1996) in Western Rajasthan Districts

| District | Percent of area irrigated area by Different Sources | | | |
|---|---|---|---|---|
| | Tube Well | Well | Tanks | Canal |
| Barmer | - | 99 | 1 | - |
| Bikaner | 1 | 3 | - | 96 |
| Churu | 6 | 94 | - | - |
| Sri Ganganagar | 1 | - | - | 99 |
| Hanumangarh | 1 | - | - | 99 |
| Jaisalmer | 4 | 8 | - | 88 |
| Jalore | 1 | 95 | - | 4 |
| Jhunjhunu | - | 99 | 1 | - |
| Jodhpur | 59 | 40 | - | 1 |
| Nagaur | 32 | 67 | 1 | - |
| Pali | 5 | 75 | 20 | - |
| Sikar | 1 | 99 | - | - |

## Materials and Methodology Used

### (a) Ground water quality data used

Data is acquired from the Ground water Department of 12 districts of the year 2007. Data for 1119 wells have been considered for present study. This data is provided with Location code (L code) with their latitude- longitude and values of water quality parameters.

### (b) Methodology

For the analysis of ground water quality Geographic Information System (GIS) is used. A GIS software package Arc GIS 9.3 is used for the Kriging estimations. Kriging is used to obtain the spatial distribution of groundwater quality parameters over the area. Spatial distribution maps or thematic maps for parameters were generated by carrying out point interpolations using Kriging. Thematic maps pertaining to EC, Na%, RSC and SAR were interpolated from the well data. Flow chart of GIS Methodology used in study is shown in Fig 2.

**Generation of Suitability Map for Irrigation -** Irrigation water quality refers to suitability of water for its agricultural use. It depends on the concentration and composition of dissolved constituents in water. Good quality water has the potential to cause maximum yield under good soil and water management practices. If total salt content increases in irrigational water, various cropping problems developed and due to this special management practices may need to apply for acceptable crop yields. The most important characteristics for quality of water used for irrigation are: a) Salinity (EC), b) Sodium percent (Na%), c) Residual Sodium Carbonate (RSC) and d) Sodium Adsorption Ratio (SAR).

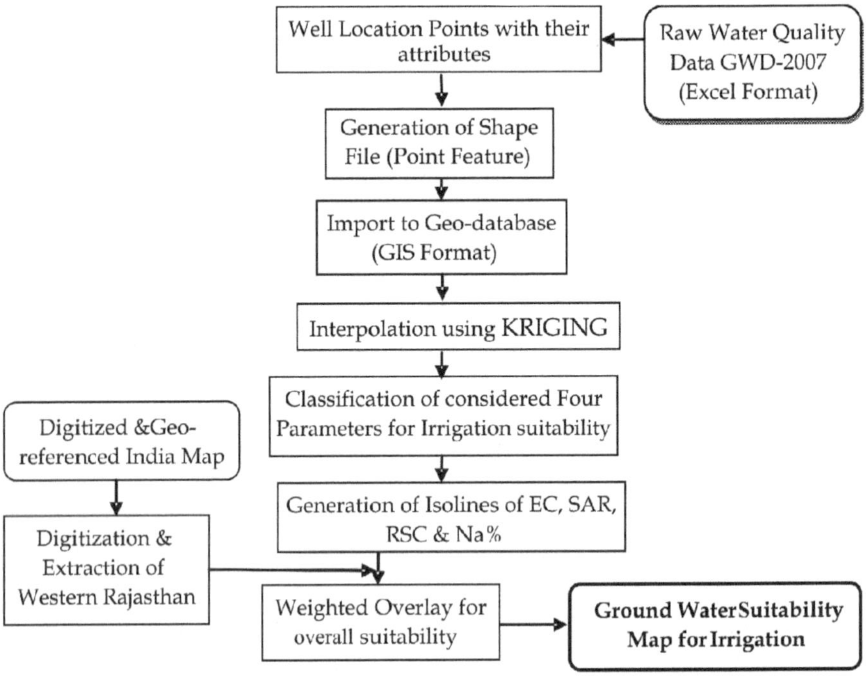


**Fig 2 :** Flow chart showing the methodology adapted for the generation of Maps

a) **Salinity Hazard** - This term means the salinity or total dissolved salts in water use for irrigation. It is measured in Electrical Conductivity (EC). EC measures salinity from all the ions dissolved in a water sample. The primary effect of high EC water on crop productivity is the inability of the plant to compete with ions in the soil solution for water (physiological drought). The higher the EC, the less water is available to plants, even though the soil may appear wet. Because plants can only absorb "pure" water, usable plant water in the soil solution decreases dramatically as EC increases. The amount of water absorbed through a crop is directly related to yield; therefore, irrigation water with high EC reduces yield potential.

b) **Sodium Percent -** The sodium in irrigation waters is usually denoted as per cent sodium.

$$\%Na = \frac{(Na^{+})\times 100}{(Na^{+1} + K^{+1} + Ca^{+2} + Mg^{+2})}$$

Where, the quantities of $Ca^{2+}$, $Mg^{2+}$, $Na^{+}$ and $K^{+}$ are expressed in meq/l.

c) **Residual Sodium Carbonate (RSC)** - Residual Sodium Carbonate (RSC) is criteria to evaluate the effects of carbonates and bicarbonates in irrigational water.

$$RSC = (CO_3^{-2} + HCO_3^{-1}) - (Ca^{+2} + Mg^{+2})$$

Its unit is meq/l. High carbonates cause calcium and magnesium ions to form insoluble minerals leaving sodium as the dominant ion in solution. Then these alkaline conditions in water could intensify sodic soil problems.

d) **Sodium Hazard or Sodicity** - Excess of sodium in water used for irrigation is known as sodium hazard and assessed using Sodium Adsorption Ratio (SAR).

$$SAR = \frac{[Na^{+1}]}{\sqrt{\frac{[Ca^{2+}] + [Mg^{2+}]}{2}}}$$

Where, $Na^{+1}$, $Ca^{+2}$ and $Mg^{+2}$ are expressed in meq/l. Calcium flocculate (hold together), while sodium disperses (pushes apart) soil particles. If calcium and magnesium are the predominant cations adsorbed on the soil exchange complex, the soil tends to be easily cultivated and has a permeable & granular structure. Dispersed soil will readily crust and have water infiltration and permeability problems, this result into reduction in crop production.

Guidelines {India} are available to evaluate quality of water for irrigation. For irrigation, water can be classified in five classes depending upon its chemical properties Table 2.

**Table 2:** Guidelines for Evaluation of Irrigation Water Quality Based on Na%, EC, SAR and RSC

| Water class | Sodium (Na) % | Electrical | SAR | RSC meq/l |
|---|---|---|---|---|
| Excellent | < 20 | < 250 | < 10 | < 1.25 |
| Good | 20 - 40 | 250 - 750 | 10 - 18 | 1.25 - 2.0 |
| Medium | 40 - 60 | 750 - 2,250 | 18 - 26 | 2.0 - 2.5 |
| Bad | 60 - 80 | 2,250 - 4,000 | > 26 | 2.5 - 3.0 |
| Very bad | > 80 | > 4,000 | > 26 | > 3.0 |

[Source-CGWB-2000]

## Result

The construction of the Integrated groundwater quality map was carried out through the overlapping of the thematic maps of (Na%, EC, SAR and RSC), which are produced as a result of Kriging interpolation (Fig 3). The spatial integration for final irrigation water quality mapping was carried out using Arc GIS Spatial Analyst extension and Weighted Overlay Tool. The reclassification was performed based on values pertaining to four parameters. This map shows that water with excellent and very bad quality is in very- very small region of western Rajasthan. Both excellent and very bad class is almost 0 % (fig 4, 5 and Table 3). However, most of the region is covered with good to medium quality of ground water. The map statistics shows that maximum area of western Rajasthan is under the category of good to medium (188171 Km$^2$) quality of water for irrigation purpose. Almost complete Bikaner, Shri Ganganager and Hanumangarh districts are having water of good quality, this fact is coincides with heavy agricultural activities in these areas.

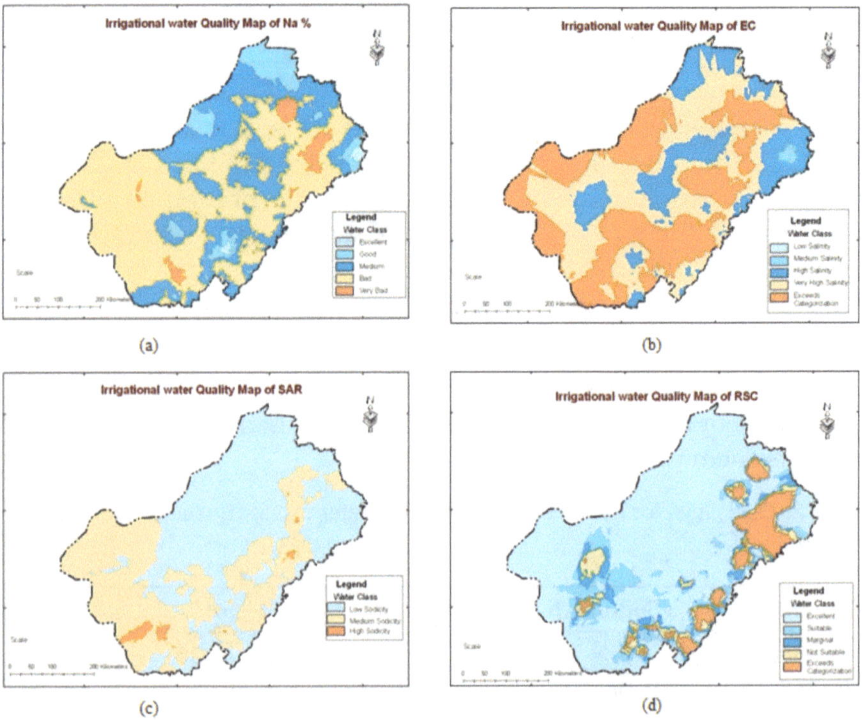


(a) (b) (c) (d)

**Figure 3 :** Ground water quality Maps for Irrigation purposes (Na%, EC, SAR and RSC [a, b, c, d])

**Table 3:** Area covered according to Categories of Integrated Irrigation Water Quality

| Water class | Area Km$^2$ | Percentage |
|---|---|---|
| Excellent | 26 | 0 |
| Good | 58109 | 28 |
| Medium | 130062 | 62 |
| Bad | 20483 | 10 |
| Very Bad | 65 | 0 |

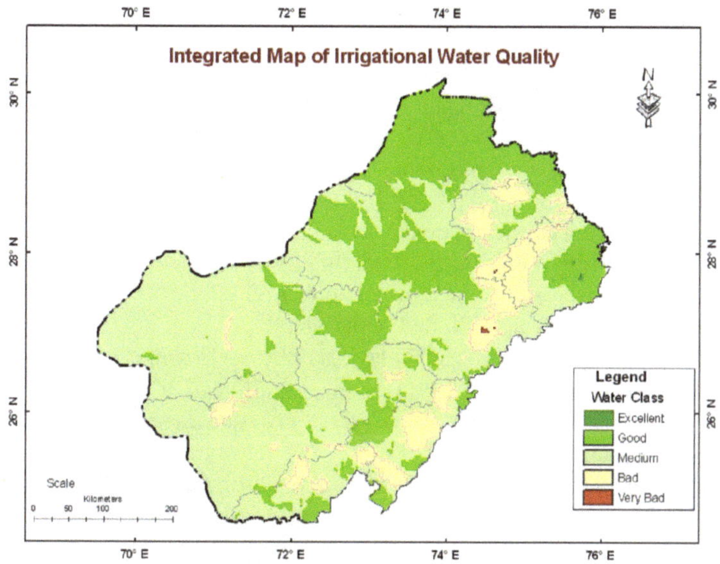


**Figure 4 :** Integrated Map of Irrigational Water Quality

**Figure 5 :** Graphical Representation of Integrated Ground Water Quality for Irrigation

## Conclusion

In western Rajasthan although ground water table is very deep and quality is very poor but here agriculture is the main occupation. People of this region adapted to such unfavorable conditions. They cultivate crops, which require less quantity of water or persist in poor quality of water. Ground water in study area is 28% (58109 km$^2$) falls under the category of good hence it is suitable for agricultural use. While remaining water is either medium to bad category which if used for agriculture, less crop production and many other problems can arise.

## References

Burrough, P. A. and McDonnell, R. A., 1998. Principles of geographical information systems. Oxford University Press, Oxford. 333pp.

Central Ground Water Department, 2000.

Goyal, S.K. and Chaudhary, B.S., 2010. GIS based study of Spatio-Temporal changes in groundwater depth and quality in Kaithal district of Haryana, *India J. Ind. Geophys. Union* ( April 2010 ) Vol.14, No.2, pp.75-87.

Johnston, K., Ver-Hoef, J. M., Krivoruchko, K. and Lucas, N., 2001. Using Arc GIS Geo-statistical Analyst. ESRI. Redlands, 300pp.

Ozha, D.D. and Golani F.M. (2005), *Journal of Env & Health*, 22 (3), 76.

Ozha, D.D. and Sharma, S.K., 2011. Mitigation of Water Quantity and Water Quality Challenges in Ground Water of Rajasthan. *Water Research & Development* Vol. 1(1) pp. 45-52; January-April, 2011.

Shankar, K., Aravindan, S. and Rajendran, S., 2010. GIS based Groundwater Quality Mapping in Paravanar River Sub-basin, Tamil Nadu, India. *International Journal* of *Geomatics and Geosciences*. Volume 1, No 3, 2010.

Uyan, M and Cay, T., 2010. Geo-statistical Methods for Mapping Groundwater Nitrate Concentrations. 3rd International Conference on Cartography and GIS 15-20 June, 2010, Nessebar, Bulgaria.

Chapter - 15

# Predicting Drought Using Time Series Analysis and Forecasting of Rainfall and Normalized Difference Vegetation Index (NDVI) Data

**Aditi Sharma, P. Dhayani and M.P. Punia**

## Abstract

*The definition of what drought is and what drought is not, has profound implications for the environment and all segments of society, yet it may be different for each. Many attempts have been made to develop a comprehensive and meaningful definition. A generic definition provides a starting point: Drought is a persistent and abnormal moisture deficiency having adverse impacts on vegetation, animals, and people. The public perceives "drought" as a serious departure from normal water conditions, a departure that requires a public response to reduce negative impacts. About two thirds of the geographic area of India receives low rainfall (less than 1000 mm), which is also characterized by uneven and erratic distributions. Out of net sown area of 140 million hectares about 68% is reported to be vulnerable to drought conditions and about 50% of such vulnerable area is classified as 'severe', where frequency of drought is almost regular. Rajasthan and Karnataka were chosen as the study area for this research work. Rainfall data from 1951-2006 was used to compute Standardized Precipitation Index (SPI) and NOAA-AVHRR NDVI for the period 1981-2006 was used to calculate Vegetation Condition Index (VCI). SPI was then used to find the drought characteristics like onset, severity, and spatial extent. Time series analysis and forecasting was done to predict drought. Rainfall and NDVI data from 1981-2005 was used for predicting the future values. Predicted values for the 2006 was compared with the actual values and 90% accuracy was achieved. From various models of time series analysis and forecasting, ARIMA model consistently gave better results. Hence, a model was developed for prediction of drought pattern from the year 2007 to 2012.*

***Keywords:*** *Drought, Data mining, Standardized Precipitation Index (SPI), Vegetation Condition Index (VCI), Time series analysis and forecasting, ARIMA model.*

## Introduction

Predicting drought depends on our ability to forecast rainfall. Forecasting is the process of estimation of unknown situation whereas prediction is a claim or statement that a particular event will occur in more certain times than a forecast.

## Time Series Analysis and Forecasting

Many methods and approaches for formulating forecasting models are available in the literature. This research exclusively deals with time series forecasting model, in particular, the Auto Regressive Integrated Moving Average (ARIMA). These models were described by Box and Jenkins (1976). The Box-Jenkins approach possesses many appealing features. It allows the manager who has only data on past years' quantities, rainfall as an example, to forecast future ones without having to search for other related time series data, for example temperature. Box- Jenkins approach also allows for the use of several time series, for example NDVI dataseries, to explain the behaviour of another series, for example rainfall, if these other time series data are correlated with a variable of interest and if there appears to be some cause for this correlation.

## Time Series Analysis

Time series analysis is often associated with the discovery and use of patterns (such as periodicity, seasonality, or cycles) and prediction of future values (specifically termed forecasting in time series context). Therefore, one may wonder what is the difference between traditional time series analysis and data mining on time series. One key difference is the large number of series involved in time series data mining. As shown in Box and Jenkins (1970, 1976) and vast volume of time series literature, traditional time series analysis and modelling tend to be based on non-automatic and trial-and-error approaches. When a large number of time series are involved, development of time series models using a non-automatic approach becomes impractical. An automated method of outlier detection in time series is an essential component of time series data mining. The adaptation of data mining on time series promises to assist the drought study in several ways.

Data mining

- Provides a method to process a large amounts of data in an automated fashion

- Provides a method of distilling vast amount of data into information that is useful in drought planning and
- Offers a consistent, reliable and quite accurate method for forecasting rainfall data over temporal sets (weekly, monthly, yearly etc.)

## General Approach to Time Series Modelling

- Plot the series & examine main features of graph, checking in particular whether there is
  - ❖ Trend
  - ❖ Seasonal component
  - ❖ Any apparent sharp changes in behaviour
  - ❖ Any outlying observations
- Remove the trend and seasonal component to get stationary residuals
  - ❖ To achieve this goal, we sometimes apply preliminary transformations to the data.
  - ❖ By estimating the components and subtracting from the data
  - ❖ By differencing the data i.e. replacing the original series
- Whichever transformation method is used, aim is to produce a stationary series, whose value we shall refer as residuals. Example for transformation: if the magnitude of fluctuations grows linearly with the level of the series, then the transformed series will have fluctuations of more constant magnitude. If some of the data are negative, add a positive constant to each data values to ensure that all values are positive before taking logarithms.
- Choose a model and fit the residuals
- Forecasting will be achieved by forecasting the residuals and then inverting the transformations described before to arrive at forecasts of the original series.

## Study Area

Rajasthan and Karnataka states were chosen as the study area for this research work. Rajasthan is the most drought prone state in North West India while Karnataka in South India suffers from occasional droughts and has second largest drought affected area in the country. Figure 1 and 2 shows the study area.

**Figure 1 :** India map highlighting Rajasthan and Karnataka

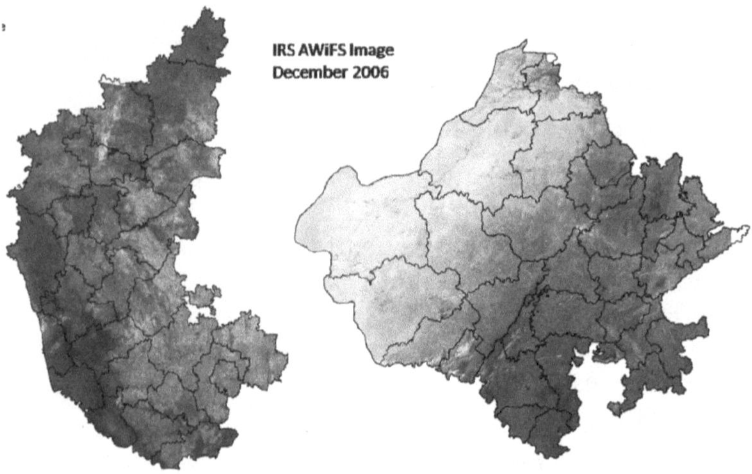


**Figure 2 :** IRS AWiFS Image (December 2006) of Karnataka & Rajasthan

## Methods

## Box-Jenkins Model

In time series analysis, the Box-Jenkins methodology, named after the statisticians George Box and Gwilym Jenkins, applies Autoregressive Integrated Moving Average (ARIMA) model to find the best fit of a time series to past values, in order to make forecasts.

The main stages in setting up a forecasting ARIMA model includes model identification, model parameters estimation and diagnostic checking for the identified model appropriateness for modelling and forecasting.

## 1. Box-Jenkins model identification

The data is examined to check for the most appropriate class of ARIMA processes through selecting the order of the consecutive and seasonal differencing required to make series stationary, as well as specifying the order of the regular and seasonal auto regressive and moving average polynomials necessary to adequately represent the time series model. The Autocorrelation Function (ACF) and the Partial Autocorrelation Function (PACF) are the most important elements of time series analysis and forecasting. The ACF measures the amount of linear dependence between observations in a time series that are separated by a lag k. The PACF plot helps to determine how many auto regressive terms are necessary to reveal one or more of the following characteristics: time lags where high correlations appear, seasonality of the series, trend either in the mean level or in the variance of the series. (Brockwell and Davis, 2002).

## 2. Model parameter estimation

The Box and Jenkins model includes autoregressive and moving average parameters as well as differencing in the formulation of the model. The three types of parameters in the model are: the autoregressive parameters (p), the number of differencing passes (d) and moving average parameters (q). Box-Jenkins model are summarized as ARIMA (p, d, q). For example, a model described as ARIMA (1, 1, 1) means that this contains 1 autoregressive (p) parameter and 1 moving average (q) parameter for the time series data after it was differenced once to attain stationarity.

### Identify p and q

Once stationarity and seasonality have been addressed, the next step is to identify the order (i.e., the p and q) of the autoregressive and moving average terms.

### Order of Autoregressive Process (p)

Specifically, for an AR (1) process, the autocorrelation function should have an exponentially decreasing appearance. However, higher-order AR processes are often a mixture of exponentially decreasing and damped sinusoidal components.

## Order of Moving-Average Process (q)

The autocorrelation function of an MA series cuts of sharply whereas as for AR series the autocorrelation function exhibits an exponential decay. The autocorrelation function of a MA (q) process becomes zero at lag q + 1 and greater, therefore we examine the autocorrelation function to see where it becomes zero. We do this by placing the 95% confidence interval for the autocorrelation function on the autocorrelation plot.

In addition to the non-seasonal ARIMA (p, d, q) model, introduced above, we could identify seasonal ARIMA (P, D, Q) parameters for our data. These parameters are: seasonal autoregressive (P), seasonal Differencing (D) and seasonal moving average (Q).

The general form of the above model describing the current value $X_t$ of a time series by its own past is:

$$(1-\phi_1 B)(1-\alpha_1 B^5)(1-B)(1-B^5)_{xt} = (1-\theta_1 B)(1-\gamma_1 B^5)e_t$$

Where:

| | | |
|---|---|---|
| $1-\phi_1 B$ | = | Non seasonal autoregressive of order 1 |
| $1-\alpha_1 B^5$ | = | Seasonal autoregressive of order 1 |
| $x_t$ | = | The current value of the time series examined |
| B | = | The backward shift operator $Bx_t = x_{t-1}$ and $B^5x_t = x_{t-5}$ |
| 1-B | = | 1st order nonseasonal difference |
| $1-B^5$ | = | Seasonal difference of order 1 |
| $1-\theta_1 B$ | = | Non seasonal moving average of order 1 |
| $1-\gamma_1 B^5$ | = | Seasonal moving average of order 1 |

After choosing the most appropriate model, the model parameters are estimated by using the least square method. In this step, values of the parameters are chosen to make the Sum of the Squared Residuals (SSR) between the real data and the estimated values as small as possible. In general, nonlinear estimation method is used to estimate the above identified parameters to maximize the likelihood (probability) of the observed series given the parameter values. The methodology uses the following criteria in parameter estimation:

- The estimation procedure stops when the change in all parameters estimate between iterations reaches a minimal change of 0.001
- The parameters estimation procedure stops when the SSR between iterations reaches a minimal change of 0.0001 (Nail and Momani, 2009)

## 3. Diagnostic checking for model appropriateness and forecasting

In diagnose checking step, the residuals from the fitted model shall be examined against adequacy. This is usually done by correlation analysis through the residual ACF plots and the goodness-of-fit test by means of Chi-square statistics $x^2$. If the residuals are correlated, then the model should be refined as in step one above. Otherwise, the autocorrelations are white noise and the model is adequate to represent our time series. Hence, the model developed incorporates the basic statiscal properties of the time series data into its parameters.

## Results and Discussion

An attempt has been made in the study to predict drought using two parameters rainfall and NDVI data. Both the datasets from the year 1981 to 2006 for both the states Karnataka and Rajasthan have been used for forecasting of rainfall and NDVI data. Both the data series from year 1981 to 2006 is fed as input to the model and forecasting of next 6 years is done i.e. rainfall and NDVI values from the year 2007 to 2012 is forecasted.

Detailed description of the process is as follows: We have used rainfall data for 4 months (June, July, August & September) from 1981 to 2006. There are 104 data points (4 months * 26 years). Out of these, we use 25 years data (100 data points) for building the model and will use last 4 data points for validating the prediction.

R statistical package has been used for time series analysis. R provides a wide variety of statistical (linear and nonlinear modelling, classical statistical tests, time-series analysis, classification, clustering) and graphical techniques, and is highly extensible.

**Step 1:** The data is first read in R using the command

read.table()

**Step 2:** Data is then plotted to see the distribution for each month separately for 25 years (plot for all districts/taluks one by one)

Plot()

After plotting of data, load the t-series library which is required for further processing. Then after load two packages i.e. quadprog package and zoo package in R.

**Step 3:** Check whether there exists any pattern in data. Here two assumptions, we use are:

Null hypothesis ($H_0$) - pattern exists in data

Alternate hypothesis ($H_1$) - there is no pattern in data.

**Step 4 :** Now, we perform standard statistical test for unit root in all four months. Unit root test means to check that whether there is any pattern in data or not. If $H_0$ proves as true, then it indicates that data is non-stationary i.e. pattern exist between the data. For removing the pattern we use differentiation .i.e. we difference the series until we get stationary residuals.

We use statistical tests to see whether series appears to be free of trend (stationary)

- Augmented Dickey-Fuller test (adf.test)
- Phillips Perron test (pp.test)

  p value should be very small.

**Step 5 :** If above test results are successful in giving evidence against Null hypothesis, then plot the

- Periodogram of series
- acf (auto correlation function)
- pacf (partial auto correlation function)

acf and pacf plot tell the time lag for two subsequent trends (reoccurrence of pattern).

We shall now develop model with different combinations for the AR and MA components for the series obtained by taking the usual difference of the seasonal difference and look for the model with minimum AIC (Acaci Information Criteria).

**Step 6 :** After getting the stationary residuals develop a model. Various models used for prediction are

- ARIMA model - essence of ARIMA is to apply ARMA model not immediately to the time series at hand, but after its preliminary transformation.
- Linear model - this model specifies linear relationship between the dependent and independent variable.
- Regression model - used for finding the best fit equation between the dependent and independent variables.

The first model is the Random walk model. Random walk is the special case of ARIMA model where all components (AR, I, MA) are 0.

Different combinations of the AR and MA components with first order integration is tried.

AR component(p) - Time series is generated by passing white noise through recursive filter. In recursive filter, output at the moment is function of output values at previous moment and input values.

MA component(q) – It represents the time series generated by passing white noise through non recursive filter. Non recursive filter is output at moment is function of only input values.

```
> log(var(y))*100+2
Rainfall
Rainfall 820.3375
```

We shall now fit the ARIMA model with the data with different combinations of the AR, and MA components with the first order integration.

```
>arima0(rainfall,order=c(0,1,0),seasonal=list(order=c(1,1,0),period=4))$aic[1] 1032.768
> arima0(rainfall,order=c(0,1,0),seasonal=list(order=c(0,1,1),period=4))$aic [1] 1031.299
> arima0(rainfall,order=c(0,1,1),seasonal=list(order=c(1,1,1),period=4))$aic [1] 976.4774
```

............and so on. After trying all the different combinations, we then look for the minimum AIC among all models.

**Step7 :** Selection of model is done on basis of

a. Acaci Information Criteria (aic criteria) – it determines how well the data supports each model, taking into account both goodness of fit (sum of squares) and number of parameters in model. We look for model with minimum aic value.

b. Standard normality checks – performed to check whether residual is normally distributed.

- Shapiro-Wilk normality test
- Anderson-Darling normality test
- Cramer-von Mises normality test
- Lilliefors (Kolmogorov-Smirnov) normality test
- Shapiro-Francia normality test

c. Independence check – to look for uncorrelated residuals. Ljung Box test is done to test if they are uncorrelated.

A model which passes all the abovementioned tests is selected. Finally, the values are predicted.

**Step8 :** Final step is to find out how accurate the model is in predicting the values. Accuracy is calculated by the following formula.

A = (100 – (mod(Actual value – Predicted value) / (Actual value))) * 100

where A is the accuracy in percentage

**Table1** : Accuracy achieved for different models implemented

| Model | Predicted values | Actual values | Month | Accuracy |
|---|---|---|---|---|
| Model 1 | 65.98 | 180.4 | June | 45.13% |
| | 196.85 | 430.9 | July | |
| | 173.48 | 157.7 | August | |
| | 130.67 | 264.4 | September | |
| Model 2 | 146.85 | 180.4 | June | 89.86% |
| | 393.76 | 430.9 | July | |
| | 190.55 | 157.7 | August | |
| | 197.49 | 264.4 | September | |
| Model 3 | 120.63 | 180.4 | June | 90.10% |
| | 390.65 | 430.9 | July | |
| | 161.54 | 157.7 | August | |
| | 259.33 | 264.4 | September | |
| Model 4 | 56.36 | 180.4 | June | 42.43% |
| | 113.76 | 430.9 | July | |
| | 170.90 | 157.7 | August | |
| | 97.49 | 264.4 | September | |
| Model5 | 153.85 | 180.4 | June | 45.37% |
| | 104.72 | 430.9 | July | |
| | 96.00 | 157.7 | August | |
| | 114.37 | 264.4 | September | |

Model 3 predicts the values with highest accuracy among all models. Similarly, rainfall and NDVI values are predicted for all stations of Rajasthan and Karnataka using model 3.

As distribution of data was similar for Rajasthan and Karnataka, same model was used to predict the values for Karnataka. ARIMA Model with parameters (2,1,2) and (2,1,1) consistently gave better results for Rajasthan and Karnataka.

It was observed that when the model was used to predict values for two simultaneous years, then accuracy of prediction for first year was 90% but for second year accuracy reduced to 73%. Hence, a feedback loop was generated, so that year by year values can be predicted and given as input to the model. It means that initially rainfall data from 1981 to 2005 was given as input to the model and rainfall value for the year 2006 is predicted. Then after, data from 1981 to 2006 is fed as input to the model, and values for the year 2007 were predicted.

After getting the predicted values of NDVI and rainfall for the years 2007 to 2012, association rule is applied on predicted values. Same set of association rules were used which was used earlier to predict drought pattern. Hence, drought pattern is extracted from the year 2007 to 2012.

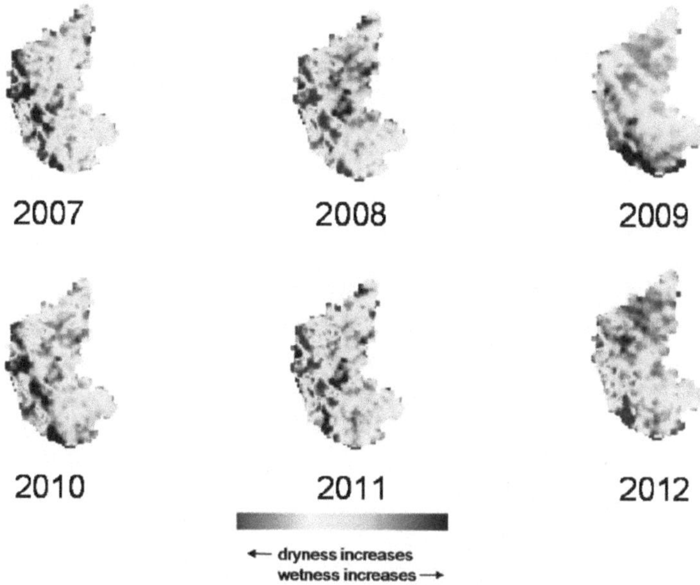


**Figure 3 :** Drought predicted for Karnataka upto year 2012

Drought pattern was predicted from the year 2007 to 2012 for Karnataka with 82.5% accuracy. (Fig 3).

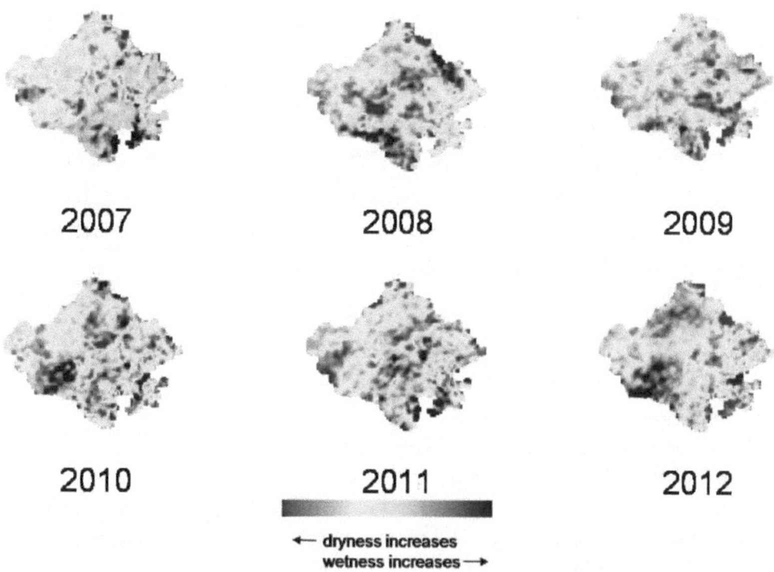


**Figure 4 :** Drought predicted for Rajasthan upto year 2012

Drought pattern was predicted from the year 2007 to 2012 as shown in figure 4. For Rajasthan, accuracy of prediction achieved was 85%.

The results obtained for the years 2007, 2008 and 2009 for Rajasthan and Karnataka states have been compared with government drought reports, rainfall data and NADAMS, project reports of NRSC. I find 90% accuracy for the years 2008 and 2009 and 75% for the year 2007 for the state of Rajasthan. Similarly, 90% accuracy for the year 2007 and 2008 and 80% for the year 2009 for the state of Karnataka.

## Conclusion

An attempt has been made in this research study to predict drought using time series analysis and forecasting. Various models like Linear Model, Regression Model and ARIMA Model were used for prediction. ARIMA Model gave better results in comparison to other models. Model developed can be used for other data, provided it should have similar data distribution plot as that of rainfall and NDVI series. ARIMA Model with parameters (2,1,2) and (2,1,1) consistently gave better results for Rajasthan and Karnataka. Association rule was applied to predicted values and drought pattern for Karnataka was extracted with 82.5% accuracy and drought pattern for Rajasthan was obtained with 85% accuracy. From the predicted drought pattern, it was observed that year 2009 and year 2011 would be drought year for Rajasthan, and for Karnataka, year 2009 and 2010 will be drought years. For Karnataka, major parts of districts like Bagalkote, Bijapur, Gulbarga, Raichur, Gadag would be affected and for Rajasthan, major portion of Churu, Bikaner, Barmer, Chittaurgarh, Jhalawar, Jodhpur, Kota, and Sikar districts would be affected. The results obtained for the years 2007, 2008 and 2009 for Rajasthan and Karnataka states have been compared with government drought reports, rainfall data and NADAMS, project reports of NRSC. It found 90% accurate for the years 2008 and 2009 and 75% for the year 2007 for the state of Rajasthan. Similarly, 90% accuracy for the year 2007 and 2008 and 80% for the year 2009 for the state of Karnataka has been obtained

## References

Brockwell, P., and Davis, R. (2002) "Introduction to Time Series and Forecasting" Springer International Edition.

Box, G.E.P. and G.M. Jenkins, 1976. Time Series Analysis: Forecasting and Control. Revised Edn., Hoden-Day, San Francisco.

Nail, P.E., Momani, M. (2009). "Time Series Analysis Model for Rainfall Data in Jordan:A Case Study for Using Time Series Analysis" *American Journal of Environmental Sciences* 5(5): 599-604.

# Chapter - 16

# Assessment and Modeling of Village Ecosytem using Earth Observation and Geospatial Technology

**Ila Agnihotri**


## Abstract

*This paper is the result of strong belief that if earth observation and geospatial techniques could be applied to a village to plan its natural resources in sustainable manner, the potential of this technique for common man is harnessed maximum. So analysis of one village natural environment, predicting its impact on humans, identifying certain possible inhabitable conditions due to existing nature & suggesting ways to utilize these conditions for sustainable human's and nature's existence is the motive behind this work.*

*A population density of 3 persons per 5 Hectare, a subcentre with a nurse, a common PHC shared with 19 more villages, rainfall is erratic and highly variable with annual average of 386 mm, having naturally gifted flora and fauna, all sustaining together represents the study area, village Melva, located at Luni block of Jodhpur district, Rajasthan.*

*For this study, quick bird imagery was interpreted visually and digitally for assessing village natural resources. Other ecosystem parameters as runoff, harvestable rainwater and land capability were modeled using derived spatial & non spatial data in GIS environment. Ecosystem study using geospatial techniques reveals that the village occupies an area of 2146 Ha, out of which 92% (1964 Ha) is meant for agriculture. The soil is sandy loam with a major percentage of sand, nearly 80%. Except built up rural land, other land belong to III (14.55%) and IV (79.39%). Land capability and Land irrigability class indicates good soil for agriculture and irrigation. The area occupied by the trees is 6.7% of the total area of the village.*

*Using Spatial Analysis and Decision Assistance (SADA), three inorganic chemicals-Chloride, Nitrate and Fluoride were evaluated against EPA toxicological and scenario parameters for finding their accumulation in humans in hazardous amount assuming dependency on groundwater. Chloride was found to be required risk management targets.*

*Smart Rainwater harvesting was proposed to be appropriate to sustain the water quality and quantity needs of the village for both-agriculture and consumption, the backbone factors for the survival of a naturally gifted self-sustaining village.*

*Keywords: Ecosystem, Melwa, Water quality, Health hazard, Self-sustaining village, SADA, GIS, Earth Observation*

## Introduction

The term ecosystem was coined in 1930 by Roy Clapham to mean the combined physical and biological components of an environment. The ecosystem concept proposes that the earth operates in a series of interrelated systems within which all components are linked, so that a change in any one component may bring about some corresponding change in other components and in the operation of whole system (Baily, 2009). Here, considering village as unit, the biotic and abiotic factors are being assessed using the techniques of earth observation and geospatial technology. Instead of natural delineation, administrative boundary of village Melwa (Rajasthan) was used for selection of ecosystem to be studied (http://socyberty.com/issues/sustainable-development-village). Assessment and modeling of village ecosystem requires village data with accuracy and spatial data with large scale, and this was made available through earth observation and geospatial techniques (Johnson, 2009).

A widely-used and accepted international definition of sustainable development is: '*development which meets the needs of the present without compromising the ability of future generations to meet their own needs*' (Brundtland Commission 1987). So in scientific terminology, to be sustainable, a village should have minimum external intervention and all its needs should suffice by intrinsic resources leading overall carbon footprint to zero. To make it feasible, a village should be treated as a micro isolated unit, here termed as village ecosystem. In other words, a village should be independent in terms of produce, healthcare, sustenance, infrastructure and nature human cohabitation without destructing each other i.e. self-sustainable.

A great step in moving our country towards developed is village level development and sustainability. Using geospatial technology with a target of village resource assessment, the components assessed here are land, soil, flora & fauna, water resources, demography, topography, climate, water quality and human health.

The study is carried on a village located in semi-arid region of the

Great Indian Desert. Being a part of desert, the basic issue of this village is also water - the quality and quantity. The village is characterized by dryness, extremes of temperature, erratic nature of rainfall. In terms of human resources, ironically, the major survival and livelihood is dependent on rainfed agriculture and drinking water needs are harnessed by village pond which is accumulated rainwater or on groundwater which is at the depth of 300-350 ft. as well as brackish. So the impact of quality and quantity of water on humans is greatly dependent on the water in available and harnessed resources-groundwater and accumulated surface water in village pond.

For this village also the ecosystem was studied for the availability of natural resources to optimize their efficiency (Bothale, Bothale, Srinivasan, & Sharma, 2002). The current practices and trends were analyzed for various ecosystem constituents. These were interlinked with the existing socioeconomic conditions (Figure 1). This helped in suggesting a plan for management of ecosystem accordingly so as to be sustainable for both the essential dimensions - environment and humans (Miller, 2009). It covers the analysis of ecosystem delineated by the village boundary along with all the fields or cadastre, analyzing its suitability for human dependency in terms of living, harvesting, drinking, irrigation, land capability and water related human health hazards, seeking alternates to escape human health hazards and village Natural resource optimization for human and environmental sustainability (Corvalan, Hale & McMichael, 2005; Sharma & Pearlman, 2009; World Bank, 2007).

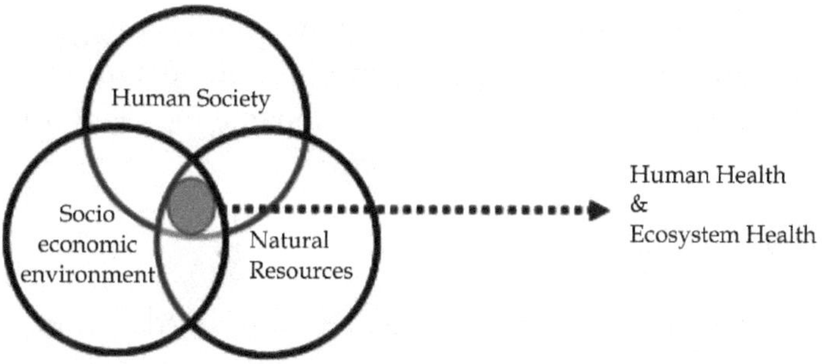


**Figure 1:** The Human Ecosystem Interlinkage Web

## Earth Observation, Geospatial Techniques and Ecosystem

Remote sensing is the science and art of obtaining information about an object, area, or phenomenon through the analysis of data acquired by a device that is not in contact with the object, area or phenomenon under investigation (Lillesand, Kiefer & Chipman, 2004). Earth Observation or remote sensing includes the process of capturing earth images by satellite sensors and processing of same images is done by the techniques of digital image interpretation. Geospatial Techniques involves the technology of geographical information system (GIS) which involves linkage of spatial data (generated by interpreted satellite image) with non-spatial data to generate more ground reality matching results (Jensen, 2004). The main interpretation phenomenon for remote sensing is visual image interpretation and quantitative analysis of same interpretation is done using GIS (Burrough & McDonnell, 2005).

Large scale quick bird imagery has been used to extract the natural resource details of micro unit of administration - a village. Well-managed water resources have helped to promote economic development, which in turn contributes to human well-being. It has been estimated that for each dollar invested in improving water and sanitation, a return of $3-34 can be expected. (Water Quality Outlook, UN GEMS/Water Programme Office). To assess water related issues, GIS environment of ArcMap and SADA (Spatial Analysis and Decision Assistance) were used. The synergy of linking health sciences with water quality indicates an immense potential in improving the infrastructure for human health and its contributors, mainly environment and water (United Nations Environment Programme Global Environment Monitoring System (GEMS)/Water Programme, 2008; Water Quality Outlook, UN GEMS). This study reflects the potential use of space science and GIS in identifying problems of a village. Rainwater harvesting was suggested as an option for this problem (UNEP Regional Land Management Unit, 2006). If this synergy of space science and human development can be replicated, it would change the way of decision making to more scientific, robust, potential and accurate (Rowe, 1961).

## Objectives

The main objectives of this study, using earth observation & geospatial techniques are:

- To assess the natural resources of a village, Melwa, Jodhpur district
- Analysis of inter-linkages between natural resources and socio-economic factors of village
- Impact on human health by water quality hazards

- Estimation of harvestable rainwater in the village
- Suggestions for a perfect strategy: Rainwater Harvesting for managing water stress

## Area to be considered for micro level study

Melwa village is appears as a hat shaped area in Jodhpur district, Rajasthan (Figure 2). The mean annual rainfall in the region is 386 mm. Monthly distribution of rainfall above 90% during May to September and rainy days are about 17 in a year. The air temperature is highest during May and June with mean maximum temperature remaining around 40°C.

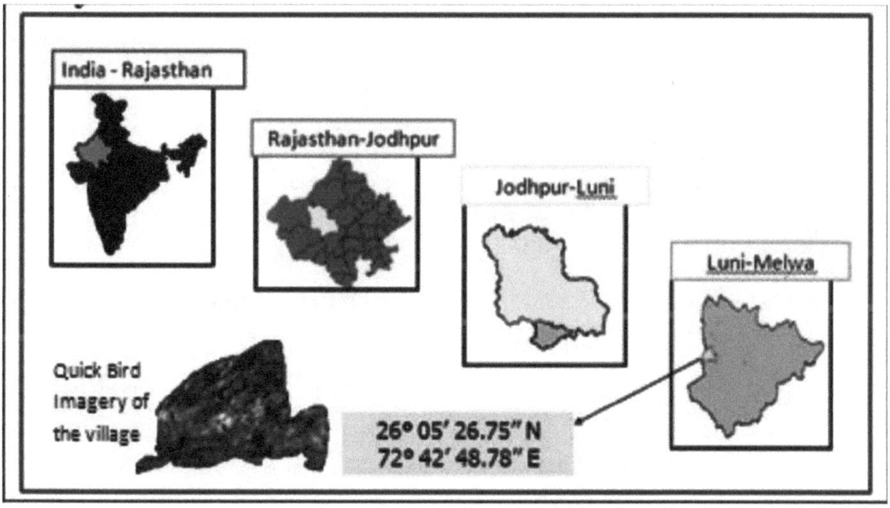


**Figure 2 :** Location of village Melwa

The economy of the area is based on rain-fed farming with pearl millet, kharif pulses (Green gram and Kidney bean), Sesame and Cluster bean crops and cattle rearing. After monsoon, the scarcity of water prompts villagers to leave their fields and search for some supplementary source of their finances for their family. Though certain government schemes as NREGA (National Rural Employment Guarantee Act), SGRY (Sampoorna Grameen Rojgar Yojna) help in preventing migration from villages, still a lot of people move out to get some form of employment. Sometimes agriculture dependents have nothing to do when there are no rains and female population is mainly indulged in carrying and fetching water to home from common water sources. Even female literacy is limited as most of their time is spent in fetching water from water sources.

For domestic water, people depend largely on a village pond, where domestic animals and wildlife also have their share. Therefore, both quantity and quality of domestic water is also an issue. The ground water is saline; it occurs 150 - 200 feet below the surface and has total dissolved solids (TDS) > 2800 ppm, which is not suitable for drinking purpose. Even large storage of water for a long time at their home leads to deplete the quality of water.

## Materials and Methods

### Data used

The following data were used to carry out the analysis:

**Primary data :** Cadastral map and Village boundary from revenue department, Quick Bird imagery, Field survey and interaction with the villagers and data from medical and health department, Jodhpur.

**Secondary data :** Census of India 2001 data, Meteorological data from CAZRI station, GWD ground water quality data.

**Derived information :** Temporal land use of 3 years, soil map (LCC and LIC), vegetation map with tree density and water resources map (runoff and harvestable water).

**Software's :** MS excel, MS access, ERDAS Imagine 9.2, ARCGIS 9.3 and SADA Software

### Methodology

The Methodology involves using Quick Bird imagery and secondary data in ERDAS and ArcMap for assessment of village ecosystem through assessing its natural resources i.e. land, soil, climate, topography, demography, fluora, fauna and water resources. This assessment was interpreted and modeled in GIS environment (ArcMap, SADA) for developing requirements for the village to be self-sustainable. The modeling consisted of land cover assessment, change detection, LCC and LIC classification, estimation of runoff, harvestable rainwater, and water quality modeling in terms of contaminant accumulation through various landuse scenarios. The steps are as follows (methodology chart in Figure 3) :

- Landcover was prepared by visual interpretation of Quick Bird imagery. This was converted into grid form. Other available Landuse of year 2005 and 2007 were also converted into grid (Map 1).
- Spatial change detection was performed by generation of an outgrid from temporal grids in ArcInfo.

- Tree distribution was generated using fuzzy classification of Quick Bird image and recoding it to two classes of area covered by trees and area not covered by trees in ERDAS Knowledge engineer.
- Soil was mapped as per its family classification, percentage of san, silt and clay, soil electrolytic conductivity, and organic carbon. ( Map 3)
- Land Capability Classification and Land Irrigability Classification were performed by the analysis of soil texture, organic carbon, soil moisture content and all the available soil parameters in soil layer. (Map 2)
- Availability of groundwater was mapped as seen in imagery. (Map 4)
- 22 watersheds were delineated by extracting the drainage from ASTER DEM. The procedure followed included delineation of flow direction, flow accumulation and then generation of drainage using CON function of Raster calculator in ArcMap. (Map 5)
- For each watershed, weighted curve numbers were generated using soil moisture and texture conditions, slope and land use. This was curve number II, which was converted to curve number I using conversion equation.
- It was further taken along with precipitation in Indian Scenerio SCS equation to calculate runoff using soil, slope and landuse of this area.
- Harvestable rainwater was estimated by calculating runoff in same watersheds, by subtracting the runoff from crop area and 2m buffer area near trees.
- This runoff and harvestable rainwater for each watershed was mapped along with cadastral boundaries. So for each fields, both runoff and harvestable rainwater was plotted on maps. (Maps 6 to 9)
- From the ecosystem assessment, landuse was combined with groundwater quality for the estimation of toxicant accumulated in the humans in various exposure scenarios of ingestion, inhalation, dermal, milk and total hazard.
- Three inorganic chemicals-Chloride, Nitrate and Fluoride were evaluated against EPA (Environmental Protection Agency) toxicological and scenario parameters.

- Well locations were plotted in ArcInfo. Attributes of groundwater quality were added.
- Then the coverage was imported in SADA (Spatial Analysis and Decision Assistance)
- SADA can use point data only for analysis, so well locations were taken as point layer and imported in SADA (.sda fie). Setup was done from the menu bar to Human Health Risk.
- Inbuilt toxicological and scenario databases were available for analysis, so Chloride, Nitrate and Fluoride were analyzed for their presence in toxicological limits on ingestion, dermal contact, through milk, vegetables and total hazard. (Maps 10 to 12).

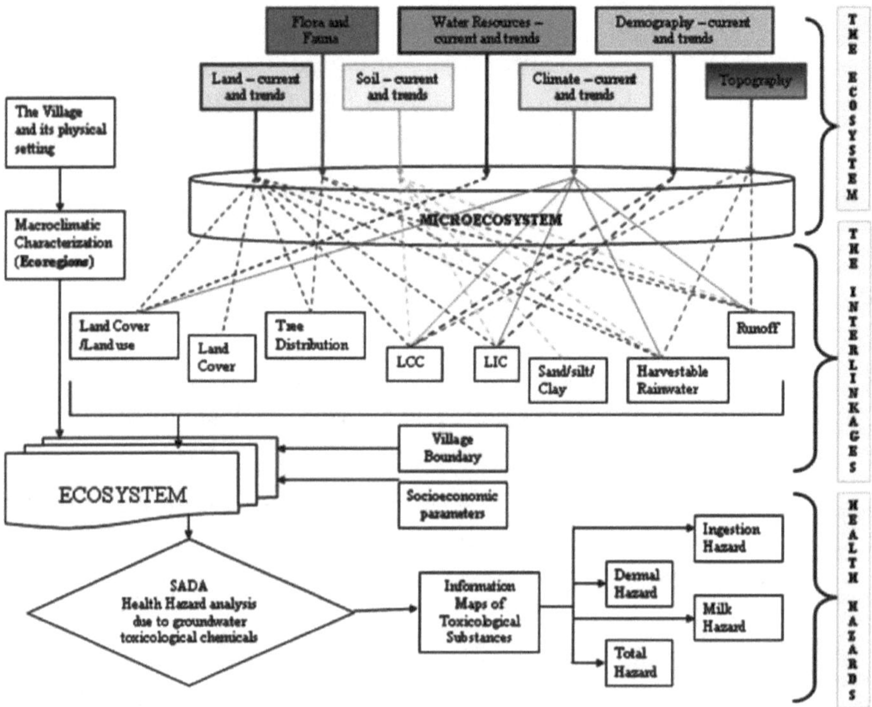


**Figure 3 :** Methodology

## Results & Discussion

Though these results represent basic village statistics, but main thing to focus is that these are obtained using recent earth observation images (Quick Bird imagery) & modern geospatial techniques as well as represented on map. These are meant to show the potential of this modern technology for a village. Results can be divided in two parts: (i) ecosystem analysis results which include village statistics and (ii) spatial representation of results i.e. maps. The results of micro ecosystem analysis are presented as follows:

1. Land - represents land use, change in land cover and implications
2. Soil - Soil type, classes, pH, suitability and implications
3. Climate - meteorological variables with trends and implications
4. Flora and Fauna - natural vegetation and animals
5. Livestock - cattle, sheep, goat and other reared species
6. Water resources - water quantity, presence of various substances in water
7. Demography - population, literacy, socioeconomic resources
8. Crops - the major crops grown in village
9. Agriculture Equipments - main agriculture equipments used in village
10. Topography - the terrain and slope, runoff, harvestable rainwater calculations
11. Ecosystem and Health - Health hazards due to chloride, fluoride and nitrate

## 1. Land - The requirement of every human

The Quick Bird extracted land cover data indicates that the area of the village is 2146 Ha (Map 1). The current land use indicates the major land cover is agriculture land, 92% area of the village i.e. 1964 ha. (Figure 4). Merely 1.9 ha of land account for water bodies (0.09%). The decrease in agricultural area and increase in area occupied by scrublands and wastelands suggest a constant increase in pressure on natural resources. Drastic regression Water body may be attributed to seasonal nature of satellite imagery used for extracting water bodies. Seasonal water bodies tend to dry in summers and therefore seen regressed. A good area (6.7%) is occupied by robust desert trees. These trees, can be assumed as nature's gift to harsh climate. These may be harvested for economic growth of the village - a step for harnessing natural resources for self sustainability.

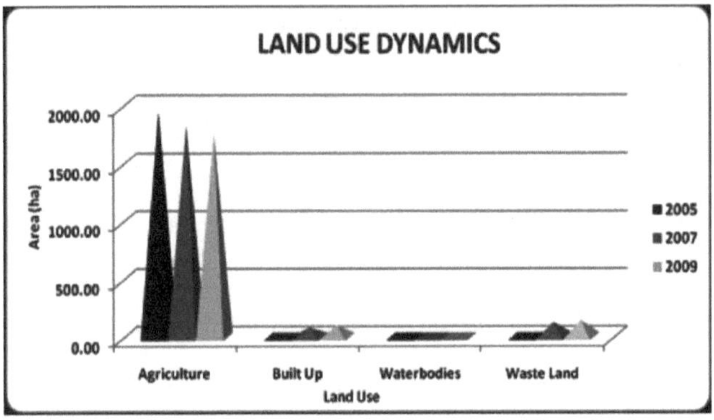


**Figure 4 :** Land Use Dynamics of the Village

The abstract of area change as implicated by change detection is given in Table 1.

**Table 1 :** Land Use Dynamics of Village in terms of percentage change from 2005 to 2009

| Classes | % Change, 2005-2007 | % Change, 2007-2009 | % Change, 2005-2009 |
|---|---|---|---|
| Agriculture Land | -5.725366612 | -3.6894 | -9.2035 |
| Built Up (Rural) | 155.0334014 | 19.0642 | 203.6534 |
| Water bodies | 446.8264752 | -87.1695 | -29.8396 |
| Waste Land | 272.3849861 | 11.6761 | 315.8650 |

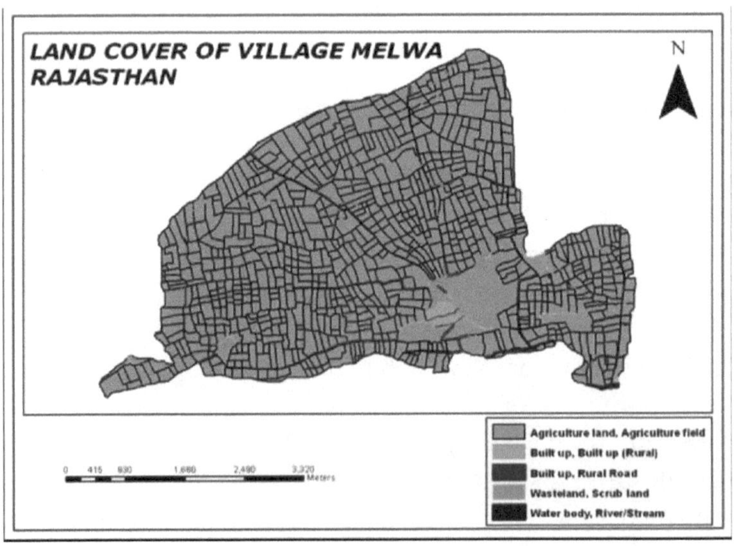


**Map 1 :** Land Cover of Village Melwa, Rajasthan 2009 (Extracted From Quick Bird Image)

## 2. Soil - The compassion of farmers

Almost 92% being agriculture, a detailed study of soil may significantly impact the current agricultural practices. The significance of various soil families and characteristics present in village Melwa, as deciphered using soil map, is as follows:

The soil in this area is mainly aridisol (98%). These soils are characterized by an ochricepipedon generally light in color and low in organic matter. They may have a horizon of accumulation of calcium carbonate (calcic), gypsum (gypsic), soluble salts (salic), or sodium (natric). The area occupied by various family textures is: 78.55 % Coarse loamy, 14.55% Fine loamy and 6.06% Loamy skeletal (Map 3).

Land Irrigability is dominated by class 3. Soil pH is mainly alkaline and 13.2% being strongly alkaline. Land Capability Classification, LCC III says that the land is *moderately good cultivable lands* having moderate erosion, slope, drainage and soil problems, whearas class IV tells that land is Fairly good cultivable lands having severe erosion, drainage, climate and soil problems and V denotes that land has limitations in terms of cultivation (Map 2).

Land suitable for irrigation as land irrigability classification is grouped under classes 1 to 4 according to their limitations. Lands not suited for irrigation are grouped under classes 5 and 6.

Soil pH is 7.9-8.4 (moderately alkaline) and 8.5-9 (Strongly alkaline). At pH 7.8 or more, calcium and magnesium are abundant. Molybdenum is also available if it is present in the soil minerals. High pH soils may have an inadequate availability of iron, manganese, copper, zinc, and especially of phosphorus and boron. Usage of pesticide must be with caution as most work under a given range of pH.

Organic carbon in the range of 5-23% represents the classification of farm lands as highly productive agriculture soils (3-6%). It consists of plant residues and living microbial biomass, detritus and humus. It accelerates decomposition of soil minerals over time, making the nutrients in the minerals available for plant uptake as well as enhances pore space through the actions of soil microorganisms. This helps to increase infiltration and reduce runoff.

Majority sandy loam soil of village suggests more water holding capacity and water movement is easier, less erodibility as compared to soils with more silt or clay content, organic matter decomposition is faster, less cation exchange capacity, potassium ion can leach in very sandy soils with low cation exchange capacity; otherwise it does not leach readily.

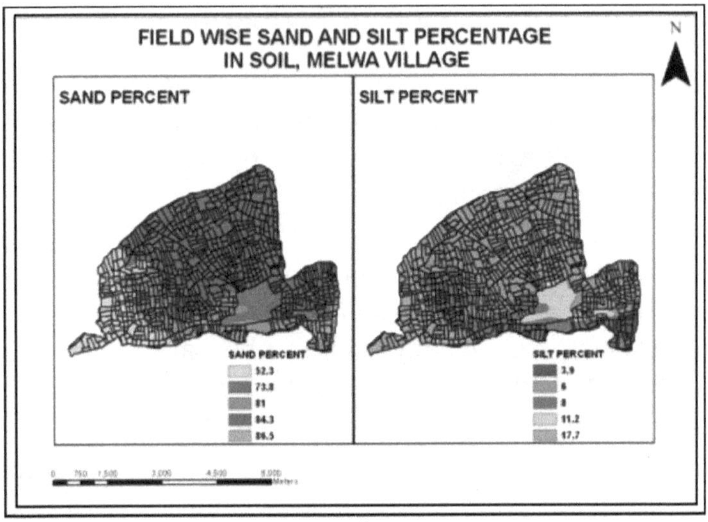


**MAP 2 :** Sand and Silt Percentage in the soil at village Melwa

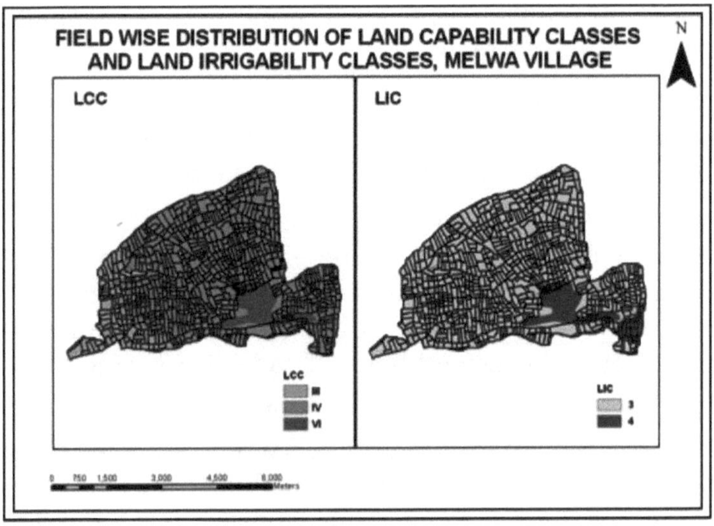


**MAP 3 :** Soil LCC and LIC at village Melwa

## 3. Climate - the backbone for survival and extremes

Highest average evapotranspiration (12.59 mm), Temperature (33.57 °C), wind speed (7.41 Km/hr) and sunshine (10.7 hrs) occurs in zaid season. Evapotranspiration was found to be 13.6 times higher than the rainfall. Potential evapotranspiration and rainfall misbalance suggests the deficit and dryness characteristics of this area. Climate is dry and during hot season winds blow in the day but nights are cool and pleasant. Analysis of climatic variables was taken into account from the nearest meteorological observatory located at CAZRI. Rainfall is accounted mainly in kharif season, which represents rainy season. Kharif has highest rainfall, represented by 165.8 mm, 72.8% of the total rainfall of this area (Figure 5).

The range of variation for all the three (as noticed in 12 years data) variables can be accounted as being within the range of ± 1 unit from previous year. So, no drastic change has been observed in terms of evapotranspiration, wind speed and sunshine of this area.

Kharif season's best rainfall availability is quiet clear from figure 6 which is nearly 4 times higher than zaid and almost 59 times higher than Rabi Rainfall. The year 2002 reflects unexpected decline in rainfall values, almost 92% decline from previous year Kharif rainfall, reflecting a meteorological drought.

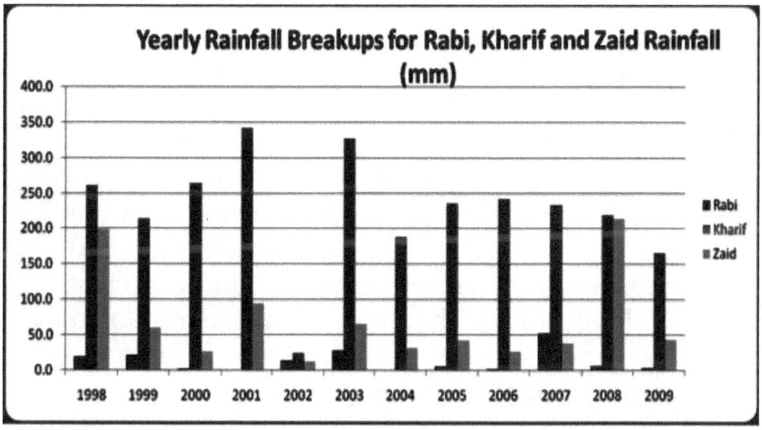


**Figure 5 :** Rainfall Variation in a year for Rabi, Kharif and Zaid seasons

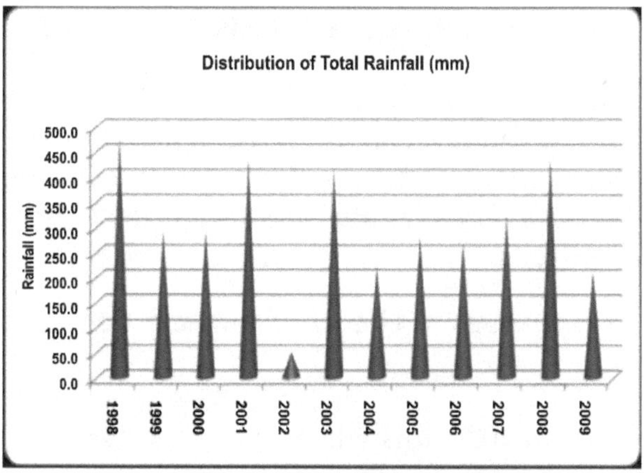


**Figure 6 :** Annual distribution of Total Rainfall (mm)

## 4. Flora and Fauna - representational species of plants and animals

### Flora

Most commonly found tree species are the ubiquitous khejri (*Prosopis cineraria*). The khejri, held sacred by the Bishnoi community of Jodhpur, is extremely drought resistant, due to its deep root system. It is a multipurpose tree - its thorny twigs are used to form barriers between fields to keep animals away from the crops, its leaves are dried and used as fodder, its fruits are eaten ripe and when unripe it is cooked and eaten as *sangri*. The utility list goes on further; the wood is used for furniture and the branches as fuel.

Other representational species are Vilayati, Kumat (*Acacia*), Babul (*Acacia arabica*), Ber (*Zizyphus*), Neem (*Azadirachta indica*).

### Fauna

Irrespective of its unfriendly terrain, western Rajasthan gives shelter to a variety of animals and birds. Antelopes and gazelles are widely available in the Jodhpur district where they are worshipped by the Bishnoi community. They are mainly responsible for the steady population of the chinkara or Indian Gazelle in this desert zone. Chinkaras are slimmer than the blackbuck and can survive without water for very long periods. They live in smaller herds and thrive mainly on wild grasses and various types of shrubs. The *nilgai* or the bluebull or bluecow are abundant on the open plains and in the foothills of the Aravalli. The fauna of the place include jackal, Jungle Cat, Indian Fox, Black Buck, Chinkara, common hare, etc.

## 5. Livestock - the reared animals for man's utility

Cattle, sheep and goat are major among livestock. A goods number of sheep's, 1468 in a village, are reflected here in the livestock census. Goats have increased by 475% since 1962 till 1997, accounting an average increase of 13.57% every year. Buffallow's are 269% more in 1997 than in 1962, an average of 7.6% every year.

## 6. Water Resources - the basic need for existance and survival

As assessed in 2004, this village comes under *critical* category of ground water resources. (Ground water Department Report). As per 2008 ground water quality data, Nitrate, Bicarbonate and Fluore are less than desirable limits (Figure 7). Major lithology is occupied by Malani Rhyolite and Granite, suggesting poor aquifers. Temporal water quality comparison as per the BIS standards, is presented in Table 2. Spatial water availability can be appreciated by Map 4.

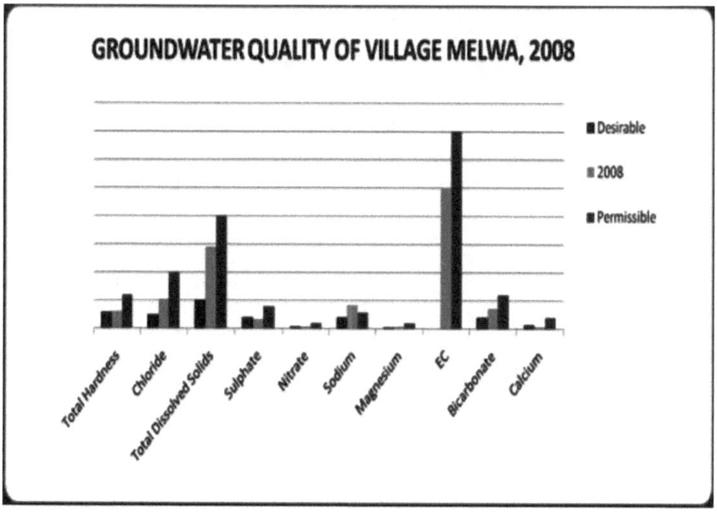


**Figure 7 :** Trends in Groundwater Quality at Village and its comparison with Reference values

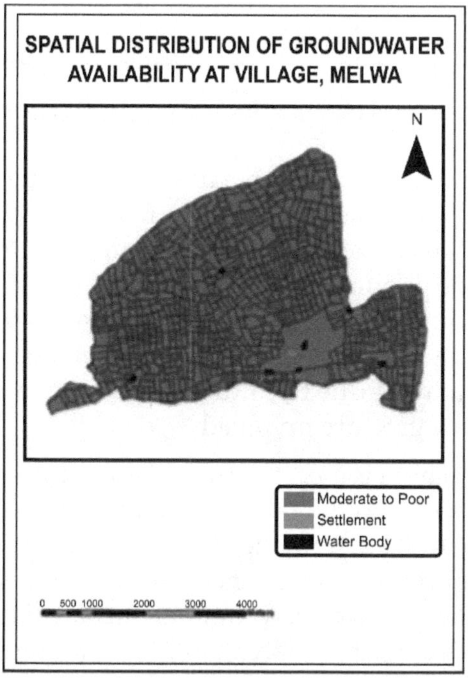


**MAP 4 :** Groundwater Availability of Village Melwa

**Table 2 :** Temporal Comparison of Water Quality

| PROPERTY | 2004 | 2005 | 2006 | 2007 | 2008 |
|---|---|---|---|---|---|
| pH | √ | √ | √ | √ | √ |
| Total Hardness | √ | | | | √ |
| Chloride | | | | | √ |
| Total Dissolved Solids | | | | | √ |
| Sulphate | | | √ | | √ |
| Nitrate | | | √ | | |
| Sodium | | | | | |
| Potassium | √ | | √ | √ | √ |
| Magnesium | √ | | | √ | √ |
| EC | | | | | √ |
| Bicarbonate | | √ | √ | √ | √ |
| Calcium | √ | √ | | | |
| Fluorie | √ | | | | |
| | Much more than permissible (>2times) | | | | |
| | More than permissible | | | | |
| | Normal | | | | |
| | Less than desirable | | | | |
| | Much less than desirable (<2 times) | | | | |
| | Not available | | | | |

## 7. Demography - the socioeconomic mirror

Population density of the area has increased by 117.6%, from 1991 to 2001 suggesting increase in pressure on natural resources of the area. The temporal analysis of male to female ratio reflects the dominance of male in each of the population categorization. Total literate to population suggests a variation and reflects the less emphasis on woman education and literacy. Literacy rates suggest less female literacy, though increasing decadal, which may be due to cultural prohibitions and water fetching activities in which girls are involved.

A primary school is a part of the village, a middle school is within a range of 5-10 km from the village and a college is also a present at a range of more than 10 km from the village. Nearest primary health centre is at Dhawa. Sources of water include - tap water, tank water and summer source as tank water. A commercial bank is present at 5-10 km, and a cooperative bank is at more than 10 km from the village. All credit societies, Agricultural, Non-Agricultural and other, are within a range of 5-10 km from the village.

## 8. Crops - the major survival way and agricultural means

The major crops of this region are mainly because of its sandy and loamy soil. Bajra (Pearl millet) is the major crop in Kharif. In Rabi Wheat, Pulse and a variety of masala like Jeera, Dhania and Red chilly are also grown.

## 9. Agriculture equipments

Plough and animal cart represent common equipments for agriculture. Other advance tools as Electrical pump set, diesel engine pump and tractor seems to be gaining importance.

## 10. Topography

The village is nearly flat with a maximum slope of 5.9 degree. Aspect represents a prominent direction of slope in North West and South East direction. The drainage derived from Aster DEM could help in delineation of small sub watersheds (Map 5). Runoff and harvestable rainwater was estimated for each sub-watershed on annual basis and for the Kharif season attached with this sheet, Drainage delineation, Watershed delineation, Runoff was estimated for each sub-catchment using SCS equation for a watershed using soil, slope and landuse of this area, harvestable rainwater was estimated for each watershed by subtracting runoff for crop area and from 2 m buffer near trees. The amount of runoff and harvestable rainwater is presented in information maps Map 6 to Map 9.

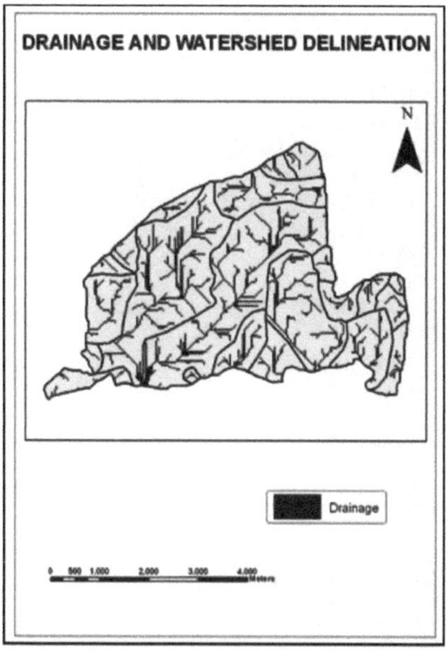


**MAP 5 :** Drainage and Watershed Map of Village Melwa

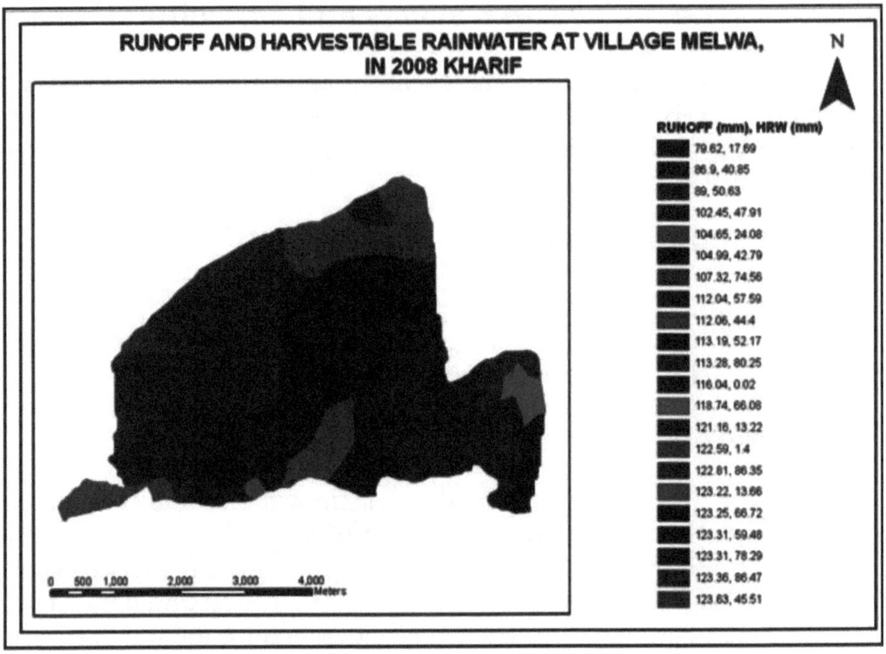


**MAP 6 :** Runoff and Harvestable Rainwater at the village in 2008

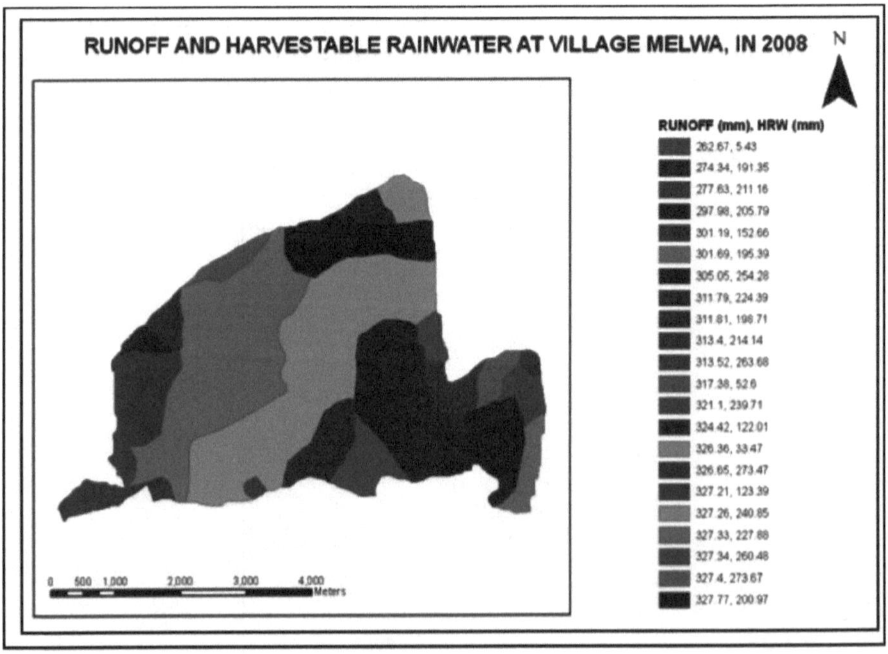


**MAP 7 :** Runoff and Harvestable Rainwater at the village in 2008-kharif

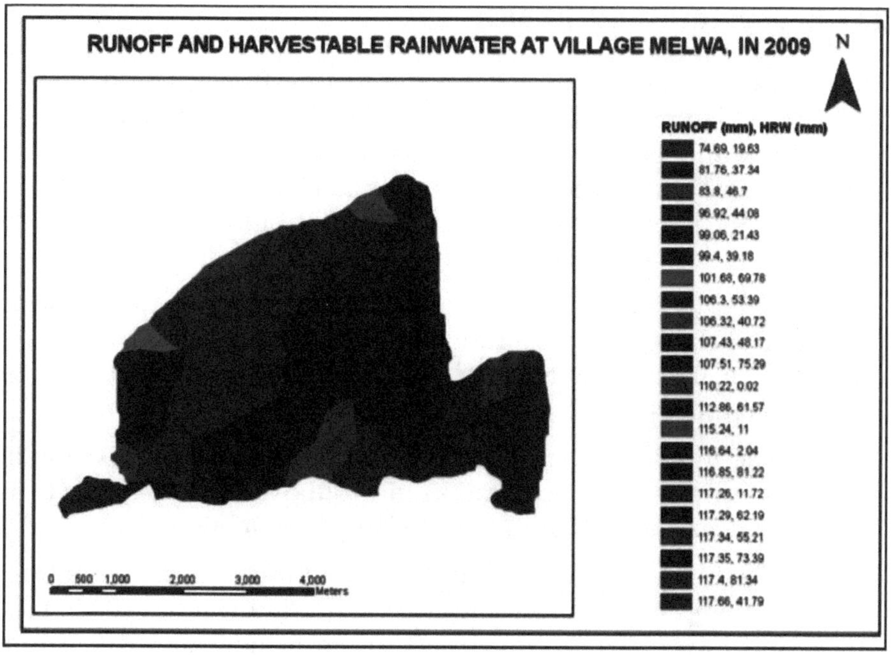


**MAP 8 :** Runoff and Harvestable Rainwater at the village in 2009

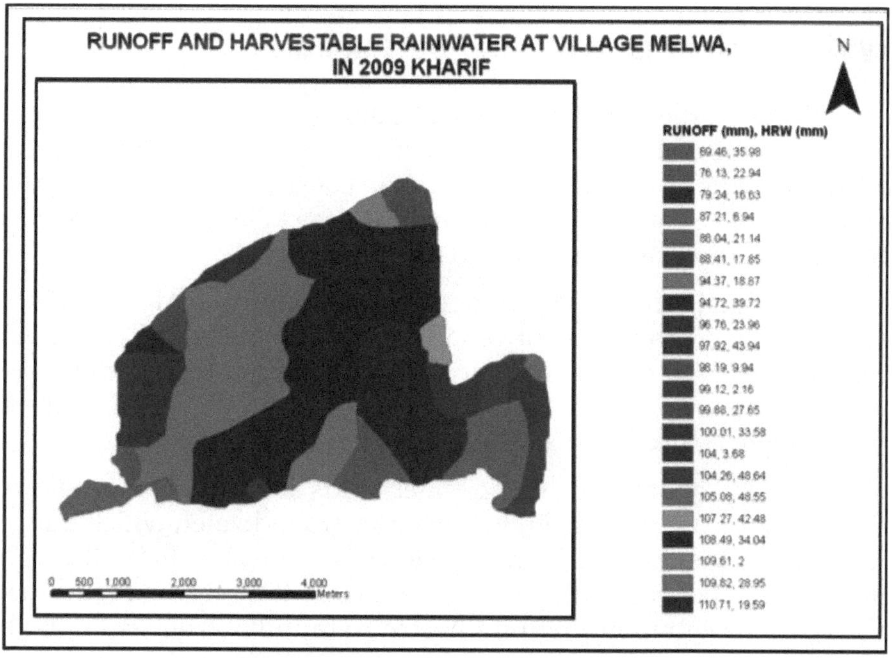


**MAP 9 :** Runoff and Harvestable Rainwater at the village in 2009-kharif

## 11. Ecosystem and health

Spatial variability and risk of chloride, fluoride and nitrate according to land use scenario was calculated using SADA using EPA toxicological and scenario database and was plotted in information maps (Environmental Protection Agency, 1989; SADA Version 5 Help, 2008; Technical documentation-SADA, 1996). The result suggests a health hazard using SADA can be summerizes as

- The ingestion, dermal, vegetable, milk and total risk due to Chlorie (with concentration in groundwater as 524.8 mg/L) is high. It has to be checked to stop the potential hazrds of chlorie on human health (Maps 10 to 12)
- The ingestion, dermal, vegetable, milk and total risk due to Fluorie (with concentration in groundwater as 0.7 mg/L) is under control. It does not need any intervention.
- The ingestion, dermal, vegetable and total risk does not exist for Nitrate (with concentration in groundwater as 40.9mg/L) till now. Milk is not found to be influenced with nitrate concentrations

## Conclusion

## What it requires to be self sustainable

The work suggests that the village, though is a part of extreme and erratic climate of the desert, still supports a very sustainable ecosystem where humans and natural resources can coexist, interact and sustain each other. The culture of inhabitants bans the killing of animals, so a rich species diversity is a legacy to the villagers.

Good soil, alkaline pH, good land capability and irrigability supports good production of certain cereals, oil seeds and crops, but coexistance of erratic, low and variable rainfall limits this support (law of minimum is applicable here also). So quantity water availability is first thing to be taken care off.

Human habitation drinking water needs and irrigation needs are dependent on accumulated rainwater and groundwater, which due to its quality degrades land. Human health is also affected due to the high presence of chemicals in toxicologic limits. In 2008 it was found that chloride concentration in groundwater were in hazardeous limits. If groundwater was ingested, or animals were given that water, or vegetables were grown using ground water, or dermal contact i.e. total hazard was found.Three inorganic chemicals-chloride, nitrate and fluoride were evaluated against EPA toxicological and scenario parameters. Chloride was found to be required risk management targets with agricultural and residential ingestion being 3.9 mg/l, recreational hazard is 7300 mg/l and industrial is 10 mg/l. Fluoride and Nitrate do no show up as a risk contaminant in PRG screen through Spatial Analysis and Decision Assistance. So groundwater quality is also an issue to be embark upon.

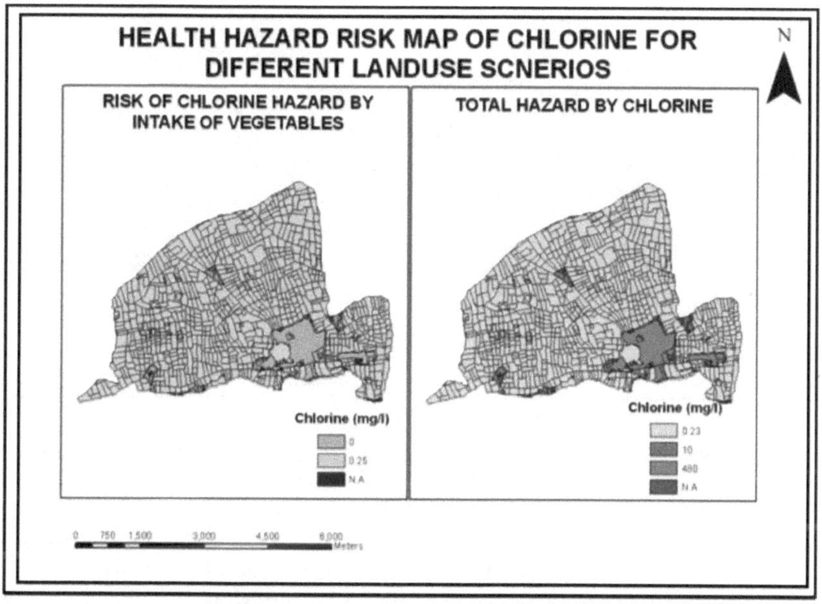


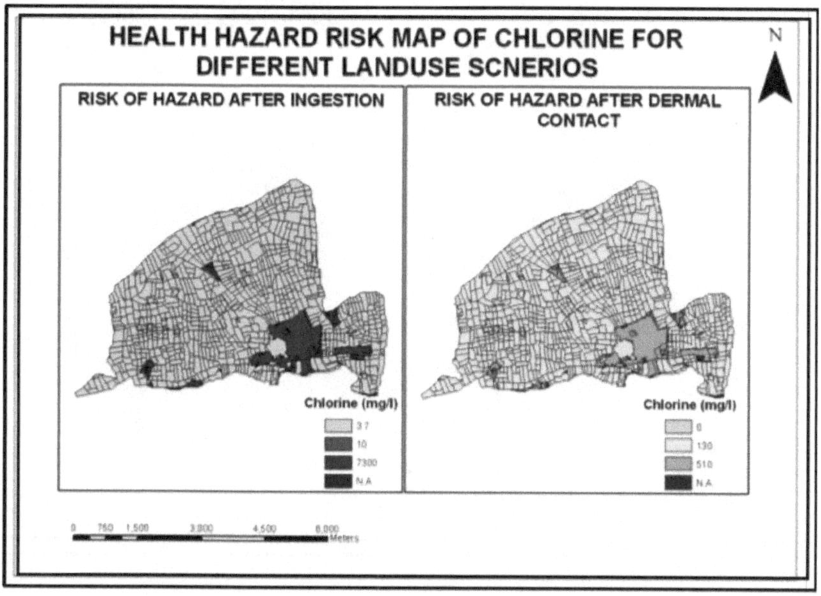


**MAP 10 :** Health hazard risk due to chlorine ingestion and dermal contact

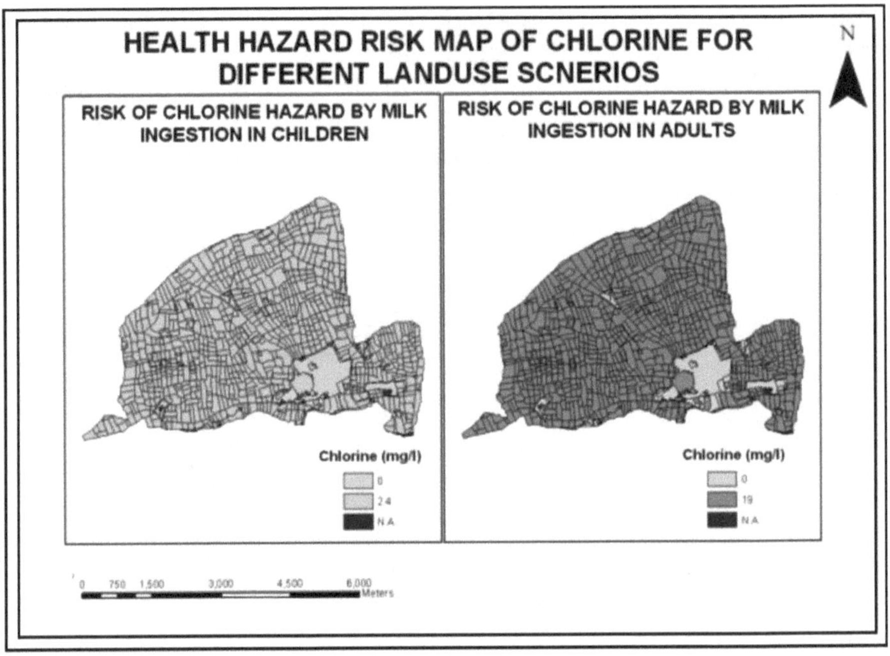


**MAP 11 :** Health hazard risk due to chlorine after milk ingestion in children & adults

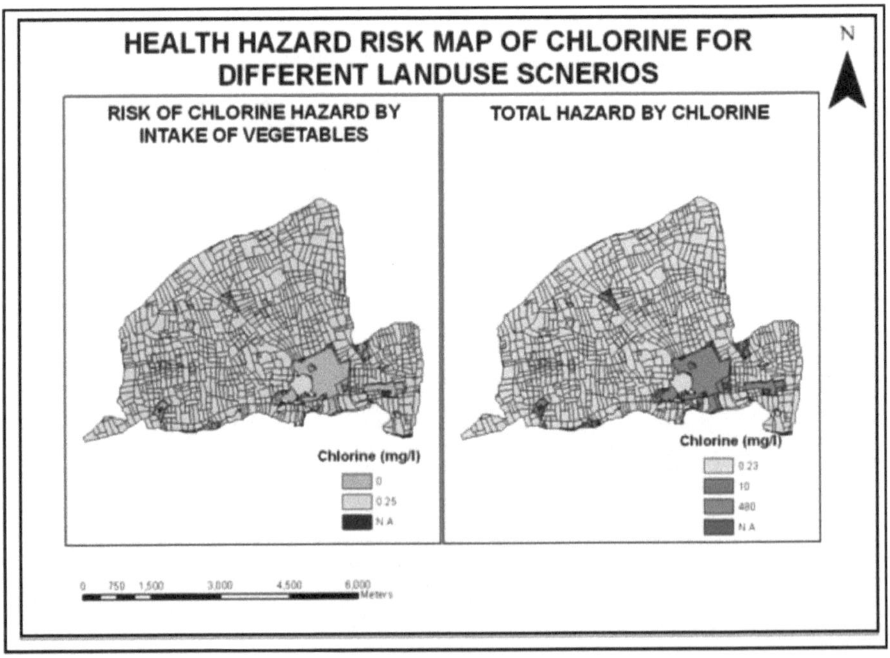


**MAP 12 :** Health hazard risk after vegetables intake and total hazard due to chlorine

Increasing population pressure demands incease in agriculture on available land. This can be achieved only if quality water will be available. Human and animal health depends on availability of quality water. So requirement of water in terms of quality and quantity is essential.

## Suggesting ways to ensure sustainability: Smart Rainwater Harvesting

To tackle these issues, Smart Rainwater Harvesting is a suggested option (Ann, 2005). Smart indicates purpose specific harvesting. Harvestable rainwater calculations spatially suggests that on an average $1/5^{th}$ of runoff can be harvested. Spatial Harvestable rainfall distribution is estimated for the entire village. To maintain different hygine condition, specific construction is suggested as cemented slopes for drinking water, slope of field can be used for harvesting water for irrigation i.e. smart rainwater harvesting. Under different hygine conditions, harvested rainwater can be useful for drinking, irrigation and livestock of the village. So an essential step to ensure the food security and sustainable livelihood at the village is Smart Rainwater Harvesting .

## Acknowledgement

Author acknowledges Regional Remote Sensing Centre-West/ Indian Space Research Organization for their unconditional support and facilitation for carrying out this study.

## References

Baily, G. R. Ecosystem Geography: From Ecoregions to Sites. Second edition. Springer. Pp-1-30, 53-55. 2009.

Bothale, RV, Bothale, VM, Srinivasan, G. and Sharma, JR. 2002. Identification of Suitable Sites for Water Harvesting Structures in Upper Betwa Watershed through WARIS. GIS Development. Accessed from http://www.gisdevelopment.net/application/nrm/water/watershed/watws0013.htm

Burrough, P A and McDonnell, R A. Principles of Geographical Information Systems, Oxford University Press, UK. 2005.

Corvalan, C., Hales, S. and McMichael, A. 2005. Ecosystems and Human Well Being. A Report of the Millennium Ecosystem Assessment, WHO.

Environmental Protection Agency (EPA). Exposure Factors Handbook. EPA/600/8-89/043. Office of Health and Environmental Assessment, Washington, D.C. 1989.

Jensen, JR. Remote Sensing of the Environment: An Earth Resource Perspective. University of South California. 2004.

Johnson, Lynn E. Geographic Information Systems in Water Resources Engineering. CRC Press, Taylor & Francis Group, FL 334872742. 2009.

Lillesand, T M, Kiefer, R W and Chipman, J W. 2004. Remote Sensing and Image Interpretation (5th ed.). John Wiley and Sons. New York, U.S.A., pp. 9-53.

Miller, D H. 2009. The Factor Of scale: ecosystem, landscape mosaic, and region. In:Robert G. Baily. Ecosystem Geography:from ecoregions to sites, Second edition. Springer. Pp-26-27.

Rowe, J S. The-level-of integration concept and ecology. *Ecology*. 42:420-427. 1961.

Phillips AA. City of Tucson - Water Harvesting Guidance Manual, Stormwater Management Section, Tucson, Arizona. 2005.

SADA Version 5 Help. 2008. University of Tennessee and the Oak Ridge National Laboratory's Environmental Restoration program, University of Tennessee.

Technical documentation- Spatial Analysis and Decision Assistance (SADA). 1996. University of Tennessee and the Oak Ridge National Laboratory's Environmental Restoration program.

Sharma, J R and Pearlman, J. 2009. Smart Rain Water Harvesting and Multiple Uses for Human Welfare, Environmental Sustainability and Adaption to Climate Change Impacts in Semi Arid Regions, Pilot project proposal for MGCS, Jodhpur

Water Quality Outlook. UN GEMS/Water Programme Office, c/o National Water Research Institute, Ontario, Canada, ISBN 95039-11-4. http://www.gemswater.org

United Nations Environment Programme Global Environment Monitoring System (GEMS)/Water Programme. 2008. Water Quality for Ecosystem and Human Health, 2nd Edition. UN GEMS/Water Programme Office, c/o National Water Research Institute, Ontario, CANADA. http://www.gemswater.org/

UNEP, Regional Land Management Unit. 2006. Rainwater Harvesting for Natural Resource Management and Sustainable Development - Strategies to eliminate poverty in the Olepolos community, Kajiado District Kenya.

World Bank. 2007. World Development Report - Agriculture for Development - Overview.WRC, South Africa: Rainwater harvesting, Technical Brief, April 2008.

http://socyberty.com/issues/sustainable-development-village/

# Chapter - 17

# Desertification Assessment Within Landuse and Degradation in Pali Development Block by Using LISS IV Data

**Mani Tyagi,**

## Abstract

*Conversion of an already arid land into a desert through indiscriminate human actions magnified by droughts. Such actions include overgrazing, repeated burning, intensive farming, and stripping of vegetation for firewood. One quarter of the earth's surface is threatened by desertification - an area of over 3.6 billion hectares. Since 1990, Six million hectares of productive land are lost every year due to land degradation. Desertification threatens the liveihoods of more than one billion people and has already made over 135 million people homeless. Desertification does not occur in linear, easily mappable patterns. Deserts advance erratically, forming patches on their borders. Areas far from natural deserts can degrade quickly to barren soil, rock, or sand through poor land management. The presence of a nearby desert has no direct relationship to desertification. Unfortunately, an area undergoind desertification is brought to public attention only after the process is well underway. Often little or no data are available to indicate the previous state of the ecosystem or the rate of degradation. Scientists still question whether desertification, as a process of global change, is permanent or how and when it can be halted or reversed. There are numbers of reasons – manmade, as well as natural for this process. The main thrust of this article is to study various types of natural factors and indicators and their severity levels, which are responsible for desertification process. Pali development block, which falls in semi arid region, is a good example, wherein natural as well as manmade factors are visibly responsible the desertification process.*

*Key words: desertification, degradation, landuse, indicators, severity level, landuse.*

## 1. Introduction

Desertification - the degradation of drylands is caused by various factors including climatic variations and human activities. This is among the most serious environmental challenges being faced by the world today. A total of over 19 million square kilometers of area is covered by the deserts. It represents almost 15 per cent of the terrestrial surface of the planet. It is currently home to more than 144 million people. In fact, poverty affects many of the people living in deserts but simultaneously, population growth and the subsequent pressure on resources (agriculture in marginal lands as well as encroachment in forest areas, excessive grazing) high animal population growth, non-scientific methods of irrigation, which develop subsequently water logging and salinity are important aspects for activation of desertification process. In 1997, the UNCCD modified definition of desertification as "land degradation in arid, semi-arid and dry sub-humid areas resulting from various factors, including climatic variations and human activities". The most commonly cited forms of unsustainable land use are over cultivation, overgrazing, deforestation, and ill-considered irrigation practices. Soils are being mined of nutrients at an alarming rate (Smaling, 1997; Van der Pol, 1992). Periodic droughts debilitate and destabilize poor societies, and contribute to land degradation by reducing vegetative cover and water supplies, triggering a desperate exploitation of the remaining resources in order to survive. There are many factors that trigger desertification, including the unpredictable effects of drought, fragile soils and geological erosion, livestock pressures, nutrient mining, growing populations, inadequate/ ambiguous property and tenure rights, landlessness and an inequitable distribution of assets, poor infrastructure and market access, neglect by policy makers and agricultural and environmental research systems, and the failure of markets to reward the supply of environmental services. Given this complexity of causal factors, an integrated approach including broad stakeholder participation is essential if peoples' livelihoods - and security needs - are to be improved. Desertification is a process of faster erosion of soil or deterioration in the quality of land accelerated due to increased human intervention or due to natural causes (Katyal and Paul Vlek, 2000).

## 2. Study Area

Pali development block of Pali district is located between 25° 27′ to 25° 55′N latitudes and 73° to 73° 34′ E longitudes, covering an area of 1320.86 sq. km with 82 villages. The block is covered by an area equivalent to 12 SOI toposheets on 1:50,000 scale. The area falls under arid agro-climatic condition with four seasons viz., winter, summer, monsoon, and post-monsoon. The temperature begins to rise from the month of March and may reach up to 46°C occasionally. High temperature may be associated with high speed dust storms. The average annual rainfall received in the block is about 472 mm, and no perennial river flows through the area. Accordingly to the census (2001) total population of Pali development Block was 307412 - Rural (119771), Urban (187641). The Scheduled castes population comprised of 16.31 per cent of total population of the block whereas 3.85 percent population was scheduled tribes (http://pali.nic.in/other1.htm).

Pali town is the main centre of industrial activity. Different types of industries include textiles, cement, chemical and leather.

## 3. Methodology

Standard methods available for measurement of desertification processes are expensive and time consuming whereas it has been demonstrated that satellite based remote sensing data offers a considerable potential.

Data used in this study include:

a. IRS 1C/1D LISS III imageries for all three season of 2000-01

b. IRS P6 LISS IV imageries of 2005

c. Survey of India topographic maps on 1:50000 scale

For the assessment of desertification indicators and their status using Remote Sensing and GIS a three step methodology has been adopted in the study and is as follows:

### 3.1. Methods for first level (LU/LC classification)

Visual as well as on-screen interpretation of IRS-1C LISS-III geocoded FCC of Rabi, kharif and summer seasons and, IRS-P6 LISS-IV (5.8m) False Colour Composite (FCC) using standard image interpretation keys like tone, texture, size, pattern, association etc. was carried out to identify broad landuse/landcover categories at first level. The land use / land cover categories (Level-1 classification) are:

1. Agriculture land
2. Forest
3. Land with scrub
4. Dune / sandy area
5. Fallow land
6. Grassland and grazing land
7. Barren rocky area
8. Waterbody /Drainage
9. Others (urban, man-made etc.)

## 3.2 Methods for second level (process of degradation)

The complex interplay of soil degradation/land degradation processes and their spatial variability were assessed according to local conditions and forms of land use. So in this step classification is based on image signature interpretation, field survey, ancillary data and maps to identify the following processes of degradation:

- Water erosion
- Salinization
- Vegetation dergradation

## 3.3 Methods for third level (severity of degradation)

Severity is calculated according to a combination of the degree of soil degradation and the percentage of the area affected. The degree to which soil is degraded has been estimated in relation to changes in agricultural suitability, in relation to reduced productivity and in some cases in relation to its biotic functions. The degree of degradation or quantifying the severity of degradation is classified according to the following scheme:

- Slight: The terrain is suitable for use in local farming systems, but with somewhat reduced agricultural productivity.
- Moderate: The terrain is still suitable for use in local farming systems, but with greatly reduced agricultural productivity.
- Severe: The terrain is not reclaimable and impossible to restore. The original biotic functions are completely destroyed.

## 4. Assessment of Desertification

## 4.1 Landuse pattern of Pali development block

Land use refers to 'man's activities and the various uses which are carried on land'. Land cover refers to 'natural vegetation, water bodies, rock/soil, etc`. Although land use is generally subjective based on the cover, yet both the terms are related and interchangeable. Land use/ land cover mapping of the area has been carried out employing IRS LISS IV imageries supported by SOI toposheets. Study area contains built-up land, agricultural land, forest, wasteland, water bodies and others land classes which comprises the land use/land cover categories as Level - I classification system.

The land use / land cover classes at Level-I classification system along with their area statistics drawn using LISS-IV data of Pali Block is shown in following Tables -

| Landuse | Grid-Code | Area (Sq.Km) |
|---|---|---|
| Agriculture Land | 1 | 600.610 |
| Built-Up Land | 2 | 33.225 |
| Fallow Land | 3 | 447.906 |
| Forest | 4 | 16.281 |
| Grazing Land | 5 | 71.450 |
| Waste Land | 6 | 126.168 |
| Water Body | 7 | 25.223 |
| Total | | 1320.864 |

## 4.2 Degradation types and severity

| Degradation Type | Code | Area(Ha) |
|---|---|---|
| Vegetation Degradation Slight | 11 | 0 |
| Vegetation Degradation Moderate | 12 | 1128.548 |
| Vegetation Degradation Severe | 13 | 1006.969 |
| Water Erosion Slight | 21 | 3057.919 |
| Water Erosion Moderate | 22 | 1809.523 |
| Water Erosion Severe | 23 | 217.4802 |
| Salinization Slight | 51 | 3263.32 |
| Salinization Moderate | 52 | 412.9086 |
| Salinization Severe | 53 | 144.1546 |
| Total | | 11040.82 |

## 4.3 Combine desertification class

| Sl.No. | Description of Desertification Class | Code | Area (Ha) |
|---|---|---|---|
| 1 | Agriculture - Water Erosion - Slight | 121 | 1213.32 |
| 2 | Agriculture - Water Erosion - Moderate | 122 | 10.28 |
| 3 | Agriculture - Salinization - Slight | 151 | 2831.07 |
| 4 | Agriculture - Salinization - Moderate | 152 | 311.46 |
| 5 | Agriculture - Salinization - Severe | 153 | 108.45 |
| 6 | Forest - Vegetation Degradation - Moderate | 312 | 762.16 |
| 7 | Forest - Vegetation Degradation - Severe | 313 | 754.76 |
| 8 | Grazing Land - Vegetation Degradation - Moderate | 412 | 366.39 |
| 9 | Grazing Land - Vegetation Degradation - Severe | 413 | 252.21 |
| 10 | Scrub Land - Water Erosion - Slight | 521 | 862.85 |
| 11 | Scrub Land - Water Erosion - Moderate | 522 | 383.28 |
| 12 | Scrub Land - Water Erosion - Severe | 523 | 85.25 |
| 13 | Scrub Land - Salinization - Slight | 551 | 432.42 |
| 14 | Scrub Land - Salinization - Moderate | 552 | 101.46 |
| 15 | Scrub Land - Salinization - Severe | 553 | 35.76 |

*contd.*

| 16 | Barren/Rocky Area - Water Erosion - Slight | 621 | 981.75 |
|---|---|---|---|
| 17 | Barren/Rocky Area - Water Erosion - Moderate | 622 | 1415.97 |
| 18 | Barren/Rocky Area - Water Erosion - Severe | 623 | 132.23 |
| | Total | | 11040.82 |

## 5. Results

It has been observed that about 110.41 ha area (8.36% geographical area of the block) was affected by various desertification processes such as water erosion, salinization and vegetation degradation. Above table shows area affected by degradation processes active in different landuse/landcover categories along with severity classes. Spatial distribution of these degraded areas is depicted in figure 1. Extent of degradation due to water erosion covered about 50.85 ha in agricultural land, scrub land and barren/rocky area. About 38.21 ha area of agricultural land and scrub land was also affected by salinization. As per Survey of India topomaps about 16.31 sq. km and 74.76 ha areas are under reserved forest and grazing land, respectively. Vegetation degradation of about 21.35 ha was confined in forest and grazing land. There are 82 villages and one town lies in the study area and magnitude of degraded areas varied in different villages. Magnitude of degradation was more only in 5 villages where it exceeded more than 20% of village area and about 60 villages were least affected. No impect of degradation is seen on the rest of the villages.

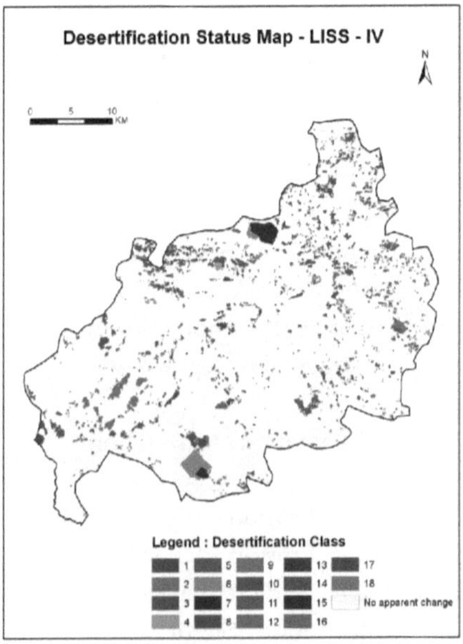


**Map no. 1 :** Desertification Status Map - LISS - IV

## References

Adu SV (1982) Desertification in West Africa. In Desertification and Soil Policy. Symposia Papers III. Transactions of the 12th International Congress of Soil Science, New Delhi, India, pp 26-43.

Ahmed YJ and Kassas M (1987) Desertification: Financial Support for the Biosphere. Hodder and Stoughton, London, UK.

Aubreville, A. (1949) Climats, Forest, et Desertification de l'Afrique Tropicale. Societe de Editions Geographiques, Maritime et Coloniales, Paris, 255 p.

Berry, L., Olson, J., Campbell, D. and F. Santibanez (2003): Case Studies: Impact & Costs of Land Degradation, Presentation at "Investing in Drylands: Lessons, Incentives and Cost of Inaction", Progress Review Workshop in Rome, June 2003.

Bishop, J., and J. Allen (1989): The on- site costs of soil erosion in Mali. The World Bank Environment Working Paper No. 21. Washington, D.C. The World Bank.

Bøjö, J. and D. Cassells (1995): Land degradation and rehabilitation in Ethiopia: A reassessment. AFTES Working Paper No. 17. Washington, D.C.: The World Bank, May

Brundtland, Gro Harlem. "The Environment, Security and Development" Sipri Yearbook 1993 - World Armaments and Disarmament. (Sipri - Stockhol International Peace Research Institute.) New York: Oxford University Press, 1993..

Colchester, Marcus. "Colonizing the Rain Forests: The Agents and Causes of Deforestation" The Struggle for Land and the Fate of the Forests. Penang, Malaysia: Zed Books Ltd., 1993.

DISMED, 2003. Information on the DISMED project is available on the web at http://dismed.eionet.europa.eu/

Drechsel, P. and L. Gyiele (1999): The economic assessment of soil nutrient depletion.- Analytical issues for framework development. Issues in Sustainable Land Management 7. IBSRAM/SWNM: Bangkok, 80 pp.

Dregne H. and M. Kassas. 1991. A New Assessment of the World Status of Desertification In : *Desertification Control Bulletin*, 20 : 6 -18.

Dregne, H.E. and N.T. Chou (1992): Global desertification dimensions and costs. In: Dregne, H.E. (Ed.): Degradation and restoration of arid lands. Lubbock: Texas Tech. University.

Dregne, H.E.1983. Desertification of Arid Lands. London: Harwood Academia Publishers.

Dregne HE (1978) Desertification: man's abuse of the land. *Journal of Soil and Water Conservation* 33: 11-14.

Dregne HE Kassas M and Rozanov (1991) A new assessment of the world status of desertification. *Desertification Control Bulletin* 19: 6-18.

Dregne, H. E. 1986. Desertification of arid lands. In Physics of desertification, ed. F. El-Baz and M. H. A. Hassan. Dordrecht, The Netherlands: Martinus, Nijhoff.

EEA, 2008. Mapping sensitivity to desertification (DISMED). Working paper on methodology. Version 2, January 2008. Prepared by: Francisco Domingues and Jaume Fons-Esteve, ETC/LUSI - Autonomous University of Barcelona.

Ekins, Paul and Hillman, M. and Hutchinson, R. The Gaia Atlas of Green Economics. New York: Anchor Books By Doubleday, 1992.

Eswaran, H., Reich, P. and F. Beinroth (2001): Global Desertification Tension Zones. In: D.E. Stott, R.H. Mohtar and G.C. Steinhardt (eds.): Sustaining the Global Farm. Selected Papers from the 10th International Soil Conservation Organization Meeting held May 24-29, 1999 at Purdue University and USDA-ARS National Soil Erosion Research Laboratory.

FGEB (Fundación Grupo Esquel Brazil) (1994): Anales del Taller Latino Americano de la Desertificación, FGEB, Fortaleza, Ceará, Brazil.

Glantz MH and Orlovsky N (1983) Desertification: A review of the concept. *Desertification Control Bulletin* 9: 15-22.

Gooland, R., and Asibey, E., and Post, J. and Dyson, M. Tropical Moist Forest Management: the urgent transition to sustainability. Washington, DC: World Bank, 1990.

IPCC (2001). Climate Change 2001: Third Assessment Report of the Intergovernmental Panel on Climate Change. IPCC, Geneva.

IPCC, 2007. Summary for Policymakers. In: Climate Change 2007: Impacts, Adaptation and Vulnerability. Contribution of Working Group II to the Fourth Assessment Report of the Intergovernmental Panel on Climate Change,

Kharin, N., R. Tateishi and H. Harahsheh (2000): A New Desertification Map of Asia. In: *Desertification Control Bulletin*, 36, 5-17.

Jagdish C. Katyal and Paul L.G. Vlek. 2000. Desertification - Concept, Causes and Amelioration, ZEF - Discussion Papers On Development Policy No. 33, Center for Development Research, Bonn, October 2000, pp. 65.

Lamprey HF (1975) Report on the Desert Encroachment Reconnaissance in Northern Sudan,21 October to 10 November 1975. Report of 1975 published as an article in 1988. In *Desertification Control Bulletin* 17: 1-7.

Mabutt JA (1984) A new global assessment of the status and trends of desertification. *Environmental Conservation* 11: 100-113.

Mainguet M (1994) Desertification-Natural Background and Human Mismanagement. Springer-Verlag, Berlin, Germany.

McIntire, J. (1994): A review of the soil conservation sector in Mexico. In: Economic and institutional analyses of soil conservation projects in Central America and the Car ibbean, ed. E. Lutz, S. Pagiola, and C. Reiche. A CATIE World Bank Project. World Bank Environment Paper 8. Washington, D.C.: The World Bank.

NAP (UNCCD National Action Programme) Turkmenistan 1996

Nelson R (1988) Dryland Management: The Desertification Problem. Environmental Department Working Paper 8. World Bank, Washington DC, USA.

Oldeman, L. R. (1998): Soil degradation: A threat to food security? Report 98/01. Wageningen, The Netherlands: International Soil Reference and Information Centre.

M.L. Parry, O.F. Canziani, J.P. Palutikof, P.J. van der Linden and C.E. Hanson, Eds., Cambridge University Press, Cambridge, UK, 7-22.

Pimentel, D., Harvey , C., Resosudarmo, P., Sinclair, K., D. Kurz, McClair, M., Crist, S., Shpritz. L., Fitton, L., Saffouri, Blair, R. (1995): Environmental and Economic Costs of Soil Erosion and Conservation Benefits. In: *Science*, 267, S.1117-1123.

Rapp A (1974) A Review of Desertification in Africa - Water, Vegetation and Man. Secretariat for International Ecology, Stockholm, Sweden.

Reich, P.F., S.T. Numbem, R.A. Almaraz and H. Eswaran (2001): Land resource stresses and desertification in Africa. In: Bridges, E.M., I.D. Hannam, L.R. Oldeman, F.W.T. Pening de Vries, S.J. Scherr, and S. Sompatpanit (eds.). Responses to Land Degradation. Proc. 2nd. International Conference on Land Degradation and Desertification, Khon Kaen, Thailand. Oxford Press, New Delhi, India.

Rozanov BG (1982) Assessing, monitoring and combating desertification. In Desertification and Soils Policy. Transactions of the 12th International Congress of Soil Science, New Delhi, India. pp 56-68.

Shafrits, Jay M. and Williams, Phil and Calinger, Ronald S. The Dictionary of 20th Century World Politics. New York: Henry Hold and Company, 1993.

Smaling EMA Nandwa SM and Janssen BH (1997) Soil fertility in Africa is at stake. In Replenishing Soil Fertility in Africa. Soil Science Society. America Special Publication # 51. American Society of Agronomy and Soil Science Society of America, Madison,Wisconsin, USA.

Solórzano, R., de Camino, R. and R. Woodward (1991): Accounts overdue: Natural resource depreciation in Costa Rica. Washington, DC: World Resources Institute.

Thomas DSG and Middleton NJ (1996) Desertification: Exploding the Myth. John Wiley & Sons Ltd., West Sussex, England

United Nations (1978) United Nations Conference on Desertification. Round-up, Plan of Action, and Resolutions, United Nations, New York, 43 p.

UNCED (1992) Earth Summit'92. The UN Conference on Environment and Development. Rio de Janeiro, Brazil

UNDP/UNSO (1997): Aridity zones and dryland population: an assessment of population levels in the world's drylands with particular reference to Africa. New York: UNDP Office to Combat Desertification and Drought (UNSO).

UNEP-UNCOD (1978) United Nations Conference on Desertification, 29 August - 9 September 1977.

UNEP (1997): World Atlas of Desertification. Second Edition. Data from GLASOD (Global Assessment of Soil Degradation) by ISRIC, UNEP, FAO (1990).

UNEP/GRID Geneva 2003, Geo Data Portal, retrievable under: http://geodata.grid.unep.ch/page.php

Warren A and Maizels JK (1977) Ecological change and desertification. UNCOD A / CONF. 74-77:1UNCCD (1994). Text of the United Nations Convention to Combat Desertification. United Nations Secretariat, New York

Wood, S., K. Sebastian and S. Scherr (2000): PAGE - Pilot Analysis of Global Agroecosystems. Washington, D.C.: World Resources Institute and International Food Policy Research Institute.

World Bank (1992): Malawi - Economic report on environmental policy, Vols. I and II. Report No. 9888-MAI. Washington D.C.: The World Bank.

World Resources Institute (WRI)(2000): World Resources 2000-2001: People and Ecosystems: The Fraying Web of Life.

Young, A. (ed.)(1994): Land degradation in South Asia: its severity, causes and effects upon the people". World Soil Resources Report 78.

**Websites**

http://www.earth-policy.org/books/out/out_table_of_contents

Action programmes: Africa. http://www.unccd.int/actionprogrammes/africa/africa.php#subregional

Climate risk management in Africa: Learning from practice. http://iriportal.ldeo.columbia.edu/portal/server.pt/gateway/PTARGS_0_2_1170_0_0_18/policy%20brief_en.pdf

Combating desertification in Africa. http://www.unccd.int/publicinfo/factsheets/showFS.php?number=11

Drought Risk and Development Policy: Discussion Paper for the UNDP-DDC/BCPR and UN - ISDR

Expert Workshop Drought Risk and Development Policy, January 31 - February 2, 2005, Nairobi.

Press Release on Joint International Conference Algiers, Dec. 17-19, 2006.

http://www.inweh.unu.edu/inweh/drylands/Algiers_news_release-Final.pdf

# Chapter - 18

# Numerical Groundwater Flow Modeling Using MODFLOW : A Case Study of Osian-Mandore Pilot area of Jodhpur District, Rajasthan

**Sneha**

## Abstract

*Groundwater flow modeling in hydrological sciences is one of the main tools used to assess the resource potential in terms of quantity and to predict the future impacts under different stresses/scenarios. This case study is prepared to present the groundwater flow model for the Osian-Mandore pilot area with 12 villages of Mandore Block and 21 village of Osian Block in Jodhpur district of Rajasthan. The spatial model parameters and spatio-temporal fluxes were acquired and processed using remote sensing, GIS and Groundwater flow modeling interfaces. MODFLOW software was used to model the groundwater flow system in three interfaces PMWIN 5.3.0, Groundwater Modeling System and Visual MODFLOW. By studying the litholog five aquifer systems with different composition and vertical spread were conceptualized. With reference to approximation of sinks and sources, the sources were recharge, and initial head and the sinks were evapotranspiration and discharge. The recharge and evapotranspiration for the study area was calculated by Penman-Monteith method and the spatio-temporal data for the same was assessed and used in the transient state model. The steady state model was recalibrated after the layer elevation corrections and input of additional data obtained. Finally, the transient groundwater flow model was developed using validated steady state model and calibrated for 5 years (2004-2008) with hydraulic head data at 18 locations of the existing piezometeres in the study area. In the study, the transient groundwater model helped to describe the flow regimes more reliably than steady. state because of the introduction of spatio-temporal fluxes constrain.*

*Keywords: Groundwater flow modeling, Remote Sensing, GIS, MODFLOW.*

# 1 Introduction

## 1.1 General

Groundwater is a distinguished component of the hydrologic cycle. Groundwater plays an important role in the supply of the ever increasing water demand (Rajapakse, 2009). Therefore an accurate estimation of groundwater resources is a prerequisite for any sustainable water management especially in water scarce (semi-) arid regions (Rajapakse, 2009). Groundwater modeling is an essential tool in the evaluation of groundwater flow and quantification of resources. It enables to predict the behavior of the groundwater system in response to future stresses due to abstractions or land cover and climate changes. Numerical groundwater flow models solve the solution of the mathematical equation and gives approximate solution to the problem, i.e., unknown variable is solved at discrete points in space (steady-state flow) and time (transient flow). The solution for the model gives the distribution of hydraulic head and describes flow. After a model is calibrated to observation values during existing conditions in the flow system, it is then used to predict future conditions with a different set of boundary conditions.

The study focuses on the use of the popular Finite Difference Method, a widely used numerical method to solve the governing partial differential equations in hydrogeology. Finite-difference model MODFLOW and its associated modular computer program simulate flow in three dimensions (HR Wallinford Report, 2002).

The MODFLOW (Modular Groundwater Flow Model) is a computer program constructed in the early 1980's and is regarded as a finite difference numerical scheme applied to the governing groundwater flow equations (Harbaugh and McDonald, 1988). To define the groundwater flow behavior the integration of proper parameters/ fluxes collected from data sources with different method is essential to select an appropriate model. The complicacies to be addressed are the spatio-temporal variability of hydraulic parameters, fluxes, recharge and evapotranspiration. For the study quantitative groundwater modeling algorithm is used which is based on the groundwater flow equations developed in combination of Darcy's law and continuity of mass equation (Anderson and Woessner, 1992).

## 1.2 Background of the research

The study area Osian-Mandore is located between north latitude 26°19′-26°41′ and east longitude 72°50′-73°01′ covering an area of 589.09 sq. km. The winter season which starts in November lasts up to February is followed by summer lasting from April to June. The minimum and maximum temperature varies from 3°C to 47°C. During April to June mean maximum temperature ranges from 38°C - 47°C and mean minimum lies between 6°C-10°C during December-January. The period from July to mid-September forms the south west monsoon season with maximum in July and August The mean annual rainfall of the area is 251.8 mm. In the last ten years (1995-2004) the annual rainfall varied between 47 and 438 mm. The average annual rainfall of the area is estimated as 318.70 mm (Census of India, 1991).

Study Area: Osian-Mandore, Jodhpur

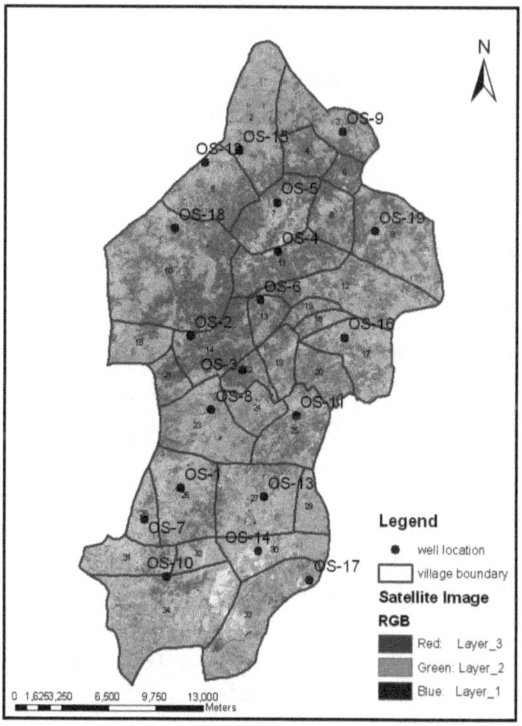


**Figure 1 :** Study area-Osian-Mandore and location of Monitoring Wells

The area around Osian-Mandore is more or less flat to slight undulating plains. Southern part of the area is comprised of isolated hills to dissected plateaus rising more than 300 meters. Although no major river is flowing through the area but there are numerous drainages originating from the hillocks in the south.

Major consumption (domestic, irrigation industry) of water is dependent on the available ground water resource. Jodhpur sandstone in the area mainly acts as a major ground water potential aquifer. This sandstone is red to pink or buffed colored, very hard and compact and at places micaceous. Lithological cross section along NW-SE direction across the Osian town (Ref. Page-486; GW Atlas of Rajasthan, 1999) shows that the thickness of the sandstone increases from southeast to north westward (Ref. Page-547; GW Atlas of Rajasthan, 1999). In the south central part, the thickness of the sandstone is found to be more than 200 meters. Except in certain pockets, the average water level in the area varies between 10-100m. Most of the tube wells discharge at the rate of 36 to 526 $m^3/d$ (e.g Binjwariya village). Osian block has been marked as over exploitable area and Mandore block is still in safer condition. Water level records in the last 10 years (2000-1009) have shown drastic changes.

## 1.3 Objective

The general objective is to prepare groundwater flow model for the study area, calibrate steady state flow and then setup and calibrate fully transient model. To carry out the study following steps were carried out:

- Review the data available about the material properties (porosity, permeability, and storativity), heads, and fluxes in the region;
- Develop a conceptual system; and
- Simulate the conceptual system developed in previous step using a mathematical model, solving numerically by finite difference method.

# 2 Review of Literature

## 2.1 Groundwater flow model

The modeling tool for groundwater depends on the nature of management, problem, decision variables, objective functions, and constraint; for which the response can be derived in the form of future spatial distribution of water levels, and/or contaminant concentrations. The results are then used to assess the impact of changes in the regime on the environment, setting/optimizing networks, and setting up groundwater protection zones. Solution to the complex groundwater system depends on accurate representation of the real world scenario to the conceptual model. The assumptions in the conceptual model typically involve the direction of flow, geometry of the aquifer, the heterogeneity or anisotropy of the sediments or bedrocks within the aquifer (Rajamanicakam et.al, 2010). Mathematical models of groundwater flow and solute transport can be solved generally with two broad approaches, analytical and numerical. Analytical solution of the mathematical equation gives exact solution to the problem, i.e., the unknown variable is solved continuously for every point in space (steady-state flow) and time (transient flow). Numerical solution of the mathematical equation gives approximate solution to the problem, i.e., the unknown variable is solved at discrete points in space (steady-state flow) and time (transient flow).The three main numerical methods that are used in hydrogeology are Finite Difference Method (FDM), Finite Element Method (FEM) and Boundary Element Method (AEM). The FEM and FDM methods are the most popular. The differential equations governing the flow (or transport) are transformed into a set of algebraic equations called Partial Differential Equations (PDE). MODFLOW is based on FDM. Two different grids are used in FDM: block-centered and mesh-centered. PDE are converted to a set of N algebraic equations involving N unknown values at these particular points. In the study block-centered grid is used to derive the FD approximations of the flow equations. Finally, a variety of solution techniques can be used to solve a set of linear algebraic equations. By combining the mathematical relation describing the physical principles of Darcy's law and mass balance equations, a general groundwater flow equation, which is a partial differential equation (PDE) is derived. There are several different forms of the general flow equation depending on whether the flow is two-dimensional or three-dimensional, isotropic or anisotropic, and transient or steady state. For transient problems, both initial and boundary conditions are needed; the flow and transport equations are solved at both discrete grid points and discrete times. Boundary conditions include things like heads at surface waters in contact with the aquifer, the location and discharge rate of a pumping well or a leaching irrigation field. For the no-flow boundary the flux boundary is specified flux of zero. The two most common types of BC are specified head and specified flux conditions.

## 2.2 Governing equations on groundwater models

### 2.2.1 Flow Equation

Modeling is translating the physical system into mathematical terms. The governing flow equation for three-dimensional saturated flow in saturated porous media is:

$$\frac{\partial}{\partial x}\left(Kx\frac{\partial y}{\partial x}\right)+\frac{\partial}{\partial y}\left(Ky\frac{\partial h}{\partial x}\right)+\frac{\partial}{\partial z}\left(Kz\frac{\partial h}{\partial z}\right)\pm Q = Ss\frac{\partial h}{\partial y}$$

*Where,*

Kx, Ky, Kz = hydraulic conductivity along the x, y, z axis which are assumed, m/sec [$LT^{-1}$]

| | | |
|---|---|---|
| h | = | piezometric head, m [L] |
| Q | = | volumetric flux per unit volume representing source/sink terms, $[T^{-1}]$ |
| Ss | = | specific storage coefficient defined as the volume of water released from storage per unit change in head per unit volume of porous material $[L^{-1}]$ |
| t | = | Time, (s) [T] |

## 2.3 Visual MODFLOW, PMWIN and GMS

Visual MODFLOW is a modeling environment for practical applications in three-dimensional groundwater flow and contaminant transport simulation (Waterloo Hydrogeologic Inc., Tutorial). Visual MODFLOW is a Waterloo Hydrologic Inc. Software product. It allows to graphically design the model grid, properties and boundary conditions, visualize the model input parameters in two or three dimensions, run the groundwater flow, pathline and contaminant transport simulations, automatically calibrate the model using Win PEST or manual methods, display and interpret the modeling results in three-dimensional space. Run the simulation (MODFLOW, MODPATH and MT3D are included), calibrate the model. The interface is divided into three separate modules, the Input module, the Run module, and the Output module. PMWIN 5.3 (Processing MODFLOW for Windows) is a graphical user-interface which supports models and programs with several other useful modeling tools (Chiang and Kinzelbach, 1998). The tools include presentation tool, field interpolator tool, result extractor tool, water budget calculator tool, and graph viewer. It supports additional packages with the original MODFLOW like, Time-Variant Specified-Head and many more. GMS (Groundwater Modeling System) is a graphical user environment by Department of Defense for performing groundwater simulations (EMRL, 2004b). The interface is developed by Environmental Modeling research Laboratory of Brigham Young University in partnership with the U.S. Army Engineer Waterways Experiment Station. It consist of a number of analysis codes like, MODFLOW, MT3DMS, RT3D, SEAM3D, MODPATH, MODAEM, SEEP2D, FEMWATER, WASH123D, UTCHEM, etc. Defining model properties can be done in a GIS interface that is independent of the simulation codes. From the GIS interface model parameters can be transferred to the appropriate grid cells/mesh elements.

## 3.0 Methodology

## 3.1 Groundwater model development process

The process of aquifer modeling was carried out in the following steps:

- Identification of the parameters characterizing the physical framework of the aquifer and stress acting on it.
- Data compilation of the relevant hydrogeological parameters.
- Interpolation/extrapolation of these parameters to characterize the entire area under study.
- Integration of the entire hydrological data to conceptualize.
- Selection of appropriate mathematical equations describing the groundwater regime.
- Calibration of the model for steady and transient condition.

The study area of approx 589.09 $km^2$ was divided into a uniform square grid of 500 m spacing, comprising 100 rows and 60 columns. The vertical expansion of the model has been divided in five aquifers. The elevation for the topmost layer has been derived from SRTM data and for the remaining layers the litholog data has been used.

## 3.2 Data collection

For the study secondary data was collected from Central Ground Water Board, and State Ground Water Department.

## 3.3 Input parameters

The recharge and evapotranspiration was given within domain boundaries. While the discharge was considered as point source distributed in the study area. *Evapotranspiration* package simulated the effects of plant transpiration and direct evaporation in removing water from the saturated groundwater regime. The *Recharge* package was used to simulate aerially distributed recharge to the groundwater system. Well location and the water level information for the stress period listed is provided and the head values was used for calibration. The water level data and the location of wells are listed in Table 1.

**Table-1** : Stress periods (based on the period of observed water level data) and the number of Observation wells

| Wells (OB-1 to OB-18) | SP-1 May 04 | SP-2 Nov 04 | SP-3 May 05 | SP-4 Nov 05 | SP-5 May 06 | SP-6 Nov 06 | SP-7 May 07 | SP-8 Nov 07 | SP-9 May 08 | SP-10 Nov 08 |
|---|---|---|---|---|---|---|---|---|---|---|

### 3.3.1 Geological Settings

By studying the litholog five aquifer systems with different composition and vertical spread were conceptualized. The multilayer aquifer system was composed of first unconfined-sandstone, second-semi-confined-shale & clay, third-confined, fourth-semi-confined-shale and fifth confined-sandstone, with the latter as the main rock type. By using the SRTM DEM the actual elevation (amsl) of the top layer was entered in the model. By using litholog data the layer thickness for all the five layers were assigned.

### 3.3.2 Wells

Based on the water level statistics given by the Groundwater Division, wells were located in the grid and the pumping rate was assigned. Uniform pumping rate were assumed throughout the year.

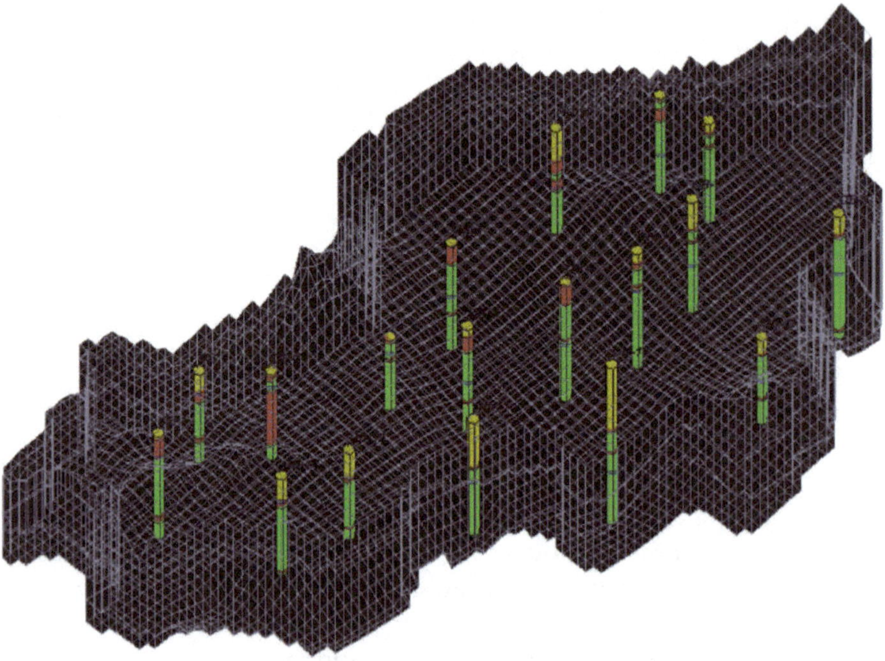

**Figure 2 :** Bore wells plotted in GMS

The calibration package of the MODFLOW saves the calculated heads at the locations of specified observation wells. This information was used to compare heads with observed heads, produce calibration statistics, and produce hydrographs at observation wells Therefore head observation wells were located in the grid. With the available literature and the CGWB report, the aquifer properties assigned in the model are listed in Table 2.

**Table-2** : Aquifer properties for the study area

| Model Properties | Unconfined | Semi-Confined | Confined | Semi-Confined | Confined |
|---|---|---|---|---|---|
| Horizontal HC m/s | 0.1 | 1e-09 | 0.000001 | 1e-13 | 0.0001 |
| Vertical HC m/s | 0.01 | 1e-10 | 0.0000001 | 1e-14 | 0.00001 |
| Effective Porosity | 0.15 | 0.44 | 0.035 | 0.005 | 0.03 |
| Storage Coefficient | 0.001 | 0.001 | 0.001 | 0.001 | 0.001 |
| Specific yield | 0.25 | 0.25 | 0.25 | 0.25 | 0.25 |
| Specific Storage m-1 | 0.0001 | 0.0001 | 0.0001 | 0.0001 | 0.0001 |

HC-Hydraulic Conductivity

### 3.3.3 Model Boundary Condition

The study area boundary is assigned as no flow boundary, because there is no influence of water table due to the denudation hills surrounding the area.

## 3.4 Model run

MODFLOW was run, calibrated and validated with 5 years data. The initial values of the parameters such as porosity, hydraulic conductivity, etc. were used for simulation. Calibration of the model is done by adjusting the input parameters. The model MODFLOW was solved using Strongly Implicit Procedure package (SIP). SIP is a method for solving a large system of simultaneous linear equations by iterations. The Model was run for transient state with 10 stress period.

## 3.5 Results and discussion

The results for the run in transient state comprises of calculated head at every node of the grid. The Head observations at the Bore well locations and the calculated value for the same is used for the calibration of the model. The results include the measure of difference between the results for head from different software and comparing them with the observed head values for each stress period. The Head values for the initial wells are matching well with the observed values. The comparison of the head values for the same well from different software came different from each other and the observed values. The wells OB-11, OB-12, OB-13, OB-14, OB-15, OB-16, and OB-17 are showing variability for stress period one in all the three software and the observed head. This can be seen in the figure 3. In stress period 2 the observation well OB-5 head value for the software is coming similar but the observed head is greater than the calculated (Fig 3). For the well OB-11 the value observed and calculated from VMOD are much closer as compared to calculated head in GMS and PMWIN. The head value for OB-16 for VMOD and GMS are closer for stress period 1.In other stress period the same head values are similar for VMOD and PMWIN. For a given general equation, there are an infinite number of possible solutions. For steady-state flow, the unique and appropriate solution is one that matches the particular boundary conditions (BC) of the conceptual model. For transient flow system and for solute transport, both initial condition and BC are required to obtain the unique solutions of head and concentrations.

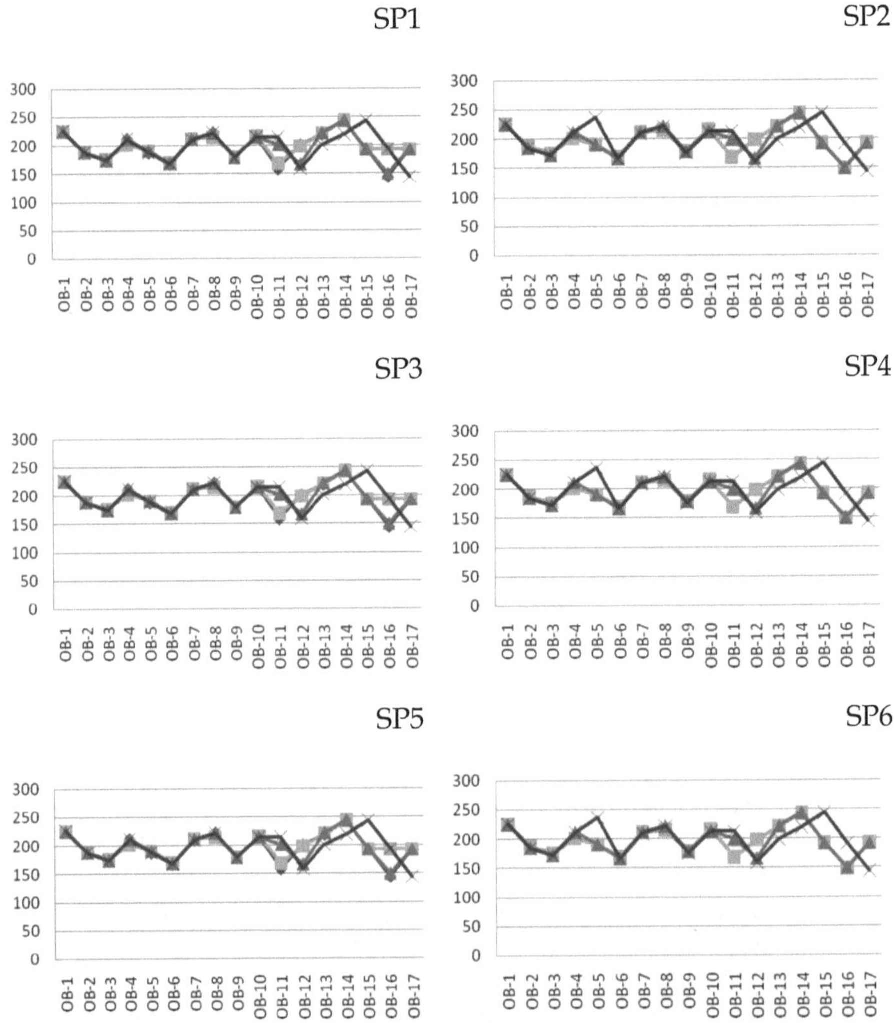


*contd.*

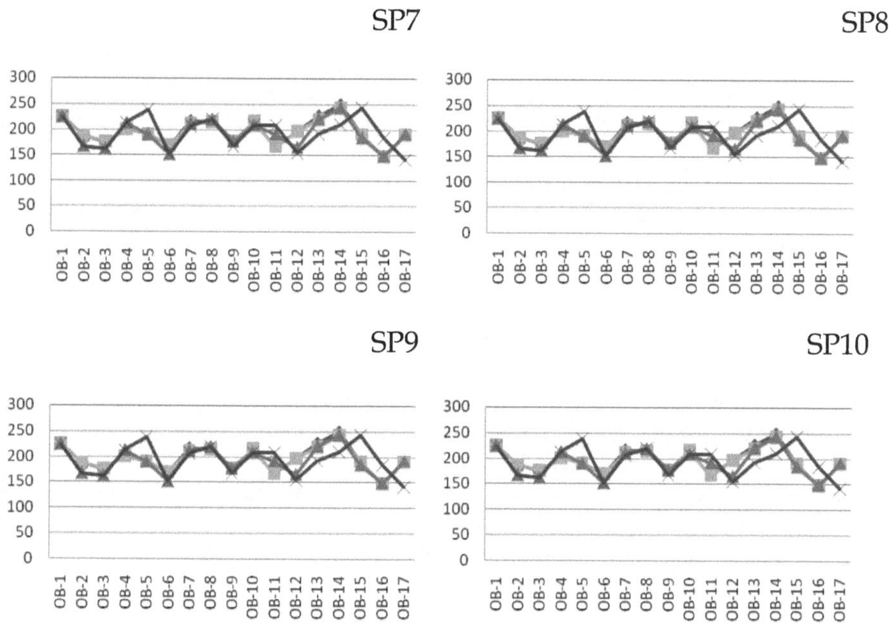


**Figure 3 :** Head Observation for Stress period 1 (SP1) to Stress Period 2 (SP2): Calculated and Observed head in three different interfaces

In flow model calibration, simulated heads and Darcy fluxes are typically compared to their observed counterparts. If a model is well calibrated, there will be some random deviations between simulated and observed data, but there will not be systematic deviations. If there are systematic deviations such as if most of the simulated heads exceed the observed heads, the calibration is poor and adjustments should be made.

When interpreting flow model results, it is important to bear in mind the simplifying assumptions that went into creating the conceptual model. Although a well-calibrated mathematical model may simulate the conceptual model with great accuracy and aesthetic appeal, it is likely to be only a crude representation of the real flow system. In some cases, there is no unique solution to the calibration problem; several different combinations of input parameters will result in models that all equally fit the observations. Problems of non-uniqueness are more prevalent when the observation data are limited. After a model is calibrated to values observed during the existing conditions in the flow system, it is then used to predict future conditions with a different set of boundary conditions.

## References

Anderson, M.P. and Woessner, W.W., 1992. Applied groundwater modeling: simulation of flow and adjective transport

Census of India, 1991

Chiang, W.H. and Kinzelbach, W., 2001. Processing MODFLOW (version 5.3), A simulation systemfor modeling groundwater flow and pollution.

EMRL (Environmental Modeling Research Laboratory), 2004b. Groundwater Modeling System Reference Manual, Version 5.0. Environmental Modeling Research Laboratory, Provo, Utah

Freee, R.A. and Cherry, J.A., 1979 Groundwater. Prentice Hall (Englewood Cliffs, New York).

Groundwater Atlas of Rajasthan, 1999

Harbaugh, A.W. and Mc Donald, M.G., 1988. A modular three-dimensional finite difference groundwater flow model MODFLOW.

HR Wallingford Report SR 598, 2002. State of art review Work Package 1

Rajamanickam, Nagam, 2010. Groundwater quality modeling of Amaravathi river basin of Karur district, Tamil Nadu, Using MODFLOW, *International Journal of Environmental Sciences* 1(1) : 94-95

Rajapakse, 2009. Numerical Groundwater Flow and Solute Transport Modelling-A case study of Sardon Catchment in Spain

Waterloo Hydrologic Inc., Visual MODFLOW Premium Demo Tutorial

# Chapter - 19

# Satellite Remote Sensing and Geospatial Technology to Support Assessment of Crop Water Information for Sustainable Water Resource Management

**Kamini Yadav**

## Abstract

*This paper sets out to describe key results and experiences obtained during the assessment of the crop response to water which varies strongly from crop to crop depending on its specific genetic characteristics and adaptation to ecological conditions. The specific crop water information for the different crops is compiled in a crop-based data base. This includes crop water parameters, yield response to water, as well as selection of economy crops. Existing natural water resources can sustain the crop production with high yield as well as economic value with minimal environment impact. Most of the crop varieties get all of its required water from rainfall- water but due to climate change there is always a demand of irrigation water from the ground water resources. Thus, there is a need to evaluate the crop water information in relation to ground water extraction as a source of irrigation. To maintain the eco-crop database and crop diversification both surface and ground water resources should be manage in a sustainable manner for the future use. In order to evaluate the crop water requirement, CROPWAT Tool has been used which in turn can be considered as the amount of extraction or discharge from the ground in the absence of surface water deposit or rainfall in semi-arid and arid regions. To have an idea of the water budget of the area SWAT tool has been used. Aquifer dynamics and properties along with the rate of infiltration and run-off determine the fluctuations in the water level during pre and post monsoon seasons. These fluctuations have been simulated through MODFLOW and analyzed in relation to agriculture water which contributes to the exploitation and depletion of ground water reserve along with the domestic and industrial use.*

*Keywords: Crop water information, Eco-crop database, CROPWAT, SWAT, Aquifer dynamics, MODFLOW.*

## Introduction

Water for public supply, irrigation and industry is needed water directly for people, whilst water for ecosystems is water indirectly for people. In the light of importance of water in society, it is essential to secure water for Ecosystem and human well-being. Management of water resources has become a major effort for governments around the world. As populations grow, the need for water for domestic, industrial, and agricultural uses increases, while resources are diminishing due to overuse and contamination. Population growth and commercial crop improvements with high yield and economic value are driving up demand for agricultural water.

Agriculture plays a vital role in the economic development of the country and continues to be the backbone of the economy. In the coming years, agriculture will face many challenges due to sharp rise in growth population. The demand for food and use in non-agricultural sectors also, like industry and urban centers will be increased which will further put greater demand on land and water resources available for agriculture. In the condition of erratic irregular pattern of rainfall, the only source from where irrigation water can be achieved is through the extraction of the ground water. The revolution of the ground water irrigation has been achieved at the stake of depletion of the available water resources. The rapid growth of groundwater use in recent years brought massive benefits to farmers, but also resulted in widespread overexploitation of ground water resources. In some basins, water tables have been falling at an alarming rate. In large areas of India and China, groundwater levels are falling by one to three meters a year. The availability of water for irrigated agriculture is decreasing day by day. Hence, there is a greater need for conservation and efficient utilization of water resources.

The water resource management has increasingly become global issue (Source: Re-engaging in Agricultural Water Management, 2005. Directions in Development, World Bank).Different decision support systems and technologies are being implemented to contribute efficiently in the sustainable water resource management. The most innovative Geospatial technology integrated with the hydrogeological parameters are the most demanding techniques in the present conditions of the depleting resources.

The use of advanced technology tools such as satellite remote sensing, geographic information system (GIS) techniques, and hydrologic modeling can greatly help to improve the economic value of the crops, their production and conservation of water resources. Aquifer characteristics can be simulated in the hydrologic models based on the input data acquired from the satellite images and processed in Geographical Information System. Models allow us to understand the data and analyze the effects of different management options. Modern hydrogeology has powerful tools for modeling water transport and flow together. Models integrating the full range of hydrological processes are emerging. Thus, on the basis of hydrologic processes crop water requirement can be estimated for any particular crop in different climatic conditions and crop combinations can be suggested to conserve the depleting resources of groundwater.

The research thrust of the study is to build the idea of the cultivation of crops suitable to the micro climate so as to conserve and manage the water resources which are being used for irrigation continuously in an unlimited amount. A decision can be made in the selection of crops which would be utilizing minimum amount of water when there are scarce resources of ground water.

It is essential to plan and implement rational policies to minimize the regional imbalances of agricultural production. The present growth in crop production has to be sustained considering ecological and economic factors. The existing as well as emerging frontiers must be examined to enhance agricultural growth on a sustainable basis. The understanding of agro-climatic relationship through crop-weather conditions, soil fertility, water use efficiency, rain water management, and appropriate cropping patterns based on regional resource potential must be emphasized. Training of farmers has to be adapted to the changing technological environment. (Source: Re-engaging in Agricultural Water Management, 2005. Directions in Development, World Bank).

## Objectives and Approach

The main objective of the study is to identify the crop water requirement of different crop varieties on the basis of their crop information which depends on the crop growing season and crop coefficients.

The second objective is to simulate the dynamics of surface and ground water by implementing the estimated crop water requirement as the discharge or extraction in the growing season of that particular crop from the ground water resource

This abstraction and infiltration is based on the soil type and land use pattern in the study area. Thus, eventually on the basis of soil, land use and watershed, water balance will be required to have an idea of the surface conditions in the area. The approach will be based on the integration of the crop information with the aquifer parameters to study the change in the water level fluctuation due to exploitation of ground water as agriculture water. For this CROPWAT tool will integrate the crop information along with the soil data, rainfall, and temperature to result in the crop water requirement. SWAT tool will integrate the soil map, land use and drainage in the area to estimate the water balance as the amount of infiltration in the form of recharge and run off in the area. Further aquifer dynamics will be analyzed in the MODFLOW to be resulted in the fluctuation of the water level between the pre-and post-monsoon season in the growing season of Rabi crops.

## Study Area

The group of villages selected as the study area in the Jodhpur district of Rajasthan has been observed to have drastic changes in the groundwater level during a period of time. Due to scanty rainfall and having no major surface water bodies (river/lake/ponds, etc.) in the area, major consumption (domestic, irrigation and industry) is completely dependent on the availability of groundwater resources. The idea behind is the identification of the crop varieties having intense impact on the water resources in the particular area. The study area (Osian-Mandore) in Jodhpur, Rajasthan, India has been shown in Figure 1.

**Figure 1 :** Study area

## Material and Methodology

### • Digital data

In the present study, digital satellite data of IRS-P6 LISS-III (23.5m resolution) of late kharif season (17.12.2003) and Rabi (3.2.2004) seasons have been used for the image interpretation for various thematic layers.

- LULC
- Soil map
- Geomorphology
- Drainage
- Collateral data
- Soil Map (Source: RRSC, Jodhpur ISRO)
- Ground water table (Source GWD, State of Rajasthan (2000-2008)
- Peizometric reading (2003-2004) from GWD, State of Rajasthan
- SRTM data (USGS)
- Meteorological data (IMD)

Tools used :

- Erdas Imagine v/s 9.3
- Arc GIS 9.3
- Arc View 3.2a (Extension SWAT)
- Processing MODFLOW
- GMS 6.0
- CROPWAT 8.0

The methodology flow chart has been given in Figure 2 which is based on the analytical tools and statistical methods and performed in three main steps. First step involve the estimation of crop water requirement which will be integrated in comparison analysis of the crop having minimum requirement of water. Such kind of eco-crop database can be maintained where different crops can be selected on the basis of amount of agriculture water used. (Source: http://fao.org, Approved Crop Calendar.) The irrigation requirement for each crop was calculated from the CROPWAT in mm per 10 day. Then it was converted to m per day, which is considered to be discharged from the agricultural area in the Osian- Mandore area that is of 469.75 sq. km. Total consumption by each crop in the agriculture area from each of the well i.e. Crop water requirement (Discharge) of different selected Crops has been given in the Figure 3.

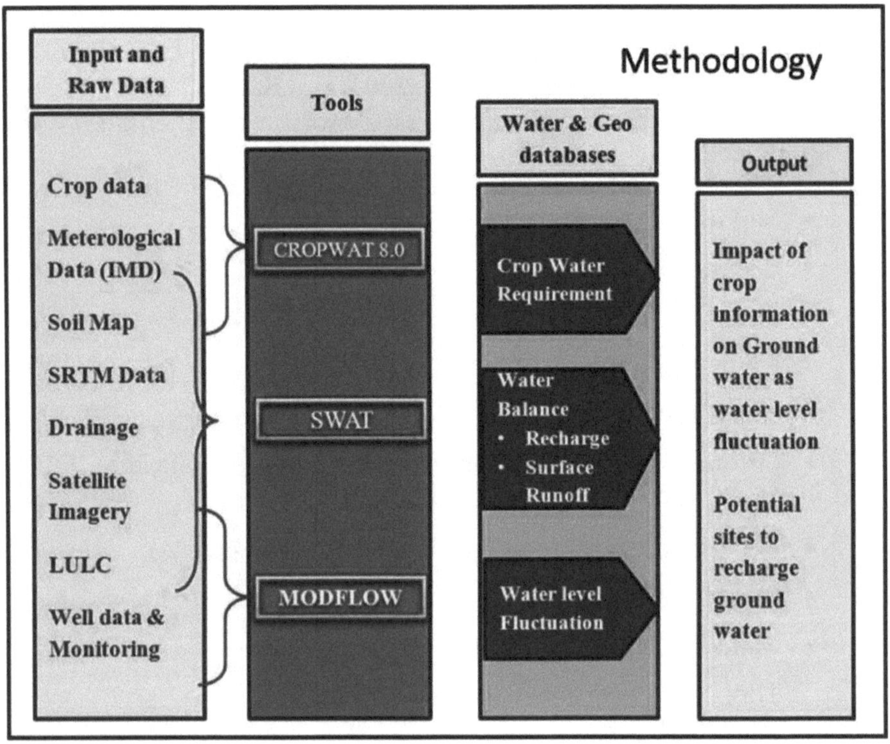


**Figure 2 :** Methodology

**Table 1 :** Irrigation requirement during growth of various crops

| Crop | Irr. Req. (mm) | Growing Period (days) | Irr. Req. (m) | Irr. Req. per day | Agri. Area ($10^4$Sq.m) | Total consumption | Discharge per well (cu m) (CWR) |
|---|---|---|---|---|---|---|---|
| Arandi | 270.9 | 110 | 0.27 | 0.0025 | 46975 | 1156866.14 | 60887.69 |
| Bajra | 297.0 | 105 | 0.30 | 0.0028 | 46975 | 1328721.43 | 69932.71 |
| Chilli | 704.2 | 135 | 0.70 | 0.0052 | 46975 | 2450355.19 | 128966.06 |
| Groundnut | 321.5 | 140 | 0.32 | 0.0023 | 46975 | 1078747.32 | 56776.17 |
| Jawar | 320.0 | 125 | 0.32 | 0.0026 | 46975 | 1202560.00 | 63292.63 |
| Moong | 270.8 | 95 | 0.27 | 0.0029 | 46975 | 1339034.74 | 70475.51 |
| Mustard | 286.9 | 110 | 0.29 | 0.0026 | 46975 | 1225193.41 | 64483.86 |
| Wheat | 494.7 | 160 | 0.49 | 0.0031 | 46975 | 1452408.28 | 76442.54 |

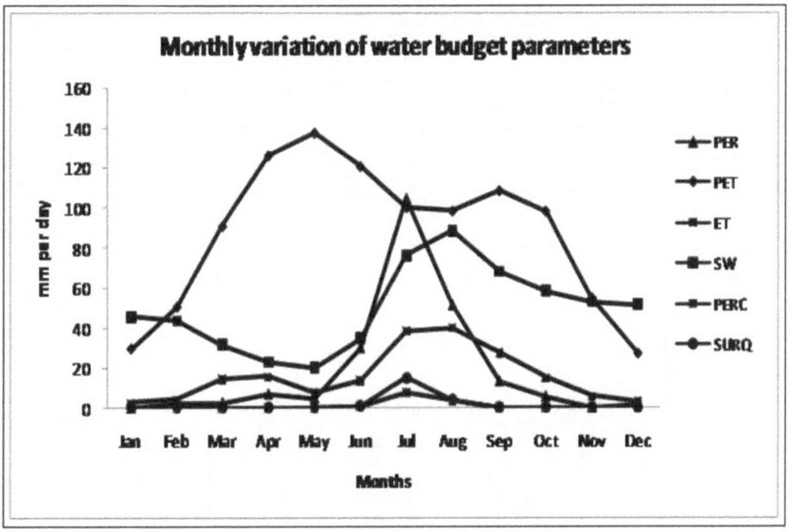


**Figure 3 :** Monthly variation of water budget parameters

On the basis of rainfall, temperature, humidity, wind speed and sunshine hours, reference evapotranspiration has been calculated using Climate and Rain module of the CROPWAT. This estimated reference evapotranspiration will be used for the estimation of crop water requirement using crop coefficient (Irrigation and drainage paper No. 56, FAO and Irrigation and Drainage Paper 24).

$$\text{CWR} = \text{KC} * \text{ETo} \qquad \text{Equation 1}$$

By using the FAO Penman- Monteith definition for ETo, one may calculate crop coefficients at research sites by relating the measured crop evapotranspiration (ETc) with the calculated ETo, i.e.

$$\text{Kc} = \frac{\text{ETc}}{\text{ETo}} \qquad \text{Equation 2}$$

Second step involve the estimation of the water budget in the area as what amount of precipitation has been infiltrated and flowed over the surface. Finally, based on discharge (crop water requirement) and recharge, fluctuation in ground water would be simulated to review the impact of crops on water resources. The equation for the estimation of water balance is:

$$\text{SWt} = SWo + \sum_{t=1}^{t}(Rday - Qsurf - Ea - Wsep - Qgw) \qquad \text{Equation 3}$$

where SWt is the final soil water content (mm $H_2O$), SW0 is the initial soil water content on day i (mm $H_2O$), t is the time (days), Rday is the amount of precipitation on day i (mm $H_2O$), Qsurf is the amount of surface runoff on day i (mm $H_2O$), Ea is the amount of evapotranspiration on day i (mm $H_2O$), Wseep is the amount of water entering the vadose zone from the soil profile on day i (mm $H_2O$), and Qgw is the amount of return flow on day i (mm $H_2O$). Monthly percolation, surface run off, evapotranspiration variation has been given in Figure 4. (Source: Soil and water assessment tool theoretical documentation).

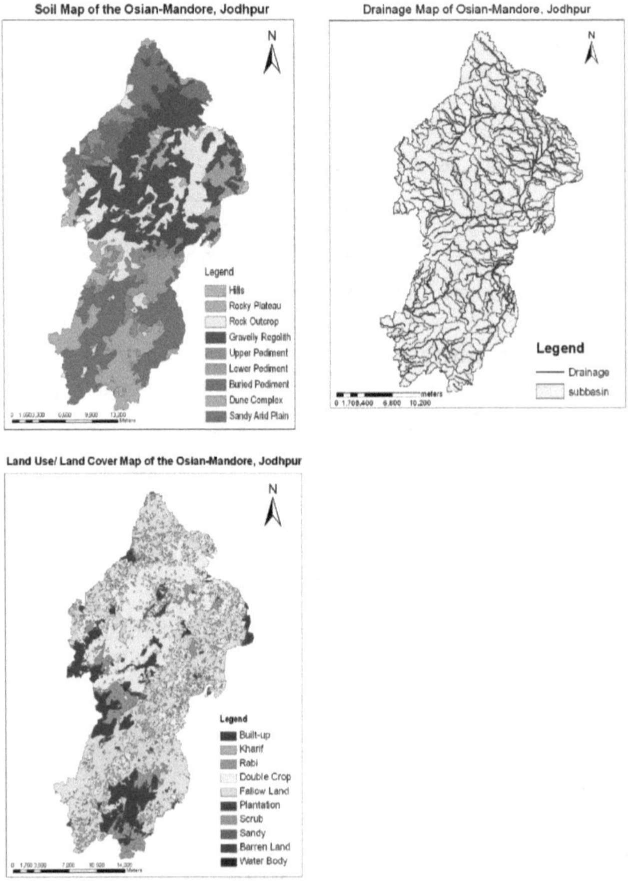


**Figure 4 :** (a) Soil map; (b) Drainage map; and (c) LU/LC map of study area

The fluctuation of water level is inter-related with the time of the precipitation. In the above graphs both rainfall pattern and fluctuations are being compared, there is change in the water level when there is rainfall in the months of May - June and again it comes down after being exploited for different uses. For each period two-dimensional representation was done on the basis of water level in that particular period. It has been observed that there is a change in the water level during 2004-2008. The volume of change either gain or loss is calculated by subtracting the Post-monsoon (December) value from that of Pre-monsoon (May). Thus, it can be inferred from the above graph that there is decreasing trend in the water level (in volume terms) till 2007 but during 2007-2008, there is slight increase in water level. The change in the cyclic behavior of the water table in the year 2007-2008 is evidenced by the construction of recharge sites during the period. The change in the water level during 2004-2008 shown in Figure 5 can be compared with the change in the rainfall pattern during 2004-2008 shown in Figure 6.

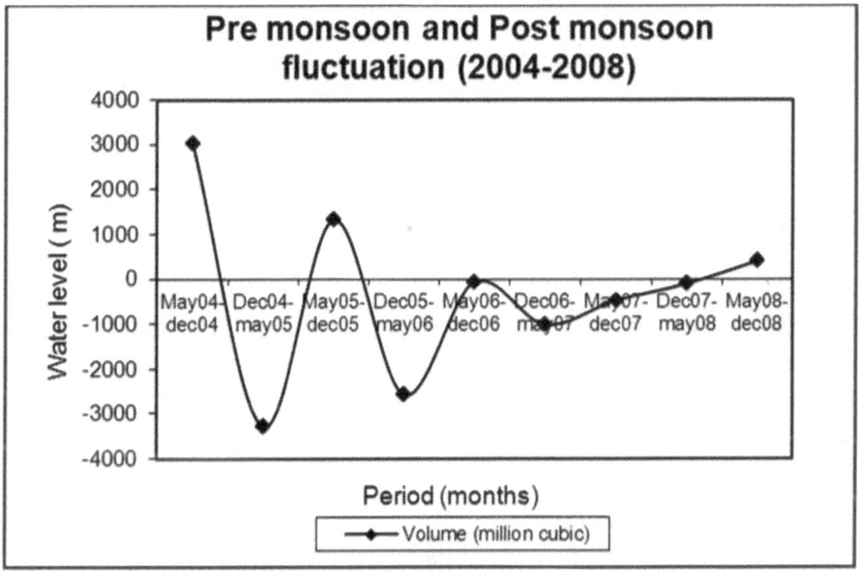


**Figure 5 :** Pre-monsoon and Post-monsoon fluctuations (2004-08)

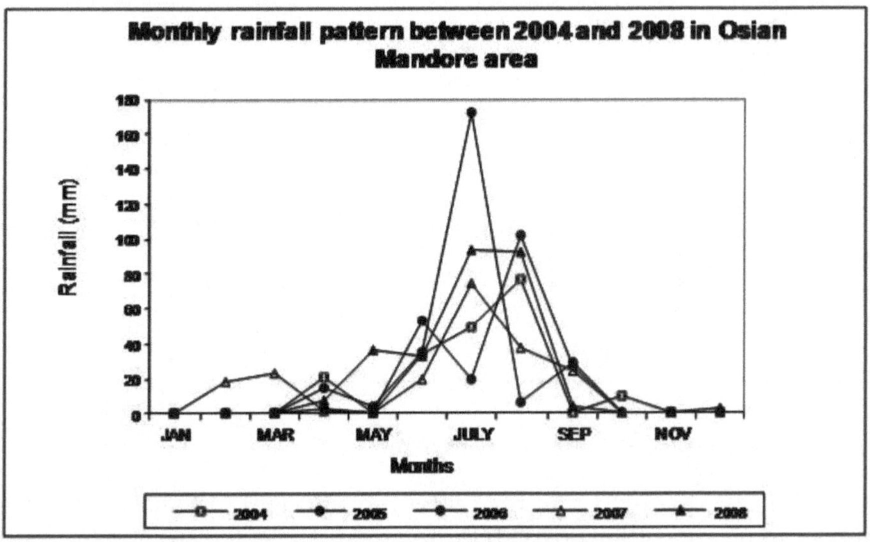


**Figure 6 :** Monthly rainfall pattern (2004-08)

The volume of water is more at the start of the monsoon season and it is extracted as forwarding step towards the end of the monsoon. Again the volume will be more after the precipitation occur and again comes down after the use in different activities. It has been observed that as we move from 2004 to 2008 the volume of water which is to be extracted from the ground water is decreasing. It means that in future ground water crisis will have to be faced. Due to the recharge structures constructed in the study area, the available amount of ground water has started to increase in the year 2007 and 2008.

On the basis of water balance and ground water fluctuation, the impact of agriculture water use can be observed and crops having less water requirement can be chosen for the cultivation which will create eco crop database.

## Result

The impact of each crop on all the wells can be shown graphically using the values of the matrix of the MODFLOW. For each crop, the discharge was loaded for its growing season only and the rest of the period of the year was considered as no discharge so that the impact of crops can be studied exclusively. The crop wise change in water level has been given as a result in Figure 7.

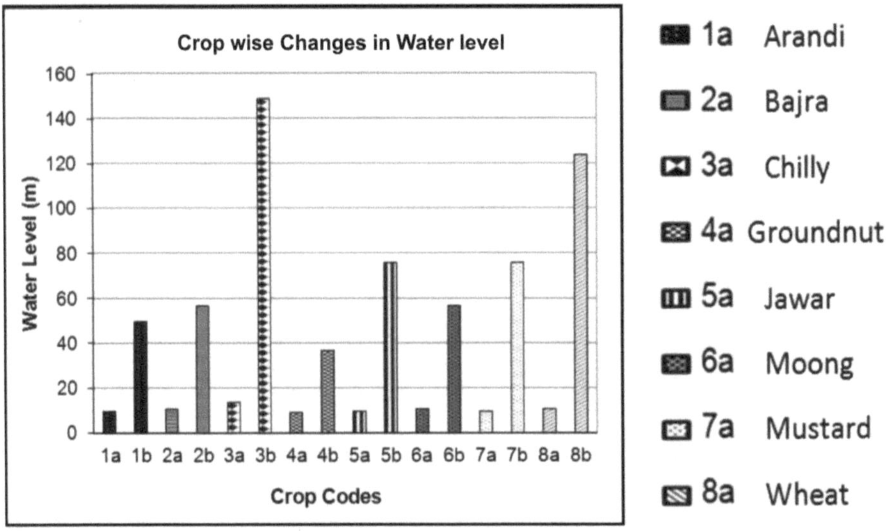


**Figure 7 :** Crop-wise changes in water level

The graph depicts that the water requirement of the crop chilly is higher than any other crop (160mm) in the selected area. Minimum amount of water is required by groundnut. In the above graph, 'a' and 'b' depicts the start and end of the growing period of the crop respectively on the horizontal axis.

The effect of all the selected crops on each of the 19 wells have been observed by using the crop water requirement estimated from the CROPWAT. Thus, in MODFLOW discharge for each crop was loaded on each of the well, which is the requirement by each well per day for the agricultural area in the Land use Land cover. The wells having dark black color are having more impact due to the agriculture water extracted from that particular well. This has been shown in figure 8 showing impact of the wheat and chilly crop water requirement on the wells. Based on the recharge values estimated in water balance, potential areas having more capacity to infiltrate precipitation into the ground can be delineated.

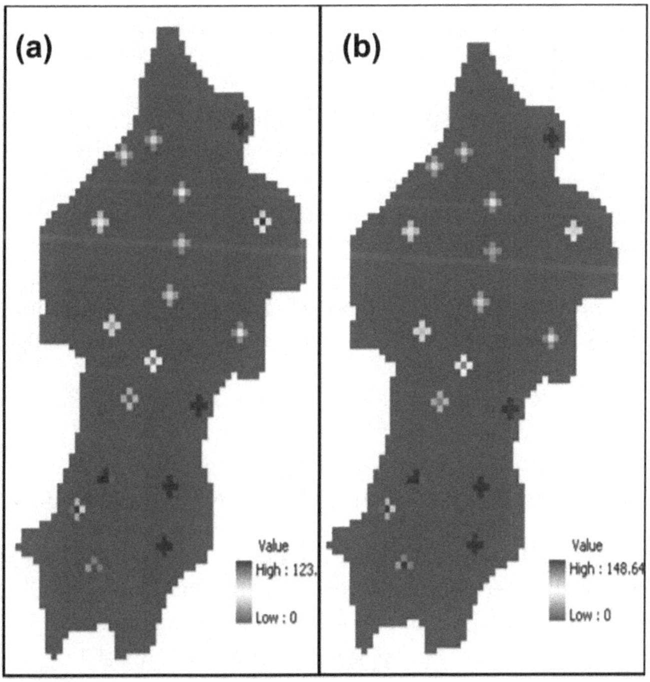


**Figure 8 :** Impact of wheat and chilly crop-water requirement on the walls in the study area

## Discussion

A large wealth of knowledge can be generated through the water resource management, farmer field based actions and approaches applicable for common people on a local scale. These kinds of interventions can have immediate benefits to small holder farmers. Based on the requirement of agriculture water and micro climatic conditions, eco-crop database will be maintained which will further aid in reengaging the agriculture water management. Thus, the commercial crops with high economic value will have less impact on the ground water extraction if the irrigation water is being utilized wisely and also groundwater resources can also be managed in a more sustainable manner.

## References


Arnold J., Neitsch S., Kiniry J., Williams J., King K. 2002. Soil and water assessment tool theoretical documentation. TWRI Report Vol. TR-191.

Chuanyan Z.and Zhongren N., 2007. Estimating water needs of maize (*Zea mays L.*) using the dual crop coefficient method in the arid region of northwestern China. *African Journal of Agricultural Research* 2(7) : 325-333

Diop M. Analysis of crop water use in Senegal with the cropwat model.

FAO (Food and Agriculture Organization), 1975. Guidelines for predicting crop water requirements. Authors, Doorenbos J & Pruitt WO, Irrigation and Drainage Paper 24. Rome, Italy.

FAO. 1998. Crop Evapotranspiration - Guidelines for computing crop water requirements. FAO Irrigation and Drainage Paper No. 56. TR No. 04-020.1998.

http://fao.org, Approved Crop Calendar.

Giorgis K., Tadege A. and Tibebe D. Estimating crop water use and simulating yield reduction for maize and sorghum in Adama and Miesso districts using the CROPWAT model.

Kassam A.and Smith M. 2001. FAO Methodologies on Crop Water Use and Crop Water Productivity. Proceedings in expert meeting on crop water productivity, Rome. Paper No. CWP-M07.

Luzio M., Srinivasan R., Arnold J., Neitsch S., 2002. Arcview Interface For Swat2000 User's Guide.

Marica A. Short description of the CROPWAT model.

World Bank 2005, Reengaging in Agricultural Water Management,. Directions in development, The World Bank.

Saradeth S., Dodo A. and Latrech D. 2010. Earth Observation in support of Management of Internationally Shared Groundwater Resources in Africa: the AQUIFER Project.

Wylie A. 2004. Confined vs. Unconfined Aquifer Representation.

Yarahmadi J., 2003. The integration of satellite images, GIS and CROPWAT model to investigation of water balance in irrigated area.

FSC
www.fsc.org
MIX
Papier | Fördert gute Waldnutzung
FSC® C083411

Zeitfracht Medien GmbH
Ferdinand-Jühlke-Straße 7
99095 Erfurt, Deutschland
produktsicherheit@kolibri360.de